The Institute of Mathematics
and its Applications
Conference Series

Volumes in the previous series were published by Academic Press to whom all enquiries should be addressed. The following and all forthcoming titles are published by Oxford University Press throughout the world.

NEW SERIES

1. *Supercomputers and parallel computation* Edited by D. J. Paddon
2. *The mathematical basis of finite element methods* Edited by David F. Griffiths
3. *Multigrid methods for integral and differential equations* Edited by D. J. Paddon and H. Holstein
4. *Turbulence and diffusion in stable environments* Edited by J. C. R. Hunt
5. *Wave propagation and scattering* Edited by B. J. Uscinski
6. *The mathematics of surfaces* Edited by J. A. Gregory
7. *Numerical methods for fluid dynamics II* Edited by K. W. Morton and M. J. Baines
8. *Analysing conflict and its resolution* Edited by P. G. Bennett
9. *The state of the art in numerical analysis* Edited by A. Iserles and M. J. D. Powell
10. *Algorithms for approximation* Edited by J. C. Mason and M. G. Cox
11. *The mathematics of surfaces II* Edited by R. R. Martin
12. *Mathematics in signal processing* Edited by T. S. Durrani, J. B. Abbiss, J. E. Hudson, R. N. Madan, J. G. McWhirter, and T. A. Moore
13. *Simulation and optimization of large systems* Edited by Andrzej J. Osiadacz
14. *Computers in mathematical research* Edited by N. M. Stephens and M. P. Thorne
15. *Stably stratified flow and dense gas dispersion* Edited by J. S. Puttock
16. *Mathematical modelling in non-destructive testing* Edited by Michael Blakemore and George A. Georgiou
17. *Numerical methods for fluid dynamics III* Edited by K. W. Morton and M. J. Baines
18. *Mathematics in oil production* Edited by Sir Sam Edwards and P. R. King
19. *Mathematics in major accident risk assessment* Edited by R. A. Cox
20. *Cryptography and coding* Edited by Henry J. Beker and F. C. Piper
21. *Mathematics in remote sensing* Edited by S. R. Brooks
22. *Applications of matrix theory* Edited by M. J. C. Gover and S. Barnett
23. *The mathematics of sufaces III* Edited by D. C. Handscomb

Continued overleaf

24. *The interface of mathematics and particle physics*
Edited by D. G. Quillen, G. B. Segal, and Tsou S. T.
25. *Computational methods in aeronautical fluid dynamics*
Edited by P. Stow
26. *Mathematics in signal processing II*
Edited by J. G. McWhirter
27. *Mathematical structures for software engineering*
Edited by Bernard de Neumann, Dan Simpson, and Gil Slater
28. *Computer modelling in the environmental sciences*
Edited by D. G. Farmer and M. J. Rycroft
29. *Statistics in medicine* Edited by F. Dunstan and J. Pickles
30. *The mathematical revolution inspired by computing*
Edited by J. H. Johnson and M. J. Loomes
31. *The mathematics of oil recovery* Edited by P. R. King
32. *Credit scoring and credit control*
Edited by L. C. Thomas, J. N. Crook, and D. B. Edelman
33. *Cryptography and coding II* Edited by Chris Mitchell
34. *The dynamics of numerics and the numerics of dynamics*
Edited by D. S. Broomhead and A. Iserles
35. *The unified computation laboratory: modelling, specifications, and tools* Edited by Charles Rattray and Robert G. Clark
36. *Mathematics in the automotive industry* Edited by James R. Smith

Mathematics in the Automotive Industry

Based on the proceedings of a conference on Mathematics in the Automotive Industry, organized by the Institute of Mathematics and its Applications and held at the University of Warwick in September 1989

Edited by
JAMES R. SMITH
Chief Engineer in Advanced Engineering,
Jaguar Engineering Centre, Coventry, UK

CLARENDON PRESS • OXFORD • 1992

Oxford University Press, Walton Street, Oxford OX2 6DP
Oxford New York Toronto
Delhi Bombay Calcutta Madras Karachi
Petaling Jaya Singapore Hong Kong Tokyo
Nairobi Dar es Salaam Cape Town
Melbourne Auckland
and associated companies in
Berlin Ibadan

Oxford is a trade mark of Oxford University Press

Published in the United States
by Oxford University Press, New York

A catalogue record for this book is available from the British Library

Library of Congress Cataloging in Publication Data
(Data available on request)

ISBN 0–19–853660–7

Printed and bound in
Great Britain by Biddles Ltd,
Guildford and King's Lynn

Preface

The papers contained in these proceedings represent those selected for publication from those given at the September '89 IMA Conference at Warwick University.

The conference was an attempt to expose typical theoretical problems in the automotive industry and the role that mathematics can play. As such it was the first event of its kind in the UK. In this regard, as it emerged, the papers have emphasized the engineering science involved, and in varying degrees of sophistication, the role of mathematical modelling. Although it was a hope of the organisers that the conference might elicit some problem or problems leading to the use or need for the development of some area or areas of applicable mathematics per se, this did not occur. All the problems discussed arose from complex engineering situations demanding numerical assessment. The techniques involved, though possibly complex and beyond the undergraduate in nature, represent methodologies known to the professionals in the various specialist areas.

In retrospect it was probably optimistic to suppose that topics of a methodological nature would arise. Anyone who has had contact with the automotive industry will have recognised that it is always crucially concerned with questions that are highly practical and of concern on relatively short time-scales. The skills required therefore are:

i) a thorough knowledge of relevant engineering science,
ii) concomitant abilities in modelling in the relevant discipline,
iii) a thorough appreciation of relevant mathematical methods, and
iv) the skills to progress the analysis through computer simulation.

It is readily appreciated that traditional UK training in classical applied mathematics (mechanics, fluid dynamics, elasticity/plasticity, electromagnetism, and latterly control theory) is a necessary condition for effective performance by a mathematician in this industry, as indeed it is in any engineering industry. It is therefore a matter of concern that our valuable heritage of classical applied mathematics is not eroded by removal from the undergraduate curriculum of rigorous and extensive instruction in relevant parts of physical science. This remark applies with especial force at the present time where it is a matter of common observation that standards of attainment in engineering science in engineering departments have fallen noticeably over the last decade.

With regard to the topics chosen for discussion, vehicle dynamics and control, noise, shape design and geometry, aerofluid dynamics, and stress analysis and optimisation, all were served by excellent review papers, with the exception of the last topic. Here, the organisers would wish to acknowledge with gratitude the contribution of C. Neale-Sturgess, Jaguar Professor of Automotive Engineering, Birmingham University who delivered a paper, at comparatively short notice, on the finite element analysis of anistropic media, when the original (overview) contributor was obliged to withdraw on a change of employment which entailed a conflict of commitments, and an inability by their former employer to provide a substitute expositor. It is of interest to note the greatest number of papers received for consideration lay in vehicle dynamics and control, closely followed by aerofluid dynamics, with about equal numbers in noise and stress analysis and optimisation, and (surprisingly) least in shape design and geometry.

The opening address by R. Bertodo set the scene by an overview of the rise of applicable mathematics, its increasing employment in engineering, and in particular, what is now the routine employment of very sophisticated technical software in, for example, the analysis of problems of structural performance, vibration and fluid flow - a theme that was amply illustrated by the conference.

On vehicle dynamics and control, L. Segel gave an invited paper that must count as something of a definitive statement on the history of vehicle dynamical analysis, post World War II. Much of it closely parallels aerospace experience and indeed Segel's own aeronautical background was undoubtedly crucial to his seminal '57 paper on vehicle dynamics delivered before the U.K. Institution of Mechanical Engineers. This paper is probably the most widely cited reference in all the literature on vehicle dynamics. The changing tools of the trade, slide rules - electro-mechanical calculators - analogue computers - hybrids - digital computers, and hand generation of equations of motion by Lagrange's method now replaced by automatic software generation and the use of algebraic manipulation codes, were all illuminatingly (and entertainingly) reviewed (as was the accompanying experimental test equipment). The following papers gave accounts of the use of control theory in problems of current interest, viz four-wheel steer (a basic linear optimal control model, and the use of a Kalman filter), the use of linear optimal control theory in the generation of control laws for active and semi-active suspensions, the control of a CVT (continuously variable transmission), and the application of an adaptive technique to dynamometer control. Whereas the first paper is a typical example of early system performance design, the second an overview of current approaches, the third and fourth give indications of the amount of work that must be done on simulation and non-linear optimisation to bring a practical system to a satisfactory state of operational definition.

On noise, A. Dowling gave a comprehensive review of the two major sources of automotive noise, viz. the engine and the tyre, concentrating on the former, and reviewed the relevant modelling of acoustical phenomena in these contexts. A particularly illuminating, simple analysis of unsteady combustion in a shell, quickly focused attention on the main issues to be addressed in engine combustion generated noise. The analysis was a good example of classical applied mathematical analysis guided by sure physical insight. A concise account of the so far limited use of Statistical Energy Analysis was given, but its potential clearly indicated. There followed a review of the calculation of radiated sound based on the Green's formula method, drawing careful distinctions between the effects of sub and supersonic surface waves. The final section was mostly devoted to the analysis of tyre noise, indicating that its mechanism is still not well understood, and some indication of flow noise. These is little doubt that these two sources of noise are very likely fruitful fields for modelling. The next paper amply illustrated that development of the Green's formula into the boundary element method can give a realistic picture of engine vibration generated noise. The next illustrated the potential of a boundary element code to discuss the effect of shielding, concurring A. Dowling's analysis that shields may decrease, but can easily increase radiated noise. The final paper illustrated the power of using relatively simple plane wave analysis in conjunction with the use of impedances (generated from three-dimensional analysis of 'standard' discontinuities) for assistance in the acoustic analysis of intake and exhaust systems.

The paper by P.E. Bezier was an extensive review of how the detailed processes by which a stylist's conception of a car is translated into a set of actual "panels" , have been changed subsequent to the appearance of the digital computer, by one who played a major part in that change. It is an example of the fact that a mathematician may be called upon to understand a whole process (and its practices and cost implications) rather than "just" a mathematical problem of the definition of curves and surfaces. For those unacquainted with this "revolution", a careful reading will show the impressive contribution that a mathematician can make to process improvement when prepared to come to grips with a real work situation. The next paper gave an attractive and straightforward means of defining surface features (pockets, channels, ribs, etc), and the necessary transition surfaces. The final paper addressed the important problem of surface data exchange, reviewing current standards (including the draft ISO, STEP proposal), the limitations of the neutral file concept, and the parametric evaluator concept.

The section on fluid dynamics received a masterly review by A.D. Gosman. Overall, this field must be considered something of a technical triumph bearing in mind the fearful complexity of most real flows.

The first paper by Johns and Sandford on in-cylinder flow in a two-stroke engine, after a concise review of the basic equations and computational methodology, showed that even with the difficulty of estimating initial and boundary conditions, predicted air-flow rate, temperature and flow characteristics are either highly accurate or realistic enough to be used as design-aids in engine development. The paper by Myhill and Sweetenham on the simulation of flame propagation and knock, reflects one aspect of the fundamental research on automotive conbustion that has been conducted by means of a "club" at Harwell for some years past. It shows that the use of lumped parameter models combined with curve-fitting of certain empirical data can lead to excellent results when the equation solver (in this case the in-house FACSIMILE code) is a "stiff" ODE solver made necessary by the differing time-scales of the complex phenomena involved.

The paper by Hawkins, Shaw and Simcox on external vehicle flow simulation shows that while remarkably good qualitative predictions of pressure distribution can now be achieved, accurate drag prediction - although improving - is still inadequate. It is clear that bluff-body aerodynamics is still a field for major research efforts at mesh refinement (achieved algorithmically), and that intriguing area of turbulence modelling.

The final paper by Veshagh et al. addresses a prime problem of passenger comfort, namely, vehicle heating and cooling. Here the approach was to synthesize into a system, a series of lumped-parameter, validated models of components, which combined with energy balances, and numerical fitment of certain empirical data, led to a model giving useful predictions. Clearly, flexibility could be improved by replacing empirical data with physically based models, but the effectiveness of simple models well-based on engineering science was clearly in evidence. One notes that this system, even when optimised for its desired performance, would have to be such that the Electrical Department would be able to obtain a "load-balance" over the operating envelope, ie. this system is a sub-system of a larger system with its own constraints.

Unfortunately, the planned overview paper on structural analysis and optimisation could not be given, and the gap was filled by C.E.N. Sturgess' paper on finite element analysis of anisotropic deformation. Anisotropic analysis is of interest to automotive engineers, not only to cope with certain kinds of composite materials, but to assist in the analysis of such as pressing or stamping operations - still very much a"black-art". The basis of the proposed method consists of relating element node force to element node deflection via a stiffness matrix itself split into three parts, a deformation stiffness matrix, a stress-increment correction matrix, and constant dilation correction matrix. Incremental stress-strain relationships are via an elastic-plastic constitutive tensor, which requires a yield criterion to distinguish between elastic and plastic regions of deformations. The core of methodology is the application of the Taylor-Bishop-Hill theory (for power dissipation) for polycrystalline aggregates to determine a yield criterion. A major computational point (other than the inevitable increases in complexity due to anisotropy) is that as deformation proceeds, changing orientation requires substantial effort in the solution of the linear equations required to redefine the coefficients in the Taylor function.

The next paper by Whalley and Chambers described how the problems arising from the demands for high mechanical integrity in automotive electronic components have resulted in sophisticated thermo-mechanical modelling of electronic components. The FE approach to temperature distribution, solder-joint response and connector performance was described, and the new interesting one-dimensional model for solder creep discussed.

The following paper by Devlukia and Davies echoed the previous one in noting the rising pressures to produce non-failing components. Fatigue is notoriously difficult to predict, and the paper describes the well tried algorithms in this area, and the steps being taken to develop a knowledge-based design support tool.

In closing the IMA would wish to express its gratitude to Rover and Jaguar Cars for the financial support provided for this conference.

J.R. Smith
Jaguar Cars

ACKNOWLEDGEMENTS

The Institute thanks the authors of the papers, the editor, Dr. R. Smith (Jaguar Cars) and also Mrs. Karren Waller, Miss Adèle Sharples and Miss Debbie Brown for preparing and typing the papers.

CONTENTS

CONTRIBUTORS

E. AERNOUDT; Department MTM, Katholieke Universiteit Leuven, de Croylaan 2, B-3030, Leuven, Belgium.

R.G. BERTODO; Rover Group Limited, Fletchamstead Highway, Canley, Coventry, CV4 9DB.

P.E. BÉZIER; 12 Avenue Gourgaud, Paris 75017, FRANCE.

N.W.M. BISHOP; Department of Civil and Structural Engineering, University of Sheffield, P.O. Box 600, Mappin Street, Sheffield, S1 4DU.

E. BRIGHT; Department of Engineering, University of Warwick, Coventry, CV4 7AL.

K.J. BURNHAM; Control Theory and Applications Group, Coventry Polytechnic, Priory Street, Coventry, CV1 5FB.

J.C. CAVENDISH; Department of Mathematics, General Motors Research Laboratories, Warren, Michigan 48090, USA.

D.J. CHAMBERS; Lucas Automotive Limited, Advanced Engineering Centre, Dog Kennel Lane, Shirley, Solihull, West Midlands B90 4JJ

J. DEVLUKIA; Rover Group Limited, Gaydon Test Centre, Banbury Road, Lighthorne, Warwick, CV35 0BL.

A.P. DOWLING; Department of Engineering, University of Cambridge, Trumpington Street, Cambridge, CB2 1PZ.

R.J. GOULT; Department of Applied Computing and Mathematics, Cranfield Institute of Technology, Cranfield, Bedford, MK43 0AL.

P. HARTLEY; School of Manufacturing and Mechanical Engineering, University of Birmingham, Edgbaston, Birmingham, B15 2TT.

I.R. HAWKINS; Computational Fluid Dynamics Services, B392 Harwell Laboratory, UKAEA, Oxfordshire, OX11 0RA.

D.J. HENWOOD; Department of Mathematical Sciences, Brighton Polytechnic, Moulsecoomb, Brighton, East Sussex, BN2 4GJ.

D.J.G. JAMES; Control Theory and Applications Group, Coventry Polytechnic, Priory Street, Coventry, CV1 5FB.

R.J.R. JOHNS; Ricardo Consulting Engineers Limited, Bridge Works, Shoreham-by-Sea, Sussex, BN4 5FG.

R.P. JONES; Department of Engineering, University of Warwick, Coventry, CV4 7AL.

P.J. KING; Control Theory and Applications Group, Coventry Polytechnic, Priory Street, Coventry, CV1 5FB.

S.M. KIRKUP; Department of Mathematics and Computer Science, University of Salford, Salford, M5 4WT.

S.P. MARIN; Department of Mathematics, General Motors Research Laboratories, Warren, Michigan 48090, USA.

A.J. MARTYR; Froude Consine Limited, Gregory's Bank, Worcester, WR3 8AD.

R. MOFFATT; Department of Engineering, University of Warwick, Coventry, CV4 7AL.

R.G. MYHILL; Computational Fluid Dynamics Services, B392 Harwell Laboratory, UKAEA, Oxfordshire, OX11 0RA.

K.S. PEAT; Department of Mathematical Sciences, University of Technology, Loughborough, Leicestershire, LE11 3TU.

I. PILLINGER; School of Manufacturing and Mechanical Engineering, University of Birmingham, Edgbaston, Birmingham, B15 2TT.

T. RAHULAN; Advanced Engineering Group, Jaguar Cars Limited, Whitley, Coventry, CV3 4LF.

M.H. SANDFORD; Ricardo Consulting Engineers Limited, Bridge Works, Shoreham-by-Sea, Sussex, BN4 5FG.

L. SEGEL; University of Michigan, Ann Arbor, Michigan 48109, USA.

R.S. SHARP; School of Mechanical Engineering, Cranfield Institute of Technology, Cranfield, Bedford, MK43 0AL.

S.R. SHARPE; Froude Consine Limited, Gregory's Bank, Worcester, WR3 8AD.

C.T. SHAW; Department of Engineering, University of Warwick, Coventry, CV4 7AL.

F. SHERRATT; Department of Engineering, University of Warwick, Coventry, CV4 7AL.

S. SIMCOX; Computational Fluid Dynamics Services, B392 Harwell Laboratory, UKAEA, Oxfordshire, OX11 0RA.

C.E.N. STURGESS; School of Manufacturing and Mechanical Engineering, University of Birmingham, Edgbaston, Birmingham, B15 2TT.

W.P. SWEETENHAM; Computational Fluid Dynamics Services, B392 Harwell Laboratory, UKAEA, Oxfordshire, OX11 0RA.

R.J. TYRRELL; Structural Analysis Department, Ricardo Consulting Engineers Plc., Bridge Works, Shoreham-by-Sea, West Sussex, BN43 5FG.

A. VAN BAEL; Department MTM, Katholieke Universiteit Leuven, de Croylaan 2, B-3030, Leuven, Belgium.

P. VAN HOUTTE; Department MTM, Katholieke Universiteit Leuven, de Croylaan 2, B-3030, Leuven, Belgium.

A. VESHAGH; Department of Engineering, University of Warwick, Coventry, CV4 7AL.

D.C. WHALLEY; Department of Sensors and Microelectronics, Lucas Automotive, Advanced Engineering Centre, Shirley, Solihull, West Midlands.

R.A. WILLIAMS; Advanced Engineering Group, Jaguar Cars Limited, Whitley, Coventry, CV3 4LF.

D.A. WILSON; Department of Electrical and Electronic Engineering, University of Leeds, Leeds, LS2 9JT.

Mathematics in the Automotive Industry

R.G. Bertodo

Rover Group Limited

Why me? - That is the question I have been asking myself ever since I was asked to open this conference. I have finally come to the conclusion that it can only be because I am probably the only remaining engineer in the Rover Group who can still find his way around a set of log tables, or do a long-division if the computer breaks down! Clearly, age still has its benefits: but then, longevity has always been a mathematician's best friend in many other ways. My formative years were adorned with hand cranked calculating machines, thermionic valves and analog computers the size of a ballroom and the power of a cheap hand calculator. Clearly, I have a natural affinity for history!

The pace of change of all levels and in every sphere of man's endeavour has been accelerating at an ever increasing rate since the beginning of time.

This turbulent scene has been particularly evident in the automotive industry during this decade, where the technology race has become a race for survival. Competing companies are attempting to climb ever higher up a steepening technological gradient. The roots of this automotive technology race are easily found. The major turning points of the industry have occurred at about 20 year intervals. The adoption of mass production techniques in pre-war north America was followed by the more recent evolution of 'Marque' differentiation in Europe and then by the productivity and quality breakthroughs engineered by Japanese producers in the 1970s. The industry is now "Globalising". In doing so, the traditional advantages and distinctive values of European design philosophies are being eroded.

Japanese producers have been the main beneficiaries of the structural changes. In an attempt to recover lost ground, European manufacturers have sought economies of scale through the consolidation of manufacturing facilities and product ranges. This "Production-Driven" strategy has contrasted sharply with Japanese thinking. They embarked on an ever-widening span of development projects, reducing product age and expanding model ranges. The subsequent European "Knee-Jerk" reaction has been the application of "Technology" as a marketing tool.

The productivity, direction and focus of technological resources have become the key determinants of competitive positioning. It is this search for ever increasing productivity of technical resources and the need for more nearly correct first time solutions which has opened up a new and important window of opportunity for the application of analytical methods in Automotive engineering. No automotive engineer will readily admit today that he has overlooked the possibility of using mathematical models in his work. In Rover, these models range from simple stress analyses on desk-top PCs to sophisticated combustion models on the BAe Cray Computer. But before I expand on some of the more recent applications, let us look back to the source of knowledge.

Three times before, have mathematicians gloried in an age of prominence. Each age offers us important lessons for the present. Babylonians, 4,000 years ago, in between hanging gardens and building towers, gave birth to algebraic equations, then disappeared in the mist of ancient history. Lesson 1: evidently, we cannot survive by mathematics alone.

Fifteen hundred years later, a one-tenth genius, nine-tenths scoundrel, named Pythagoras opened the Greek era. This was to last 300 years and drown in the overflow from Archimedes' bath. The Greeks' elevation of mathematics back into respectability, allowed Euclid to make major contributions to two-dimensional geometry, through the observation of mechanics; advances which are still exploited today in advanced surfacing of car bodies. Euclid's fixation with prime numbers proved less successful, but his successors, in expanding his series to the 65, 087 digit 391, 581 x $2^{216,192}$- 1 have recently developed algorithms and computer processing techniques for rapid multiplications. Such methods are invaluable to scientists and engineers, as demonstrated in the use of fast transforms to explore complex vibrations in vehicle structures. Lesson 2: mathematics must be used creatively.

The Greek world was a "Flat Earth" environment and their work rapidly degenerated into a stultified game to be played within imposed, rigid and unchangeable rules. Archimedes, in frustration, turned to designing catapults to sink Roman galleys. The immensely practical Romans brought down the curtain on Archimedes and mathematics for 800 years. Lesson 3: evidently, communication plays an important part in the success of mathematics, analytical methods and mathematicians.

Galileo gave birth to the golden age of mathematics in renaissance Europe by the swing of a church lantern in the 1500s - an age free of the Greek constraints, full of abstract thinking and rich in imaginative thought. It gave rise to a 200 year spell littered with great mathematicians, with household names and embellished by an abundance of creativity. The Swiss Bernoullis, that mathematical act of eight performers spanning three generations: experts in calculus, catenaries and fluid dynamics; the founders of mathematical physics and used in the solution of many of today's problems. Boole, inventor of logical thought, Boolean Algebra and posthumous sorter of today's electronic circuits and Rover 800 wiring harnesses. Euler, the algorithm. Fourier, solver of complex wave forms. Precocious Gauss; and Jacobi, failed politician, but master of elliptical functions. Lagrange; Laplace, superb developers of the work of the outstanding genius Newton, inventor of calculus, theoretical mechanics, optics and universal gravitation and Pascal, geometor . Caley brought this rich era to an end with the introduction of matrix algebra. It took a full 100 years before an application could be found for Caley's contribution. It is a sobering thought that if he had persevered and found an application for his findings, matrices might be known as Caley's today. Lesson 4: researchers are soon forgotten, but their work evidently lives on.

What of the past 150 years?

The traditional public perception of men of numbers is that of unworldly, harmless, slovenly, eccentric dreamers - an image brought about by the role mathematicians have taken onto themselves since the early 1800s. Like a researcher, secure in old age through his past achievements, becomes detached from reality, tends to lose interest in practical applications of his work and turns more and more to the ephemeral realms so has mathematical theory - turning from the adventure of the body to the adventure of the mind and on to the purest realm of thought, where no unpleasant consequences can follow from error, mistake or failure. Hardly the warriors to pitch against the virile and aggresive eastern industrial force. Yet it must be so.

Mathematics has longevity on its side: The achievements of the past are forever present. Algorithms, relational databases, finite elements and mathematical simulation have become part of the daily engineering vocabulary. Fluid and vehicle dynamics, stress analysis, surface design and noise reduction all have received the mathematical treatment in an attempt to reduce cost, increase precision and reduce lead time.

We are entering the age of applied mathematics, where the theoretical work of the past can and must be exploited to succeed in the future. Ever more powerful computers are making this a more practical task if we have the will to abandon the search for ever more elegant solutions in favour of the more risky and less glamorous role of ingenious applicator of known solutions.

Simultaneous engineering, expert systems and unified information technology are the fashion, but what of integrated analytical and predictive systems for total vehicles? Have we, and are we doing enough to exploit the work of the past for a more effective future? Are we creating the competitive edge for a post-1992 environment? These are the questions I would like you to bear in mind during the course of this conference. Personally, I believe we have made a good start, but we need to progress even faster if we are to keep pace with customer demands and economic needs.

Ruminations of a Reclusive Dynamicist

L. Segel

University of Michigan

Abstract Some historical reflections are presented with respect to the mathematical tools available to the dynamicist beginning with the World War II era and continuing to the present period. An example of the old methodologies versus the new is given, with reference to the task of deriving the equations of motion of a two-mass motor car.

1 Introduction

In selecting a title for these remarks, I took my cue from the title of a speech given by Eli Sternberg on the occasion of his receiving the Timoshenko Medal of the American Society of Mechanical Engineers in 1985. I ran across his title in an Applied Mechanics Division newsletter and it hit me as being peculiarly suited to my situation if I were to substitute "dynamicist" for "elastician".

Consider the following: "ruminate" means "to muse upon, contemplate over and over, ponder over"; "rumination" is the act or process of ruminating". "Reclusive", on the other hand, means "marked by seclusion or retirement". Given that I have retired, in a formal sense, and I wish to muse upon the extent to which the solution of mathematical problems, in a vehicle dynamics context, has changed over the years, the selected title seems highly appropriate. Clearly, I am not a historian. Nevertheless, my objective is to present some technical material in a decidely historical perspective.

2 From Aircraft to Automotive Stability and Control Issues

To a large degree, my ruminations derive from personal experience. Accordingly, I note that after obtaining an undergraduate degree in aeronautical engineering, I obtained a position as a junior engineer in the Flight Research Department of the Cornell Aeronautical Laboratory (CAL) in Buffalo, New York. My first assignment, as a young man practising engineering, consisted of a classified project being conducted for the U.S. Navy

Bureau of Aeronautics. The objective was to develop a device for measuring the attitude of an aircraft with respect to its surrounding airmass. I mention this project primarily because it provided an opportunity to become sufficiently involved in the theoretical problem of calculating the flow alignment ahead of an aircraft that I was able to write a thesis serving to fulfil the requirements for a master's degree in mechanical engineering.

My next major assignment put me squarely into the dynamics arena. In the late forties and early fifties, the Flight Research Department at CAL was heavily involved in measuring the dynamic response of aircraft caused by displacement of the control surfaces. The objective was to demonstrate that measured aircraft stability and control properties are predictable with the aid of aerodynamic theory, wind tunnel measurements, and linearized equations of motion. The world was entering the jet age and it was exciting to be the sole author of a report entitled "Dynamic Lateral Stability Flight Tests of an F-80 Airplane by the Forced Oscillation and Step Function Response Methods" [24]. A related project resulted in my preparing a follow-on report entitled "Investigation of Discrepancies Between Flight-Measured and Calculated F-80A Derivatives and Responses–Including a Study of Aeroelastic Effects" [25].

The date is now September 1952 and shortly thereafter my superiors asked whether I would be willing to take on the direction of a research project entitled "Automobile Stability and Control". I said "yes" because I was interested in challenges and not because I had any love affair with the motor car, which was a rather common state of affairs among the average American male.

The situation was as follows: my department head had convinced two very senior engineers at General Motors – namely, Maurice Olley and Robert Schilling – that the methods applicable to the study of aircraft stability and control were also applicable to studying the stability and control of the motor car. A research contract (having a starting date of 24 December, 1952) was signed between GM and CAL which contained the three objectives shown in Figure 1. Work towards achieving these objectives began in 1953, with additional research objectives being added subsequently, resulting in an effort that lasted for ten years.

3 Computational Tools Available in the Fifties and the Sixties

I turn now to the tools which were typically available to perform the mathematical operations of interest to the vehicle dynamicist. My ultimate objective is to show the extent to which the field has changed from the years following the war to the world as we know it today.

We are all very familiar with the current rapid evolution of the digital

G.M.-C.A.L. CONTRACT OBJECTIVES

1. The establishment of valid equations of vehicle motion
2. Comparisons of overall vehicle behavior with predictions based on said equations
3. A review of dynamic theory and experimental data, a re-check of said equations, and the establishment of firm analytical equations and constants

Figure 1. The Objectives of the Research Contract Between GM and CAL

computer. We have perhaps forgotten the extent to which the analyst of dynamic systems and the control-systems engineer had to use devices and methods in the forties and the fifties that are crude and slow by today's standards.

In the late forties and early fifties, the card-programmed digital computer was generally not available or familiar to the average engineer. Millions of calculations were performed on the electro-mechanical computer which sat on the top of a desk. Every mathematical operation which is typically performed on a digital computer today in micro-seconds, or less, was done slowly and laboriously on these mechanical devices, whenever accuracies greater than those obtainable with a slide rule were required. The work was monotonous and dull, and male chauvinists that we were, we typically hired hordes of young women to perform these routine tasks. Consider also that whereas today we gather experimental data in the form of electrical signals which are instantaneously converted to digital form by analog to digital (A-D) devices, it was not always thus. In the forties and fifties, young women laboriously measured, at successive intervals of time, the height of an analog trace on oscillograph paper to produce a digital file in meaningful engineering units.

In retrospect, we can marvel that engineers were willing to go to such great lengths to perform various mathematical operations such as are required, for example to perform a harmonic analysis of a repetitive or non-repetitive signal. Since a harmonic analysis capability is essential if one plans to compare dynamic theory with experimental results in the frequency domain, the device pictured in Figure 2 proved to be extremely

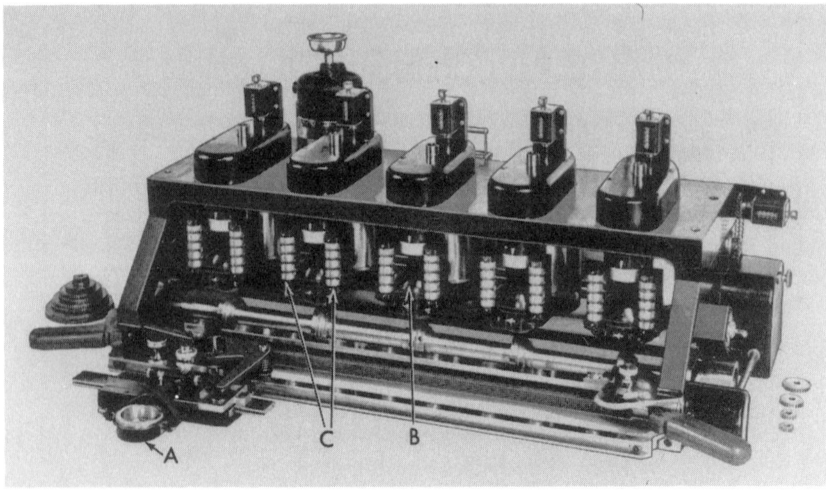

Figure 2. Rolling Sphere Type Mechanical Harmonic Analyzer

useful to the GM-CAL research project. Note that this device contains five spheres, which, on rolling on a flat surface, become integrating devices comparable to the mechanical integrators used by Vannevar Bush in his mechanical differential analyzer, more popularly known as an "analog computer". A human operator followed the analog trace of an input or output signal by controlling the vertical motion of a pair of crosshairs while a motor slowly drove the crosshairs horizontally whenever the operator closed a switch. A single traverse of the signal trace yielded five harmonic amplitudes and phase angles. On changing the gears associated with each sphere and repeating the operation, five more harmonics could be determined, and so on.

Consider now the situation which prevailed in the early fifties. In the absence of either digital or analog computers, it was a major undertaking to integrate either linear or nonlinear differential equations in order to obtain time histories which could be compared with actual measurements. On the other hand, it was relatively straightforward to evaluate the transfer function, i.e., the frequency response, of a system described by linear differential equations, even though considerable time had to be spent operating a mechanical calculating machine. Consequently, there was substantial motivation for making the small disturbance assumption in order to (1) maximize the likelihood of linear behaviour and (2) exploit the mathematical advantages accruing from the use of linear equations. Thus, the reasons for making comparisons of theory and experiment in the frequency domain were substantial and the existence of the Corradi Harmonic Analyzer made the automotive stability and control research undertaking considerably simpler than it would have been without it.

I cannot recall the extent to which my associates, at that point in time, debated the pros and cons of using analog or digital computers to perform the mathematical operations involved in describing or predicting the dynamic behaviour of aircraft or automobiles. I do, however, recall authoring a memorandum in which I pointed out the merits of using the card-programmed calculator (made by IBM) as a means of increasing our productivity as researchers. I also recall that in 1951, I used the IBM computer (in residence at CAL) to calculate the responses of the X-1A aircraft to step displacements of the various control surfaces, However, we were not integrating the equations of motion. Rather, we were using the computer to find the roots of the characteristic equation and to evaluate the closed-form solutions obtained with the aid of the operational calculus.

To the best of my recollection, if you were working with differential equations in the mid-fifties that were higher than fourth-order and/or also wished to include some nonlinear functions, then the scale was tipped in favour of the analog computer. Consequently when efforts were initiated to analyze and predict the directional behaviour of a motor car which included a steering system lumped into two masses subjected to dry (Coulomb) friction forces, the use of an analog computer appeared essential. On the other hand, by ignoring the Coulomb friction and by invoking the small disturbance assumptions, the equations of motion became linear. In this instance, the eigenvalues of the motor car were obtained by the usual matrix calculations that can easily be performed on a digital computer.

Clearly, the analog computer had considerable appeal for the dynamicist at that point in time. For example, it was not convenient for an engineer to interface directly with the digital computers then available. At that time, the concept of "software packages" did not exist, although scientific subroutines probably were being developed. Even if the engineer knew how to write a digital computer program with the aid of a computing language such as FORTRAN, he/she frequently had to rely on additional personnel for the specialized expertise required to operate these rather esoteric devices. The analog computer, on the other hand, was easy for the engineer to understand. He, or she, could, on their own, wire the patchboard and thereby connect all the summing and integrating amplifiers to produce a circuit which corresponded directly with the equations they wished to solve. This process not only accomplished your objective – it was fun. Within seconds you could change a property of a tire-vehicle system and see what difference it made.

Let me now take a moment to present an example of how an analog computer was used in 1961 to address a vehicle ride issue in a rather novel way [15]. The question being asked was whether a train of coupled twoaxled vehicles would yield a better ride than a single vehicle element and thereby be able to travel at higher speeds when the ride disturbance is cross-country terrain. The problem immediately arose as to how to represent a random profile comparable to cross-country terrain and how to generate an electrical signal proportional to the height/elevation of the terrain, which signal, delayed in time, would represent the disturbance input to each of the axles following the lead axle. This problem was solved in the manner pictured in Figure 3. The pictured device is a profile generator and consists of six linear photocells and an opaque disk arranged as shown. The opaque disk rotates at an angular velocity scaled to the assumed constant forward velocity of the vehicle system. Note that the radius of the disk controls the amount of light reaching the photocell as a function of disk-rotation angle. The resistance of each photocell varies as a function of the light flux reaching the cell and the voltage impressed across the cell, thereby yielding a signal proportional to the elevation of the ground. Note that considerable flexibility exists to adjust the wheelbase of each vehicle and the spacing between vehicle units. The opaque disk, as prepared, represents an actual section of proving ground terrain, with a length sufficient to contain the longest wave lengths of interest.

Clearly, such a device provides what is needed to study the ride behaviour of a multiaxle, multi-unit vehicle by means of an analog computer. In view of the large variation in ground elevation, the vehicle model was highly nonlinear in that tires could lose contact with the ground and suspension travel was finite, leading to bump stop contact as forward velocity was assumed to increase. The findings of the study will not be reviewed here. The point being made is that we can now achieve the purpose of this optical-mechanical device in a completely electronic manner.

During the late fifties and early sixties, the analog computer reigned supreme among those researchers who were investigating the ride and handling characteristics of the motor vehicle. Within this time period, the transistor began to replace the vacuum tube, making it feasible to build analog computers with operational amplifiers in the hundreds rather than in the tens. However, these developments came about too late to be of assistance in the GM-CAL project endeavour in which a socalled "variable-stability and response" automobile was designed and built in the mid-fifties. The objective was to create a device which would permit investigators to conduct systematic investigations of the manner in which the stability and control properties of an automobile interrelate with the skills and capabilities of the human controller. At that point in time, it was possible to separate variable stability and response systems into two categories. One category was called the "variable feedback" system, as diagrammed in Figure 4. The second category was known as a "model reference" system. (See Figure 5.) Both types of systems require motion sensors and feedback control mechanisms. However, the model reference system has much to recommend it

Figure 3. Terrain Profile Generator

in that it is much simpler to implement and to use in a man-vehicle study. In the latter case, the response of the vehicle is measured and compared with a "desired" response, with the vehicular controls being actuated to null any difference or error that exists. The "desired" response is generated by applying the operator's input to an electronic analog of a vehicle, as implemented by means of an analog computer. The computer model is adjusted to produce the dynamic behaviour that we wish to simulate. A high gain loop is placed around the vehicle to modify its transfer function towards unity, to the end that the vehicle is able to duplicate the model with minimal error.

Despite the substantial advantages of the model reference technique, the methodology defined in Figure 4 was adopted [26] in 1956-7 to create a variable stability and response motor car, since the electronic state of the art, at that time, produced analog computers with formidable weight and space requirements, thereby precluding their installation into a motor car. (Nevertheless, 11 years later, it became practical to build a "variable braking" car [23] based on the model reference principle, as is indicated in Figure 6, which is reproduced from Reference [5].)

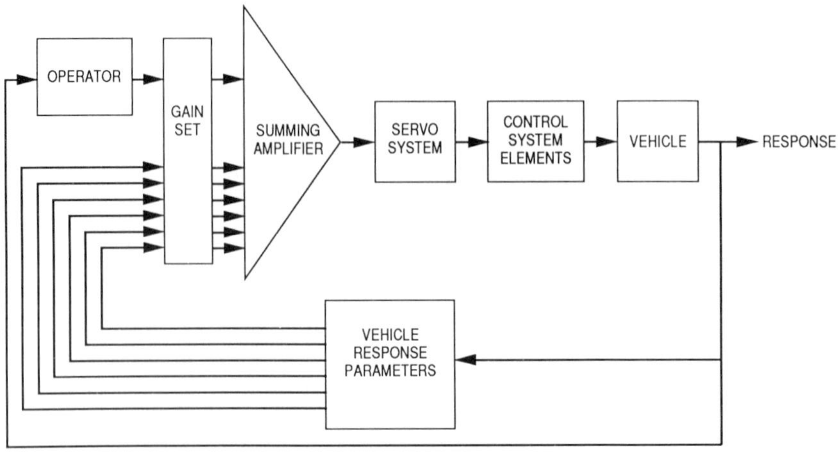

Figure 4. Basic Control Loop of a Variable Feedback Installation

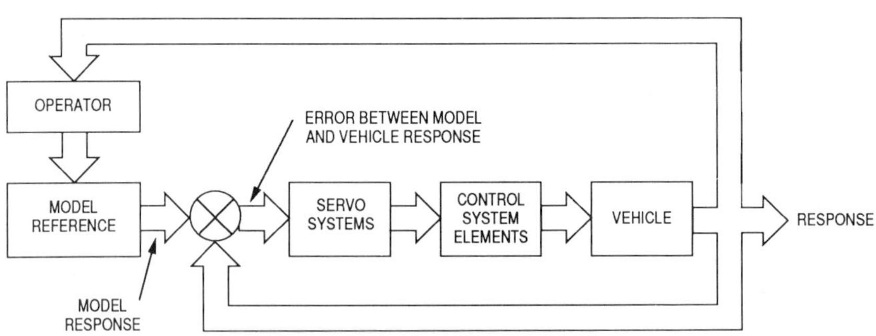

Figure 5. Basic Control Loop of a Model Reference Installation

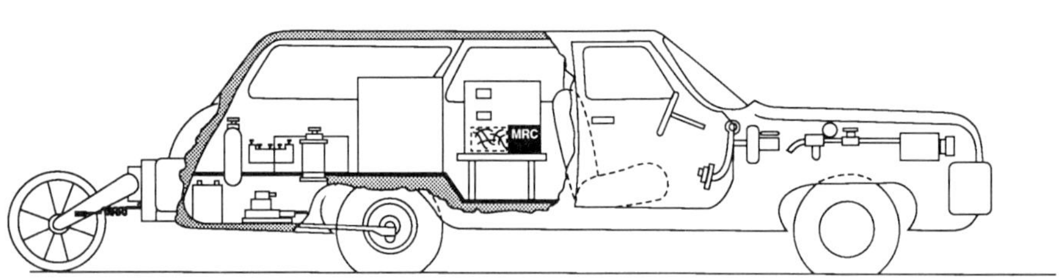

Figure 6. A Variable Braking Vehicle Containing a Model Reference Computer

Figure 7. Hybrid Computer Facility at the University of Michigan

4 Developments in the Late Sixties and in the Seventies

Notwithstanding the rapid advances achieved during the fifties and sixties in the computation power and speed of digital computers, the analog computer continued to be an important tool in vehicle dynamics studies during the late sixties and early seventies. The evolution of the hybrid computer, a combination of digital logic and analog circuitry, was, in large measure, responsible. In Figure 7, we see the hybrid computer which was used at the Transportation Research Institute (formerly the Highway Safety Research Institute of The University of Michigan) to simulate the dynamic behaviour of a wide variety of motor vehicles.

Deserving of first mention is the research in which the spin velocity of each wheel of a four-wheeled motor car is considered to be a degree of freedom of a tire-vehicle system. Figure 8 diagrams the simulation employed in this study [2], which, in retrospect, can be viewed as being a rather pioneering endeavour. Several comments are in order. First, the objective of the study was to determine how the lateral and longitudinal traction properties of the tire influence the behaviour of the motor vehicle during combined cornering and braking manoeuvres. This objective requires that a tire-vehicle simulation account for the influence of combined lateral and longitudinal slip on the shear forces produced at the tire-road interface. As indicated in Figure 8, both the braking/driving forces (F_{xi}), created at the road surface, and the braking/driving torques (T_{wi}), developed by the driveline or the braking system, act on the wheels to produce a spin acceleration and, ultimately, a spin velocity (Ω_i). The spin velocity and the translational velocity of each wheel determine the socalled "slip" of each wheel, which "slip" determines the lateral and longitudinal shear forces created at each tire. Consequently, the calculations implied by the blocks labelled "wheel rotation dynamics", "tire kinematics", and "tire mechanics" had to be performed separately for each of four wheels.

Even though the hybrid computer pictured in Figure 7 possessed a sizeable number of operational amplifiers (e.g., 150), it did not have computing capacity, sufficient to make computations for all wheels simultaneously. Accordingly, the digital logic features of this computer were used to interrupt the overall computation at discrete time intervals in order to perform the computations associated with each tire in a sequential manner. After updating the forces created by each tire, the simulation was permitted to run a short time before stopping to perform another set of tire calculations.

The tire-vehicle system simulation diagrammed in Figure 8 was used in two ways. First, it permitted investigators to establish the equilibrium turning conditions achieved in constant-speed, constant-radius turns. (In effect, the analog computer was providing solutions of the nonlinear equations of equilibrium). The simulation was also used to determine the dynamic behaviour of the motor car in response to transient inputs of steering displacement, brake torque, and driving torque. It should be observed that the first mentioned use of the simulation corresponded, in many respects, to a study which had been conducted in the late fifties [17], but which did not get published in the open literature until 1963 [20].

Let me now diverge from the general continuity of these remarks to go back in time and reflect on how steady turning, as described by a set of nonlinear equations of equilibrium, was investigated using a completely different solution technique. In this earlier work, the objective was to solve the nonlinear equations of equilibrium applicable to steady turning as results from a fixed steer displacement and a fixed level of drive thrust. The equations are nonlinear because the forces and moments generated at each of four tires are nonlinear functions of normal load, lateral slip and longitudinal slip. In addition, the mechanical and kinematic properties of the front and rear suspensions are nonlinear functions of the jounce/rebound displacements. Since a tire model featuring the combined influence of lateral and longitudinal slip did not exist in 1959, socalled "friction circle"

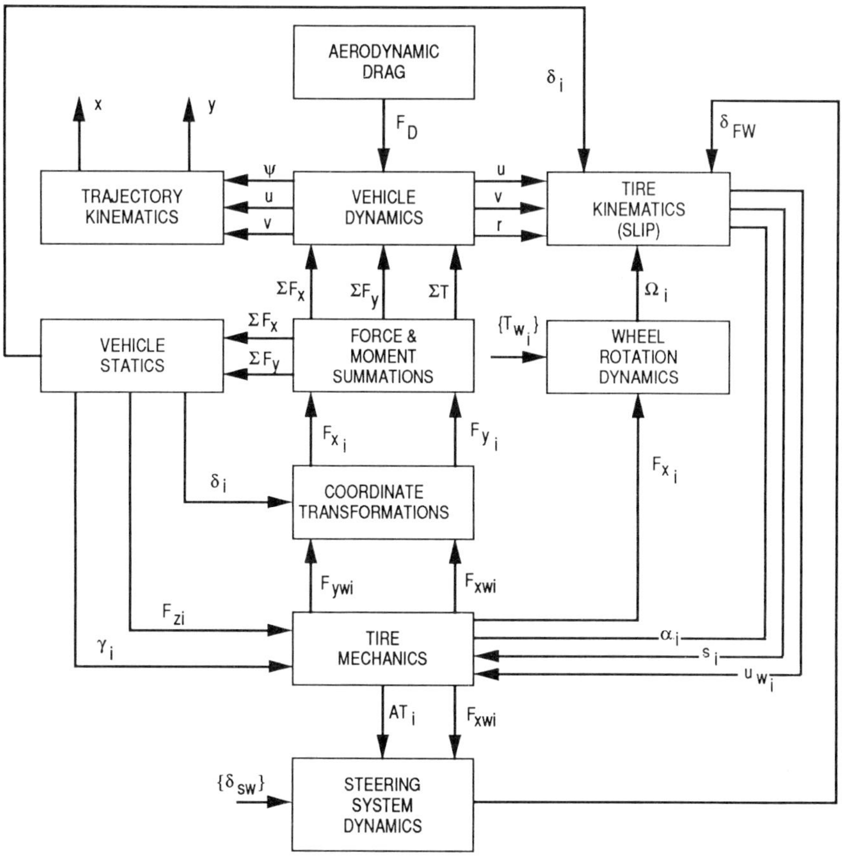

Figure 8. Block Diagram of a Simulation Model Mechanized on a Hybrid Computer

theory was utilized to account for the reductions in lateral force that result when the tire is forced to generate increasing levels of longitudinal force. As already mentioned, the objective was to determine the extent to which the nonlinear properties of the tire-vehicle system cause equilibrium turning to depart from linear predictions as drive thrust is increased to cause high levels of lateral acceleration. For a given steer angle, increased drive thrust will eventually cause the lateral acceleration to exceed 0.3g, a level which experiments had demonstrated previously [27] to constitute an upper bound (more or less) beyond which a passenger car will begin to behave in a nonlinear manner.

In this study of steady turning, the motor car was modelled as a system of rigid bodies free to translate, yaw, and roll. Specifically, the total vehicle was free to (1) translate in the lateral and longitudinal directions, and (2) yaw about a vertical axis. The sprung mass was free to roll about a roll axis established by the kinematic properties of the front and rear suspensions, and the front- and rear-unsprung masses were free to roll about a longitudinal axis located at ground level. The resulting six nonlinear equations of motion were reduced to six nonlinear equations of equilibrium and solved, using the Newton-Raphson procedure. The initial values of each response variable (other than longitudinal velocity), required to start the iterative solution procedure, were obtained by solving a linear set of equations applicable to a five-degree-of-freedom motor car. A highly informative finding proved to be the results shown in Figure 9, which is a plot of path curvature (1/R) as a function of velocity at various fixed levels of front-wheel displacement. The figure shows that the results of the linearized analysis closely approximate the results yielded by the nonlinear analysis at lateral accelerations below 0.3 g. Clearly, we have obtained a mathematical finding which agrees exceedingly well with empirical findings. It can also be observed that, at moderate levels of lateral acceleration, the nonlinear "vehicle" is more understeer than the linear "vehicle". This finding is to be expected in view of the larger lateral transfer of load experienced by the front tires, as caused by the roll stabilizer bar assumed to be connected to the front wheels. The calculations performed for the nonlinear vehicle also show that the vehicle will become oversteer above certain levels of drive thrust, when large steer displacements (e.g., $\delta = 8$ or 16 degrees) are applied.

My personal recollections are that considerable time was spent in both writing and troubleshooting the computer code required to implement the Newton-Raphson solution procedure. (The computer being used in 1959 was an IBM 704). As indicated earlier, similar steady-state solutions were obtained ten years later [2] by permitting an analog simulation to reach a state of equilibrium, following the application of a specific drive thrust, in combination with a specific steer displacement. It could be argued that the analog method was better suited to the task at hand, but one should also realize that the finite capacity of the analog computer limited the scope and complexity of the vehicle model under study.

It should also be noted that whereas the inclusion of the wheel rotation degree of freedom has little, or no, impact on an analog simulation, it has significant impact on a digital simulation because it introduces a very "stiff" subsystem into an otherwise rather "soft" system. This state of affairs led to the idea [1] of analytically solving the equations summarizing the moments about the spin axes and thereby eliminating the need to integrate the total set of equations with a step as small as would be required to integrate the stiff equations corresponding to each of the rolling wheels.

The above described studies, involving the use of hybrid and large-scale

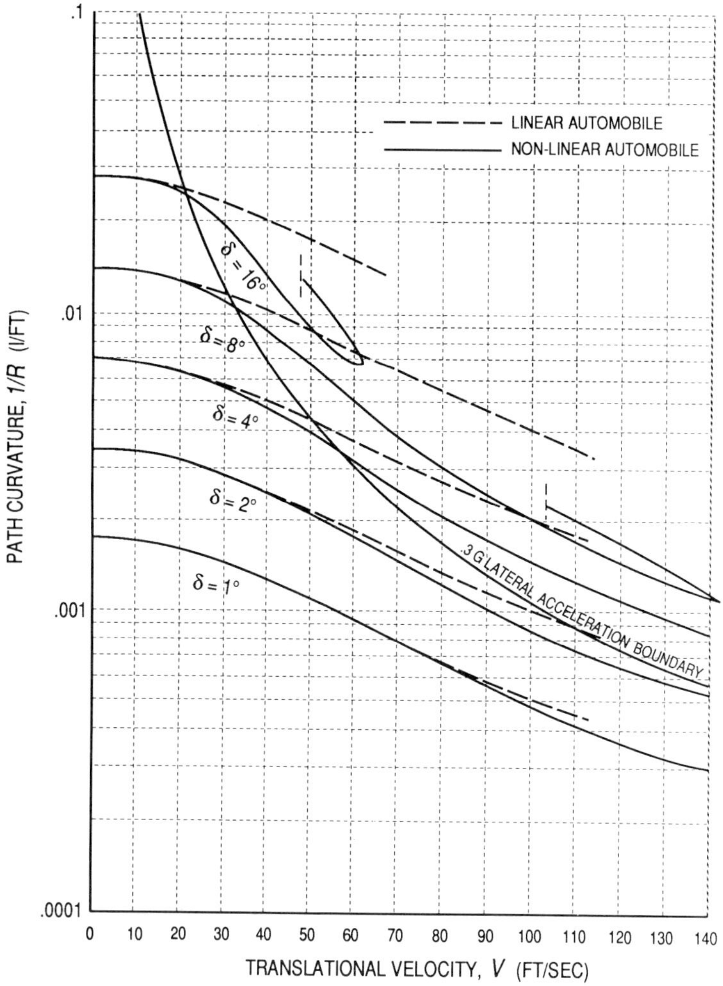

Figure 9. Path Curvature as a Function of Translational Velocity

digital computers are representative examples of studies performed to examine the dynamic behaviour of the motor car. During the sixties and seventies, analog and digital computers were used to study the directional dynamics of motorcycles [29,3,31,16] and articulated vehicles [10,4,13,12] with considerable success. An elaborate computer code was also developed to describe the combined directional and vertical plane dynamics of the motor car. Whereas the original motivation for developing the passenger car code known as HVOSM [14] was the prediction of vehicle trajectories following impact with highway guardrails (and comparable structures found on highways), this code has frequently been used to conduct studies concerned exclusively with the behaviour of the motor car. (An example of such an application is the study performed in 1982 to explain why a U.S. Army vehicle [the M151] exhibited a sudden tendency to roll over after attaining/exceeding a certain lateral acceleration level [28].) The code is still distributed and maintained by the Federal Highway Administration in the U.S., and a version [22] exists in which the wheel rotation degrees of freedom were added in order to compute the longitudinal slip that is experienced by each tire during combined steering and braking manoeuvres.

5 The Current Era

The eighties have brought substantial changes in how dynamicists and control system analysts/designers perform their every-day tasks. The computing power existing in a typical desk top computer or work station has revolutionized the way in which engineers and applied mathematicians think about problems/questions and getting solutions/answers. To be forthright, I have not been part of this revolution – it really belongs to a younger generation. However, I have been able to see that there is a wide divergence of opinion as to the best way to approach a dynamics problem. For example, there are those who swear by the efficacy of Bond graphs [8,19,18] and would not think of deriving the equations of a dynamic system in any other way. There are also those who have become convinced that commercially available multibody dynamics programs are the most efficacious way of developing a simulation of dynamic machine elements or other mechanical systems, such as the motor car [11,30]. The computing inefficiencies of these codes lead to very long computation times, thus prompting others to argue that specialized computer codes, written to represent specific vehicle categories in specific situations, represent a more prudent choice. Finally, there now exist, or are under development, symbolic dynamics codes. In effect, this type of software [9] creates special purpose codes automatically, thereby eliminating a need for an analyst to formulate the equations of motion applying to a specific system.

For example, this latter task can be accomplished by applying the methods of analytical mechanics (i.e., Lagrange's equations) to a mechanical system defined by individual mass elements which are connected in a specific manner. Deriving the Lagrangian of this system can be a very demanding task, which provides considerable opportunity for making algebraic errors. Deriving the equations of motion also provides opportunities for making errors in the partial differentiations that must be performed. The symbolic dynamics codes, on the other hand, not only generate the desired equations of motion, but also produce and compile the FORTRAN code which integrates and solves the equations. A code (AUTOSIM) [21], presently under development, does the above and also provides plots of the time histories computed in response to specified initial conditions or to a specified input(s). Thus, the application of classical methods (i.e., Lagrangian methods) and a symbolic dynamics code to a specific analysis task should serve to contrast the old versus the new. Such a comparison will be made to conclude this rumination.

The current era has also seen an increasing application of modern control-system theory to automotive design and development issues. This statement applies particularly to the growing interest in the use of active suspensions and active steering systems on motor vehicles. Whereas the objective of building very high speed, ground transportation systems provided the primary motivation for considering the use of active suspensions [7], the recent evolution in electronic, hydraulic, and sensor technologies has also resulted in considerable active-suspension research and development in the highway vehicle context. In the case of high-speed ground transportation systems, wherein the supporting guideway becomes a major part of the total system's cost, there is ample justification for considering active suspensions as a means of offsetting the costs of building a higher quality guideway. The system design problem is one of trading off guideway costs against the costs of the "rolling stock" in a manner that minimizes the cost of the total system. Consequently, it is understandable that a major part of the active-suspension literature is concerned with the ride dynamics vehicles which do not use pneumatic tires.

In the automotive context, on the other hand, the "bottom line" question becomes that of determining whether the gains in ride quality are sufficient to justify the additional power consumption costs as well as the initial capital costs. Thus, we see a strong trend to develop suspensions which are semiactive. Such suspensions do not require a continual supply of energy. Rather, they have features which enable them to be modified as a function of the prevailing operating conditions.

In the case of fully-active suspensions, investigators typically employ optimal control theory as a means of determining the extent to which active suspensions yield a finite improvement in ride quality. From a mathematical point of view, it is fortunate that the traversal of a paved road surface constitutes a small disturbance phenomenon, thereby justifying the use of linear models to represent actively suspended motor cars being exposed to a random excitation process. Nonetheless, the digital computation chores involved in the application of optimal control thoery are substantial. It follows that the extent to which active-suspension systems are currently being explored reflects a highly interactive synergism between the applicable mathematics and the computing tools available to reduce the applicable equations to numerical findings.

Notwithstanding the above statement, it must be observed that inventors of active suspensions generally are not stymied by the absence of rigorous analysis tools or the absence of suitable computing equipment. Inventors have the ability to reduce complex physical phenomena to fundamental issues of physics. For example, Hanna and Osbon [6] were able to discern that a hydraulic actuator which is controlled so as to make the sprung mass an effectively larger mass yields a basic improvement in ride quality. This principle was reduced to working hardware in 1961, but history has shown that even though active suspensions are physically realizable, they do not necessarily survive in the market place.

6 An Example: Derivation of the Equations of Motion Applicable to a Two-Mass Motor Car

Consider a motor car designed with a independent front suspension and a beam-axle rear suspension. In this case, the front-wheel masses roll when the sprung mass rolls. Thus, the vehicle is characterized by a rolling mass (m_r), consisting of the sprung mass and the front wheels, and a nonrolling mass (m_{nr}), consisting mainly of the rear axle and the rear wheels. These two masses are linked together such that the nonrolling mass is constrained to translate over a horizontal halfspace and the rolling mass is constrained to rotate about an axis fixed in the nonrolling mass. On recognizing that the geometry of the front and rear suspensions create an instantaneous centre about which rolling occurs, one can postulate the existence of a roll axis passing through these two roll centres. Further, on assuming that this two-mass system is subjected only to small disturbances from rectilinear motion, the roll axis, so defined, becomes an axis fixed with respect to the nonrolling mass.

The above assumptions permit us to observe that only four generalized coordinates are needed to locate and orient this two-mass system with respect to inertial space. Specifically, three coordinates (X,Y, and Ψ) are needed to locate the nonrolling mass with respect to an inertial set of coordinates. (See Figure 10, a view of a four-wheeled vehicle in plan.) Only a single angular-displacement coordinate (ϕ), representing rotation about the fixed roll axis is needed to define the position of the rolling mass relative to the nonrolling mass.

It proves convenient to fix an orthogonal axis system (x,y,z) in the nonrolling mass as indicated in Figure 11. Note that the z axis is vertical and passes through the centre of mass of the two-mass system defined when the roll angle (ϕ) is zero. The vehicle is assumed to be symmetrical, right to left, yielding a plane of symmetry in which the x axis is located. This x axis is arbitrarily assumed to pass through the nonrolling mass centre, which because of the assumptions made earlier regarding the type of rear suspension being employed, is also the location of the rear roll centre. Thus,

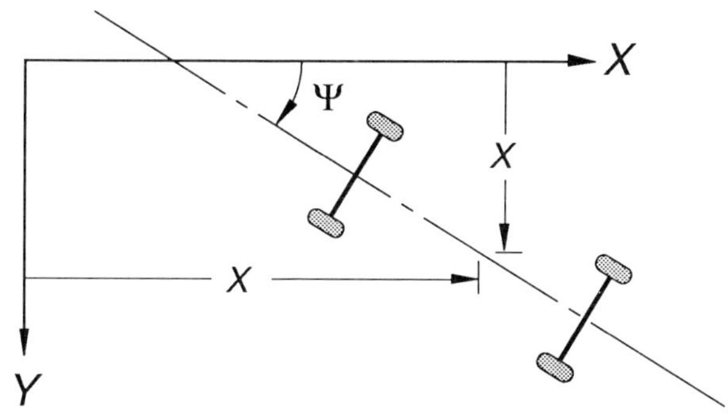

Figure 10. Definition of Generalized Coordinates Which Locate a Vehicle in Plan

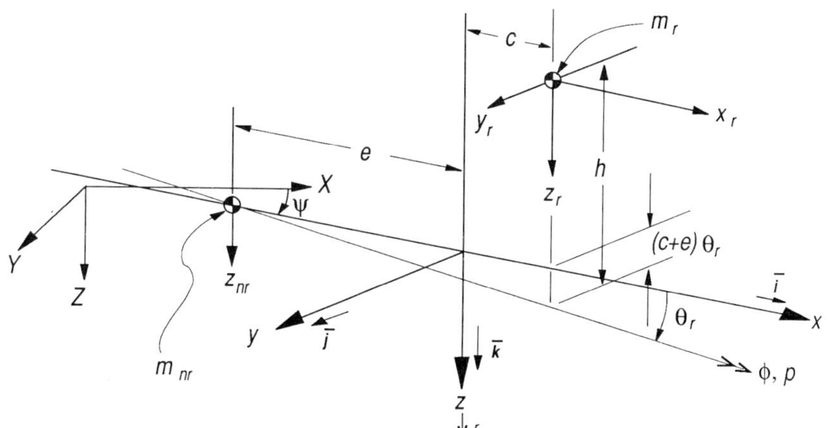

Figure 11. Placement of an Orthogonal Axis System in a Two-Mass Vehicle

Figure 11 shows the socalled "roll axis" passing through the nonrolling mass centre. Further, the roll axis is inclined downward towards the front of the vehicle, since the roll centre of a typical independent suspension is usually located at, or near, ground level. With the z axis assumed to be positive downward, the positive y axis is oriented to the right as seen by an observer looking forward in the positive x direction.

If c and e are distances measured along the x axis between the z axis and the centres of mass of the rolling and nonrolling bodies, respectively, it follows that

$$m_{nr} \bullet e = m_r \bullet c \tag{6.1}$$

Further, it should be noted that the downward inclination angle (Θ_r) of the roll axis is a small angle, such that

$$sin\Theta_r = \Theta_r; cos\Theta_r = 1 \tag{6.2}$$

Also in accordance with SAE recommended practice, let us call the time rates of change of Ψ and ϕ (i.e., $\dot{\Psi}$ and $\dot{\phi}$) r and p, respectively. Let us further designate the origin of the xyz axis system as point o. The translational velocity of point o ($\bar{V}o$) is a vector lying in the x-y plane. Specifically,

$$\bar{V}o = u\bar{i} + v\bar{j} \tag{6.3}$$

where

$$\begin{aligned} u &= \dot{X}cos\Psi + \dot{Y}sin\Psi \\ v &= \dot{X}sin\Psi + \dot{Y}cos\Psi \end{aligned} \tag{6.4}$$

Application of the equations of Lagrange to the assumed two-mass system, constrained as indicated, leads to four nonlinear equations of motions of motion which shall not be presented here. On further assuming that

(1) the component of $\overline{V}o$ in the x direction (u) is large and the component in the y direction (v) is small

(2) the x component of $\overline{V}o$ is given by

$$u = u_0 + u'$$

where

u_0 is a large velocity, constant with time, and

u' is a small disturbance velocity, variable with time

(3) the velocities u', v, r, and p and displacement ϕ are sufficiently small that their squares and products may be neglected,

the derived equations of motion become linear. Namely, we find that

$$\sum F_x = \dot{u}' \tag{6.5}$$

$$\sum F_y = m(\dot{v} + u_0 r) + m_r h\dot{p} \tag{6.6}$$

$$\sum M_z = I_z\dot{r} + I_{xz}\dot{p} \tag{6.7}$$

$$\sum M_x = m_r h(\dot{v} + u_O r) I_x \dot{p} + I_{xz}\dot{r} \tag{6.8}$$

where

$$m = m_r + m_{nr}$$

$$I_z = I_{zzr} + I_{zznr} + m_r c^2 + m_{nr} e^2$$

$$I_{xz} = m_r hc - I_{xzr} + \theta_r I_{zzr}$$

$$I_x = I_{xxr} + m_r h^2 - 2\theta_r I_{xzr} + \theta_r^2 I_{zzr}$$

and

I_{zzr}, I_{zznr} are the moments of inertia in yaw about the vertical z axes passing through the centres of mass of the rolling and nonrolling masses, respectively

I_{xzr} is the product of inertia of the rolling mass defined with respect to an axis system having its origin located at its mass centre, with these axes parallel to the xyz axis system

I_{xxr} is the moment of inertia in roll about a horizontal axis (x_r) passing through the centre of mass of the rolling mass.

Note that the longitudinal acceleration term, $\dot{u}'$, appears only in the first equation of the four equations of motion. If it is assumed that $\dot{u}' = 0$, namely that $u = u_o = a$ constant, we can ignore the first equation and conclude that the remaining three equations apply to a two-mass motor vehicle which is moving with a constant velocity, U_o, along its longitudinal x axis. The remaining three equations are a coupled set and constitute the linearized equations of motion of a two-mass motor car constrained to translate and rotate as outlined above. To apply these equations one must express the external side force ($\sum F_y$), yawing moment ($\sum M_z$), and rolling moment ($\sum M_x$) as linear functions of v, r, ϕ and p and the angular steer displacement (δ) of the front wheels. Experience [27] has shown that these equations do a most adequate job of predicting the directional response of a motor car to front-wheel displacement, provided one has valid information on the relevant stiffnesses of the pneumatic tires, the geometry of the vehicle, the kinematic properties of the front and rear suspensions, and the inertial properties of the masses involved. With respect to this latter point, note that the inertia matrix of this coupled two-mass system is symmetric and is given by:

$$[I]\{\ddot{q}\} = \begin{bmatrix} m & 0 & m_r h \\ 0 & I_z & I_{xz} \\ m_r h & I_{xz} & I_x \end{bmatrix} \begin{Bmatrix} \dot{v} \\ \dot{r} \\ \dot{p} \end{Bmatrix}$$

To derive equations (6.5), (6.6), (6.7), and (6.8), operations involving vector and scalar algebra and partial and ordinary differentiation had to be performed. Such tasks appear to be a reasonable undertaking when the number of generalized coordinates is not too large, say three or four. However, anyone who has applied the methods of analytical mechanics to a system which has a rather large number of degrees of freedom will highly value any scheme in which the analyst is relieved of performing the laborious mathematical operations, necessary as they may be. Thus, it is appropriate to consider what an analyst has to do, if he or she opts to use a symbolic dynamics program.

In order to demonstrate the ease with which equations can be derived with a symbolic dynamics program, let us use AUTOSIM [21] to derive the inertia matrix of the same two-mass motor car treated in the above example. Figure 12 shows the input statements required to achieve this particular objective. In effect, five statements are required to (1) attach reference frames to each of the two masses, (2) specify that the nonrolling body can rotate only in yaw about a z axis parallel to the z axis of an inertial reference frame, (3) define the orientation of an axis fixed in the nonrolling reference frame about which the second body rolls, (4) specify that the mass centre of the nonrolling body is free to translate in the horizontal plane, (5) specify that the $x - z$ plane of the rolling body is a plane of symmetry, (6) declare that the sideslipping and yawing velocities are small quantities, and (7) declare that the longitudinal velocity of the nonrolling body is a constant. When additional information is provided to define the external forces and moments (including the force of gravity) acting on the system, a complete set of equations are produced, in a format that can be compiled and subsequently integrated. Here, however, we shall direct our attention exclusively to the inertia matrix yielded by this procedure.

It should be noted that AUTOSIM does not permit the user to locate an axis system in an arbitrary fashion, for reasons of simplifying the final result. Rather, one is restricted to locating the origin of the axis system fixed in the rolling body on the axis about which the body rolls. The input statements shown in Figure 12 specify that the origin of the axis system fixed in the rolling mass is located at the origin of the axis system fixed in the nonrolling mass. In other words, the origins of both axis systems are located at the centre of mass of the nonrolling body. Whereas in the previous example, the sideslipping velocity, v, was defined as the lateral velocity of the nonrolling mass at a point below the total mass centre, in

```
AUTOSIM Input Statements
(small thetar ixz)

(add-body n nrb :name "nonrolling body" :translate (1 2)
     :parent-rot-axis 3 :inertia matrix #(0 0 Izznr))

(add-body nrb rb :name "rolling body"
     :parent-rot-axis #(! "cos(thetar)" 0 ! "sin(thetar)")
     :body-rot-axes (1) :small-angles (t)
     :inertia matrix #2a((Ixx 0 Ixz) (0 Iyy 0) (Ixz 0 Izzr))
     :cm-coords #(! "cc+e"  0  ! "-h"))

(small (dot (rot nrb) '[n3]) dot (vel nrb0) '[nrb2]))

(set-speed-constant (u 1) speed)
```

Figure 12. AUTOSIM Input Statements Defining a Two-Mass Motor Car

this latter example, the sideslipping velocity is defined as the lateral velocity of the nonrolling mass centre, as located a distance, e, behind the vertical axis passing through the mass centre of the total vehicle. Consequently, the inertia matrix generated by AUTOSIM will differ from that defined earlier. Specifically, one finds that the input statements shown in Figure 12 will generate the following inertia matrix, using the terminology defined earlier, viz:

$$[I]\{\ddot{q}\} = \begin{bmatrix} m & m_r(e+c+h\theta_r) & m_r h \\ m_r(e+c+h\theta_r) & I_z & I_{xz} \\ m_r h & I_{xz} & I_x \end{bmatrix} \begin{Bmatrix} \dot{v} \\ \dot{r} \\ \dot{p} \end{Bmatrix}$$

where

$$I_z = I_{zznr} + I_{zzr} + m_r(e+c+h\theta_r)^2$$

$$I_{xz} = I_{xzr} + I_{xxr}\theta_r + m_r h(e+c+h\theta_r)$$

$$I_x = I_{xxr} + m_r h^2$$

We observe that there are differences in the two results, which are to be expected in view of the two different axis systems involved. Figure 13, shows an AUTOSIM- generated plot of yawing velocity in response

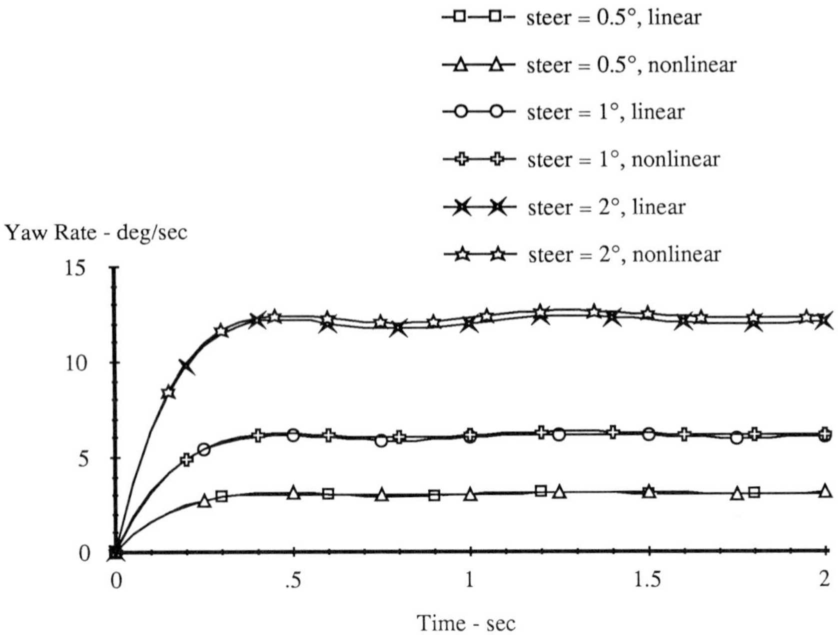

Figure 13. Comparison of the Yaw Rate Response Yielded by Linear and "Nonlinear" Equations as Generated by AUTOSIM

to a step displacement of front-wheel steer angle. Note that the linear equations produced by the AUTOSIM input statements presented earlier produce results that do not depart significantly from the results predicted when the small angle assumptions are removed. In this latter case, the equations become nonlinear, even though we have continued to assume that the tires behave in a linear manner.

References

1. Bernard, J.E., (1973) Some Time-Saving Methods for the Digital Simulation of Highway Vehicles, Simulation, 21, 161-165.

2. Dugoff, H. et al (1970), An Analysis of Tire Traction Properties and Their Influence on Vehicle Dynamic Performance, International Automobile Safety Conference Compendium, 341-366, SAE, New York.

3. Eaton, D.T. and Segel, L., (1973), Lateral Dynamics of the Uncontrolled Motorcycle, Proceedings, Second International Congress on Automobile Safety, Paper No. 73021, 22p.

4. Fancher, P.S., (1975), Prediction of Braking and Directional Responses of Commercial Vehicles, Proceedings of a Symposium on Commercial Vehicle Braking and Handling, Univ. of Mich. Report No. UM-HSRI-PF 75-6, 439-462.

5. Fisher, D.K., (1970), The Dynamic Characteristics of Vehicle Braking Systems, Unpublished Ph.D. dissertation submitted to The University of Michigan, 318 p.

6. Hanna, C.R. and Osbon, W.O., (1961), Vehicle Ride Stabilization by Inertia Control, Westinghouse Research Laboratories Scientific Paper 314-T000-PI.

7. Hedrick, J.K. and Wormley, D.N., (1975), Active Suspensions for Ground Transport Vehicles - A State of the Art Review, Mechanics of Ground Transportation Systems, 21-39, Amer. Soc. of Mech. Engineers.

8. Karnopp, D., (1976), Bond Graphs for Vehicle Dynamics, Vehicle System Dynamics, 5, 171-184.

9. Kortum, W. and Schielen, W, (1985), General Purpose Vehicle System Dynamics Software Based on Multibody Formalisms, Vehicle System Dynamics, 14, 229-263.

10. Leucht, P.M., (1970), The Directional Dynamics of the Commercial Tractor-Semitrailer During Braking, International Automobile Safety Conference Compendium, 242-254, SAE, New York.

11. Loos, H. and Dodlbacher, C., (1986), A Mathematical "Prototype" of the Vehicle to Describe Vehicle Handling Behaviour, Proceedings 9th IAVSD Symposium, VSD Journal, 15, 320-341, Swets & Zeitlinger, Lisse.

12. MacAdam, C.C. et al, (1980), A Computerized Model for Simulating the Braking and Steering Dynamics of Trucks, Tractor-Semitrailers, Doubles and Triples Combinations-User's Manual, Phase IV, Univ. of Mich. Report No. UM-HSRI-80-58, 355 p.

13. Mallikarjunarao, C. and Fancher, P.S., (1978), Analysis of the Directional Response Characteristics of Double-Tankers, SAE Paper No. 781064.

14. McHenry, R.R. and DeLeys, N.J., (1970), Automobile Dynamics - A Computer Simulation of Three Dimensional Motions for Use in Studies of Braking Systems and of the Driving Task, CAL Report No. VJ-2251-V-7, 173 p plus Appendices.

15. Metcalf, W.H., (1961), The Ride Behaviour of a Multi-Element Vehicle Traversing Cross County Terrain, CAL Report No. YM-1424-V-16, 114p.

16. Pacejka, H.B. and Koenen, C., (1980), Vibrational Modes of Single-Track Vehicles in Curves, Proceedings, The Dynamics of Vehicles on Roads and Tracks, 379-393, Swets & Zeitlinger, Lisse.

17. Pacejka, H.B. and Comeau, R.E., (1959), An Analysis of the High-Acceleration, Steady-State Turning Behaviour of an Automotive Vehicle, CAL Report No. YC-857-F-27, 80p.

18. Pacejka, H.B., (1986), Bond Graphs in Vehicle Dynamics, Third Seminar on Advanced Vehicle System Dynamics, Supplement to Vehicle System Dynamics, 16, 263-287, Swets & Zeitlinger, Lisse.

19. Pacejka, H.B., (1985), Modelling Complex Systems Using Bond Graphs, Journal of the Franklin Institute, 319, 67-81.

20. Radt, H.S. and Pacejka, H.B., (1963), Analysis of the Steady-State Turning Behaviour of an Automobile, Proceedings, Control of Vehicles During Braking and Cornering, Institution of Mechanical Engineers. 66-79.

21. Sayers, M.W., (1989), Automated by Symbolic Computation Formulation of Efficient Vehicle Simulation Codes, by Symbolic Computation Proceedings, 11th IAVSD Symposium on Dynamics of Vehicles on Roads and Tracks, 474-487, Swets & Zeitlinger, Lisse.

22. Segal, D.J., (1976), Highway-Vehicle-Object Simulation Model, Volumes 1-4, U.S. Federal Highway Administration Reports No. FHWA-RD-76-162, 163, 164, and 165.

23. Segel, L. and Murphy, R.W., (1968), The Variable Braking Vehicle: Concept and Design, Proceedings, First International Conference on Vehicle Mechanics, 276-299, Swets & Zeitlinger, Amsterdam.

24. Segel, L., (1950), Dynamic Lateral Stability Flight Tests of an F-80 Airplane by the Forced Oscillation and Step Function Response Methods, U.S. Air Force Technical Report 5995.

25. Segel, L., (1952), Investigation of Discrepancies Between Flight-Measured and Calculated F80-A Derivatives and Responses – Including a Study of Aeroelastic Effects, CAL Report No. TB-495-F-15, 70p.

26. Segel, L., (1965), The Variable Stability Automobile: Concept and Design, SAE Transactions, 217-226, (Paper No. 650658).

27. Segel, L., (1956-7), Theoretical Prediction and Experimental Substantiation of the Response of the Automobile to Steering Control, Proceedings, Automobile Division, Institution of Mechanical Engineers, 310-330.

28. Sharp, R.S. and Segel, L., (1983), Digital Simulation of Rollover of a Military Vehicle, Road Vehicle Handling, 23-29, Mechanical Engineering Publications Ltd., London.

29. Sharp, R., (1971), The Stability and Control of Motorcycles, Journal of Mechanical Engineering Science, 13, 316-329.

30. Trom, J.S. et al, (1987), Modelling of a Mid-Size Passenger Car Using a Multibody Dynamics Program, Jour. of Mechanisms, Transmissions, and Automation in Design, 109, 518-523.

31. Weir, D.H., (1972), Motorcycle Handling Dynamics and Rider Control and the Effect of Design Configuration on Response and Performance, Unpublished Ph.D. dissertation submitted to the Univ. of Calfornia at Los Angeles, 152p.

Actively Controlled Four Wheel Steering

R.A. Williams and T. Rahulan

Jaguar Cars Limited

Abstract A great deal of interest has been shown in recent years in the benefits, to vehicle handling, of steering both front and rear wheels of a motor car. This paper attempts to assess some of those benefits with the aid of a simple vehicle handling model. A range of systems is considered which vary in complexity from simple passive systems, where the rear steering angle is scheduled as a function of the front steering angle, to fully active systems where the handling and gust response can be made to be independent of the passive dynamics of the vehicle.

1 Introduction

In recent years a great deal of interest has been generated in the possible benefits, to vehicle handling and high speed stability, of steering both front and rear wheels of a motor car [1,4,5,9]. This has led three manufacturers to introduce cars with four wheel steering into the market place already, with other manufacturers no doubt close behind. Systems with four levels of sophistication can be identified and will be referred to as type A, type B, type C and type D. These are defined as:-

TypeA - Systems in which the steering angles of the front and rear wheels are a fixed function of steering wheel angle. These systems will be almost invariably under the direct mechanical control of the driver [5].

TypeB - Systems in which the front steering angle is a function of steering wheel angle, and the rear steering angle is a speed dependent function of steering wheel angle. In these systems the driver will control the front wheels directly via a conventional steering system. The rear steering mechanism will be mechanically coupled to the front steering mechanism via a variable ratio gear box, whose ratio can be changed as a function of speed [8, 10].

TypeC - Systems in which the front steering angle is a function of steering wheel angle; and the rear steering angle is actively controlled as a function of steering wheel angle, and the dynamic motion of the vehicle (forward speed, lateral acceleration, yaw rate, suspension derivatives etc) [6].

TypeD - Systems in which both front and rear steering angles are actively controlled as a function of steering wheel angle and a variety of vehicle dynamic parameters [3].

The paper reviews these four types of system and attempts to assess the benefits.

2 Theoretical Model

The two degree of freedom plan view "bicycle" handling model described in [7] has been used to assess the four-wheel steering options. The model is shown schematically in Fig 1 and the state equations given in equation 2.1.

$$\begin{bmatrix} \dot{v} \\ \dot{r} \end{bmatrix} = \begin{bmatrix} \frac{-(2C_{af}+2C_{ar})}{MV_x} & -[V_x + \frac{(21_a C_{af} - 21_b C_{ar})}{MV_x}] \\ \frac{-(21_a C_{af} - 21_b C_{ar})}{IV_x} & \frac{-(21_a^2 C_{af} + 21_b^2 C_{ar}}{IV_x} \end{bmatrix} \begin{bmatrix} v \\ r \end{bmatrix}$$

$$+ \begin{bmatrix} \frac{2C_{af}}{M} & \frac{2C_{ar}}{M} \\ \frac{21_a C_{af}}{I} & \frac{-21_b C_{ar}}{I} \end{bmatrix} \begin{bmatrix} \delta_f \\ \delta_r \end{bmatrix} + \begin{bmatrix} A_1 C_Y \\ A_1 C_\Psi \end{bmatrix} V_1^2 \qquad (2.1)$$

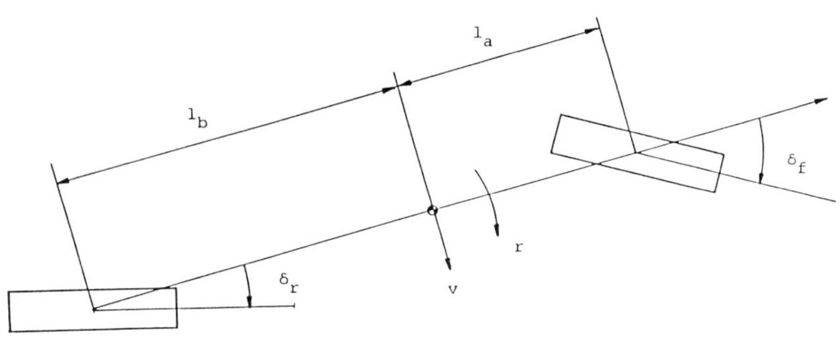

Figure 1. Bicycle Handling Model

Provision was made in the model for both front and rear wheel steering inputs. Aerodynamic inputs to the vehicle were also included, in the equations of motion, in order to investigate the possibility of using active four wheel steering to reduce the cross-wind sensitivity of the vehicle. Parameters used in the paper are typical of a large luxury motor car, the values are given with the list of symbols in section 8.

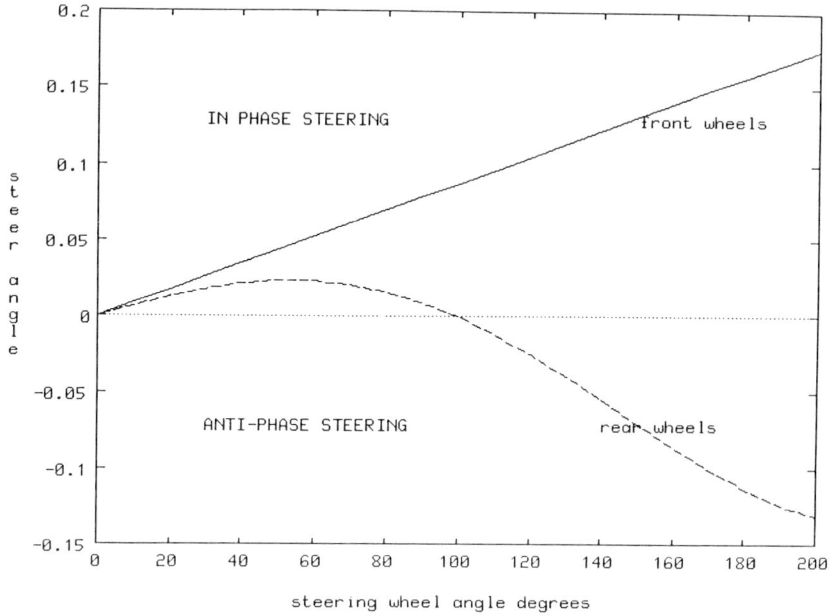

Figure 2. Rear Steering Angle as a Function of Front Steering Angle

3 Passive Systems

The systems which have been introduced so far are essentially passive systems in which the front and rear wheels are mechanically coupled to the steering wheel. In one case the rear steering angle is a fixed, non-linear, function of the front steer angle (Fig. 2), where the wheels are steered in phase for small steering angles and in anti-phase for large steering angles. In the second case the gear ratio between the front and rear wheels is varied with speed so that the wheels are in anti-phase at low speeds and in phase at high speed (Fig. 3).

A standard method of assessing the steady state handling characteristics of a motor car is to perform a steering pad test. In this test the vehicle is driven at increasing speed around a circle of fixed radius. Steering wheel angle is then plotted against lateral acceleration. Fig. 4 shows that the steering wheel angle has to be increased with speed for the standard, 2 wheel steer, vehicle. This is a condition known as understeer. Vehicle manufacturers normally aim for some degree of understeer as the opposite condition, where steering angle decreases with lateral acceleration (oversteer) is more difficult to control, and can become unstable at high speed. The simulation shown in Fig. 4 indicates that the 4 wheel steering philos-

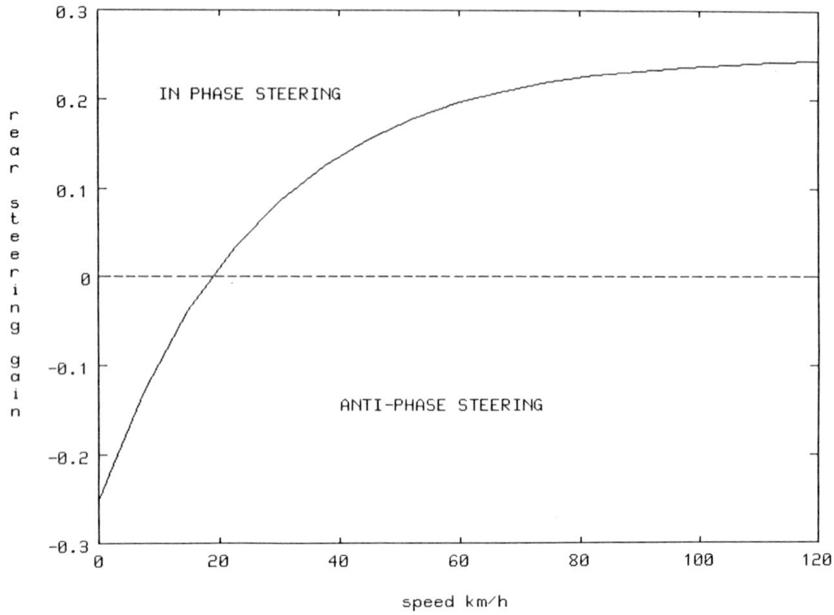

Figure 3. Rear Steering Gain as a Function of Speed

ophy, shown in Fig. 2, will reduce the amount of understeer experienced by the driver, so that the car will appear to be more responsive, while the philosophy shown in Fig. 3 will increase the degree of understeer making the vehicle feel less responsive. Note, however, that eigenvalues of the vehicle will not be affected so that the basic vehicle stability is not changed by the four wheel steering system.

3.1 Anti-phase Steering

Anti-phase steering is used to aid low speed manoeuvring, as the turning circle of the car is greatly reduced [5,9,10].

Problems can occur with anti-phase steering, such as when the car is parked next to a curb. To move away from the curb the front wheels point away from the curb and the rear wheels towards it so that they will try to mount the curb as the car moves off. A way around this problem, with a speed sensing system is to make the rear wheel steering inoperative at very low speeds.

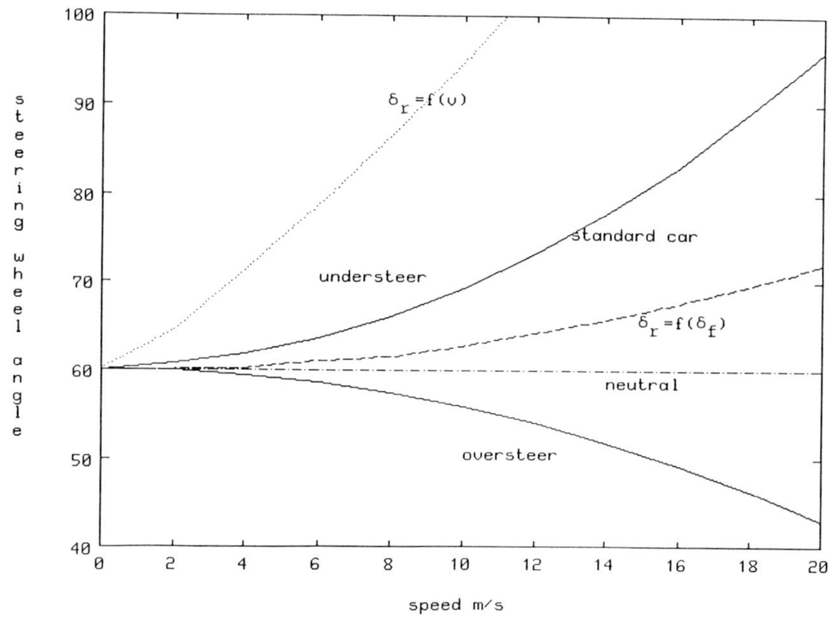

Figure 4. 50m Steering Pad Test

3.2 In-Phase Steering

In-phase steering is used to provide improved lateral response at high speeds. With the systems shown in Figs. 2 & 3 the steering gain (the rate of change of lateral acceleration with steering angle), is reduced at speed, directly in the case of the system shown in Fig.3 and indirectly in the case of the system shown in Fig. 2, by virtue of the fact that only small steering angles are used at high speeds. This reduces the vehicles sensitivity to steering inputs.

In-phase steering is particularly beneficial during evasion/lane change manoeuvres. Because a 4-wheel steer vehicle has a lower yaw velocity than a 2-wheel steer vehicle performing the same manoeuvre lower tyre forces are generated, Fig. 5, enabling more severe manoeuvres to be performed. Subjective evaluations have also shown in-phase steering to be beneficial during these manoeuvres [1].

4 Active Rear Wheel Steering

Active rear wheel steering encompasses type C systems. In these systems the steering angle of the rear wheels is determined by a servo-mechanism.

Inputs to the servo will be a combination of steering wheel angle and vehicle dynamics parameters.

4.1 Linear Optimal Control

In order to make a preliminary estimate of the effects of actively controlling the rear wheels a linear optimal regulator [2] was designed for the model given in equation 2.1.

Given the state equations

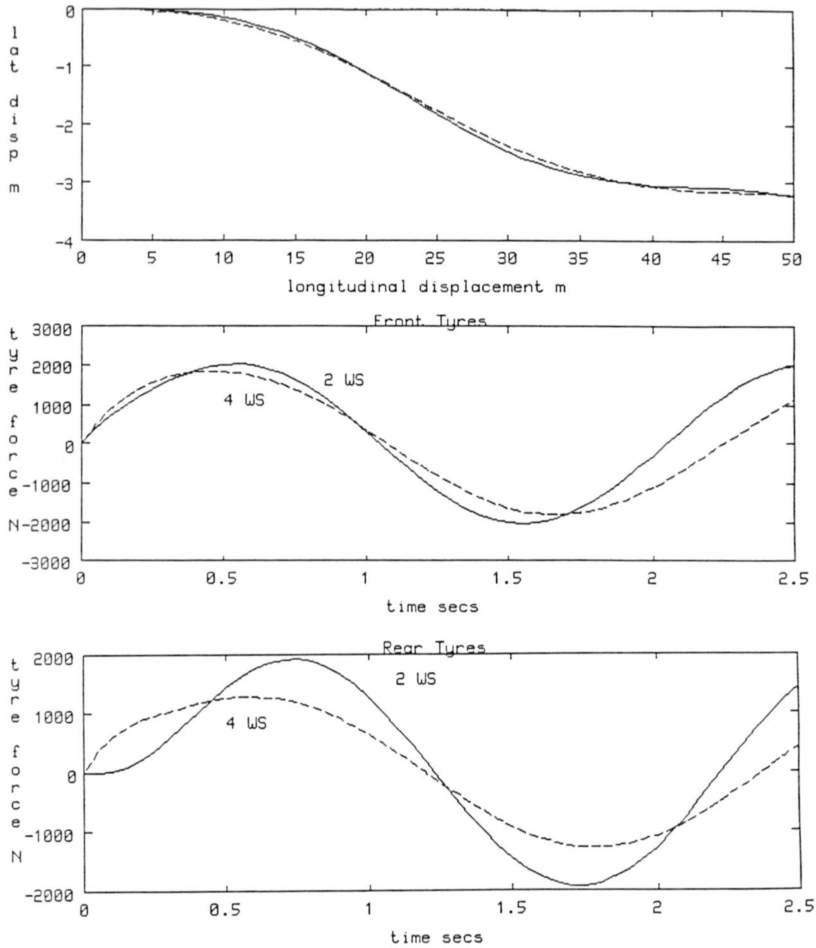

Figure 5. Slalom Manoeuvre

$$\underline{\dot{x}} = \underline{A}\,\underline{x} + \underline{B}\,\underline{u} \tag{4.1}$$

the optimal feed back matrix $\underline{K}$ is one which minimises a quadratic performance index, J, where,

$$J = \int_o^\infty (\underline{x}^T \underline{Q}\underline{x} + \underline{u}^T \underline{R}\,\underline{u})dt \tag{4.2}$$

then

$$\underline{u} = -\underline{K}\,\underline{x} \tag{4.3}$$

and

$$\underline{K} = \underline{R}^{-1}\underline{B}^T\underline{S} \tag{4.4}$$

$\underline{S}$ is the steady state solution of the Riccati equation

$$\underline{0} = \underline{S}\,\underline{A} + \underline{A}^T\underline{S} - \underline{S}\,\underline{B}\,\underline{R}^{-1}\underline{B}^T\underline{S} + \underline{Q} \tag{4.5}$$

The control input $\underline{u}$ in this case is the rear wheel steering angle (δ_r in equation 2.1).

In view of the simple structure of the control system the input weighting of

$$\underline{R} = [1] \tag{4.6}$$

was used.

The state weighting matrix $\underline{Q}$ is given by,

$$\underline{Q} = \begin{bmatrix} a\,0 \\ 0\,b \end{bmatrix} \tag{4.7}$$

The passive car, described in section 2, has two eigenvalues, at a speed of 45 m/s,

1. -3.016 4.998i
2. -3.016 -4.998i

An optimal regulator was designed around this model, using equations 4.1 to 4.7. Suitable values of a and b were found, by iteration, to be $a = b = 1 \times 10^{-3}$. Larger values increased the speed of the transient response to a level where it would probably be unacceptable, and smaller values made it tend towards the passive car. The feedback matrix, K, computed using these values is,

$$K = [0.0169, -0.0986] \tag{4.8}$$

and

$$\delta_r = -0.0169v + 0.0986r \tag{4.9}$$

The closed loop eigenvalues of this system are,

1. -6.636 7.506i

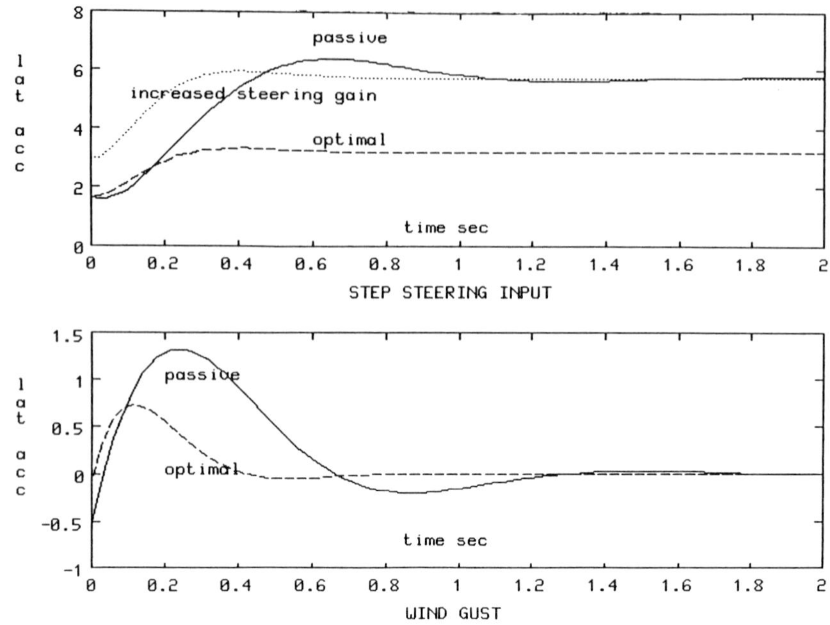

Figure 6. Linear Optimal Control of Rear Wheels

2. -6.636 -7.506i

These produce a faster transient response, with less overshoot, than the passive car (Fig. 6), when a step steering input is made to the front wheels (δ_f).

Note however that the controller has yaw rate feedback which reduces the steady state lateral acceleration, as it produces a rear wheel steering angle in phase with the front wheel steering angle. To overcome this effect, and achieve an acceptable steering response, it will be necessary to increase the steering gain. Alternatively if, in addition to the state feedback, the rear steering angle is made a function of the front steering angle, so that equation 4.9 becomes:-

$$\delta_r = -0.0169v + 0.0986r + G_r\delta_f \tag{4.10}$$

and the rear steering gain

$$G_r = -0.0986r \,/\, \delta_f \tag{4.11}$$

the rear steering angle will be zero in a steady turn if the system is stable.

A beneficial effect of yaw rate and side-slip feedback introduced by the optimal regulator is to reduce the vehicle's sensitivity to cross winds. Fig. 6 shows the effect of a 10 m/s side wind impulse on a car travelling at 45 m/s.

At high speed, most passenger cars tend to develop aero-dynamic lift, unless suitable spoilers are fitted. This causes the effective cornering stiffness of the rear tyres to be reduced.

If the rear cornering stiffness is reduced to 40 kN/rad the eigenvalues of the passive vehicle become,

1. -4.003
2. -0.327

The smaller eigenvalue causes the car to take a long time to settle into a steady turn.

Solving equations 4.1 to 4.7, with this reduced rear cornering stiffness, produces closed loop eigenvalues,

1. -5.658 4.668i
2. -5.658 -4.668i

which gives a transient response similar to that shown in Fig. 6.

Further reductions in cornering stiffness cause one of the eigenvalues of the passive car to become positive, so that the car is directionally unstable. Applying a regulator in this case will restore stability to the system. Note, however, that this improvement in stability is not accompanied by an increase in cornering power so that the maximum cornering speed is not increased.

In summary, with active rear wheel steering it is possible to:-

1. Increase the speed of the transient response.
2. Increase crosswind stability.
3. Stabilise an otherwise unstable vehicle.

in addition to the benefits of in phase and anti-phase steering given in section 3.

4.2 Optimal Observer (Kalman Filter) Design

When designing control systems it is often necessary to use, in the feedback paths, states of the system which cannot be measured directly. The problem can be overcome by means of an observer, which re-creates the states of a system from the measurements which can be made [2]. An optimal controller with a state observer is shown schematically in Fig. 7.

Given the state equations,

$$\underline{\dot{x}} = \underline{A}\,\underline{x} + \underline{B}\,\underline{u} + \underline{G}\,\underline{w} \tag{4.12}$$

where $\underline{G}$ is the noise input matrix and $\underline{w}$ the state excitation noise, and the measurement equations,

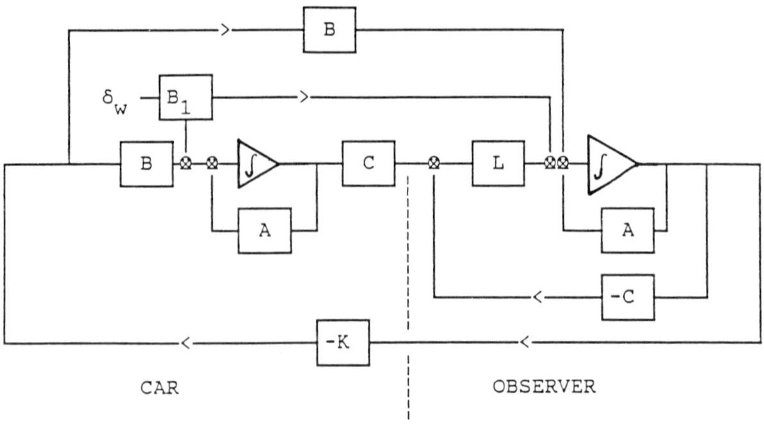

Figure 7. Optimal Control System with Observer Feedback

$$\underline{y} = \underline{C}\,\underline{x} + \underline{D}\,\underline{u} + \underline{\mathbf{v}} \tag{4.13}$$

where $\underline{\mathbf{v}}$ is the observation or measurement noise. A feedback matrix $\underline{L}$ can be found to give an optimal estimate $\underline{(\hat{x})}$ of the state vector $\underline{x}$.

$$\dot{\underline{\hat{x}}} = \underline{A}\,\underline{\hat{x}} + \underline{B}\,\underline{u} + \underline{L}(\underline{y} - \underline{C}\,\underline{\hat{x}} - \underline{D}\,\underline{u}) \tag{4.14}$$

The process noise and measurement noise have covariance,

$$\mathbf{E}[\underline{w}] = \mathbf{E}[\underline{\mathbf{v}}] = 0, \mathbf{E}[\underline{w}\underline{w}^T] = \underline{W}, \mathbf{E}[\underline{\mathbf{v}}\underline{\mathbf{v}}^T] = \underline{V} \tag{4.15}$$

The gain matrix $\underline{L}$ is given by,

$$\underline{L} = \underline{P}\,\underline{C}^T\underline{V}^{-1} \tag{4.16}$$

where $\underline{P}$ is the solution to the observer Riccati equation

$$0 = \underline{A}\,\underline{P} + \underline{P}\,\underline{A}^T + \underline{G}\,\underline{W}\,\underline{G}^T - \underline{P}\,\underline{C}^T\underline{V}^{-1}\underline{C}\,\underline{P} \tag{4.17}$$

4.3 Lateral Vehicle Dynamics Observer

An observer was designed to estimate the states of the simple handling model given in section 2 so that

$$\underline{A} = \begin{bmatrix} \frac{-(2C_{af}+2C_{ar})}{MV_x} & -[V_x \frac{+(21_a C_{af}-21_b C_{ar})}{MV_x}] \\ \frac{-(21_a C_{af}-21_b C_{ar})}{IV_x} & \frac{-(21_a^2 C_{af}+21_b^2 C_{ar})}{IV_x} \end{bmatrix} \tag{4.18}$$

$$\underline{B} = \begin{bmatrix} \frac{2C_{af}}{M} \\ \frac{21_a C_{af}}{I} \end{bmatrix} \tag{4.19}$$

As the measured parameters were lateral acceleration and yaw rate,

$$\underline{C} = \begin{bmatrix} \frac{-(2C_{af}+2C_{ar})}{MV_x} & -[V_x + \frac{(21_a C_{af}-21_b C_{ar})}{MV_x}] \\ 0 & 1 \end{bmatrix} \tag{4.20}$$

$$\underline{D} = \begin{bmatrix} \frac{2C_{af}}{M} \\ 0 \end{bmatrix} \tag{4.21}$$

In the absence of experimental data it was assumed that the process noise vector consists of two elements and an estimate of the gain matrix $\underline{L}$ was made by setting,

$$\underline{G} = \underline{C}^T \tag{4.22}$$

as this will give $\underline{L}$ as a function of the signal to noise ratio of the system. Let

$$\underline{V} = \begin{bmatrix} 1\,0 \\ 0\,1 \end{bmatrix} \tag{4.23}$$

$$\underline{W} = \begin{bmatrix} a\,0 \\ 0\,b \end{bmatrix} \tag{4.24}$$

For the observer to function effectively some of the closed loop poles of the observer should be sited well away from the poles of the system which is being observed, in this case the passive system poles are,

1. -3.016 4.998i
2. -3.016 -4.998i

Arbitrary values of $a = b = 10$ were taken for a first estimate of the observer. These gave closed loop observer poles,

1. -24.46 8.845i
2. -24.46 -8.845i

Simulations showed that the difference between the observed and actual lateral velocity and yaw rate are very small when the vehicle is subjected to a step steer input.

Reducing $a = b = 1$ gave closed loop eigenvalues,

1. -11.23 9.406i
2. -11.23 -9.406i

the observed and actual step steering responses were still excellent.

In order to investigate the possibility of measuring yaw rate only values of $a = 0$ and $b = 1$ were used. The tracking was still very good and the closed loop eigenvalues,

1. -7.937 8.825i
2. -7.937 -8.825i

Similarly if $a = 1$ and $b = 0$, so that the only measurement is lateral acceleration, the tracking also remains good and the eigenvalues become,

1. -10.05 8.016i
2. -10.05 -8.016i

While these results indicate that it should be possible to use an observer to generate lateral velocity and yaw rate from a lateral accelerometer or a yaw rate gyro a more exhaustive study is required on a more complete model of the vehicle.

5 Active Four Wheel Steering

Section 4 demonstrated how active rear wheel steering can be used to improve the transient handling response of a vehicle, while improving the immunity to external inputs, such as wind gusts. The problem with controlling the rear wheels alone is that the two functions are interrelated, which limits the control action which can be applied. In order to apply the control functions independently it is necessary to apply an active control system to the front and rear wheels (Fig. 8).

Taking the right hand side of Fig. 8, the sideslip velocity and yaw rate feedback, with gains $G_y G_\Psi$, are provided as steering inputs to the vehicle. These reduce the response of the car to wind gusts. Optimal values of the control gains G_y and G_Ψ can be computed using equations 4.1 to 4.7. Large values of G_y and G_Ψ provide good immunity to external inputs and lead to small values of the lateral and yaw errors ε_y and ε_Ψ, so that the vehicle lateral and yaw rates will follow the rates commanded by the vehicle model, on the left hand side of Fig. 8.

The vehicle model used to compute the step steering response shown in Fig. 9 had the same dynamic characteristics as the passive car used for the earlier part of the study. However, there is no reason why alternative characteristics cannot be selected to override underlying deficiencies in the car's handling, or even include a driver selectable sport/touring mode.

The lateral wind gust response for the controlled and passive car shown in Fig. 9 was computed for a 10 m/s impulse to a vehicle travelling at a speed of 45 m/s. The immunity to wind gusts shown in Fig. 9 will also apply to other unwanted handling characteristics, such as the oversteer which occurs with some rear wheel drive cars when the drivers foot is suddenly lifted of the throttle during a cornering manoeuvre.

In the practical implementation of the controller, the car is not com-

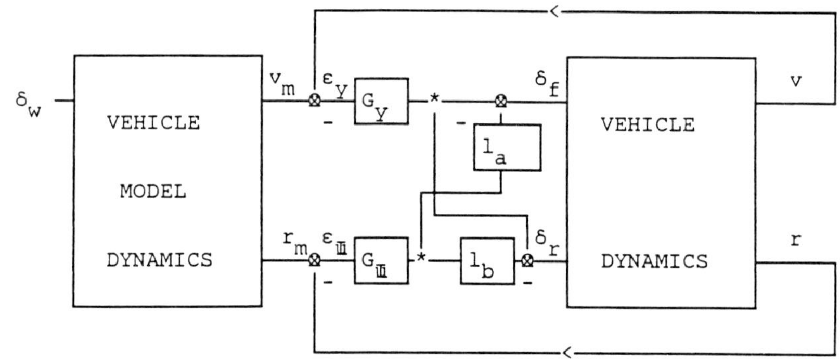

Figure 8. Full Dynamic Control System

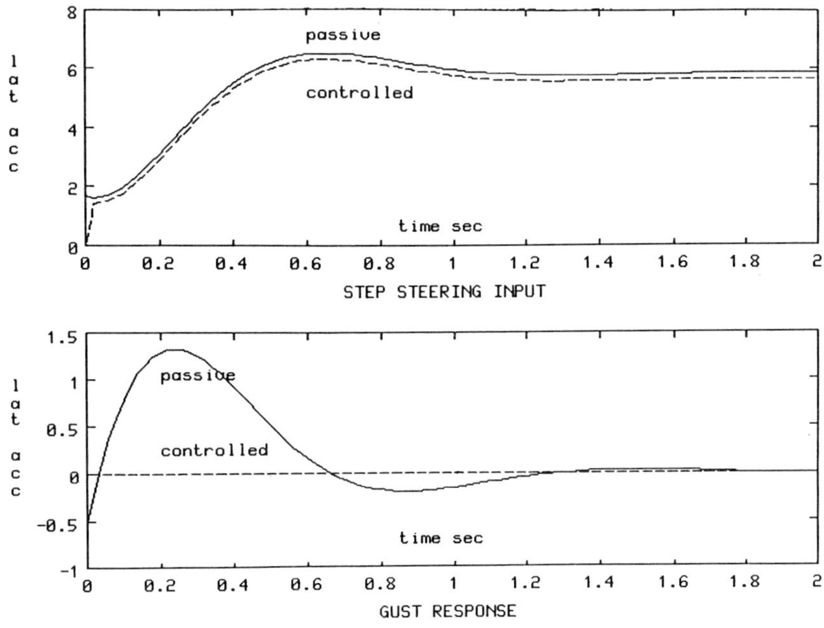

Figure 9. Full Dynamic Control System

pletely drive by wire, as the steering wheel is mechanically coupled to the front wheels and limited authority active control superimposed on top of normal steering inputs. It is also necessary to use an observer, as described in section 4, to generate the yaw rate and side slip velocities as it is not cost effective to measure them directly. The vehicle model is also replaced by an observer to ensure that the vehicle and model are behaving in a similar manner.

The controller is capable of detecting when the car is approaching the control limits, as the error signals ε_y and ε_Ψ will become large. This will enable a warning to be given to drivers that they are in danger of exceeding the limitations of the vehicle for the prevailing conditions, or that a defect has occurred - such as loss of a tyre pressure.

6 Conclusions

It has been shown that four-wheel steering systems are capable of modifying a car's handling characteristics over a very wide range. There is no doubt that this can be used to enhance the car's dynamic qualities, with benefits in vehicle safety. At present it is difficult to assess the attributes which should be built into such a system, particularly with a full dynamic control system such as the one described in section 5, as it is possible to introduce characteristics which are outside current experience. It is quite likely, therefore, that a long period of assimilation will be required with four wheel steering systems before it is known how they can be used to the greatest effect.

References

1. Furukawa, Y., Nakaya, H., (1986), Effects of Steering Response Characteristics on Control Performance of the Driver - Vehicle System, Int. Journal of Vehicle Design, Special Issue on Vehicle Safety, 7, 5/6, 262-278.

2. Kwakernaak, H., Sivan, R., (1972), "Linear Optimal Control Systems" ISBN 0-471-51110-2, Wiley - Interscience.

3. Lugner, P., (1985), Some Investigations on Computer Aided Steering, Proc. 9th IAVSD Symp., ISBN 90 265 0710 0, 353-366.

4. Mouri, H., Kuroki, J., Sugasawa, F., Irie, N., (1988), Handling and Stability Improvement achieved with Four-Wheel Steering, Proc. I.Mech.E., ISBN 0 85298 661 0, 191-197.

5. Sano, S., Furukawa, Y., Shiraishi, S., (1986), Four Wheel Steering System with Rear Wheel Steer Angle Controlled as a Function of Steering Wheel Angle, S.A.E. 860625.

6. Sato, H., Hirota, A., Yanagisawa, H., Fukushima, T., (1983), Dynamic Characteristics of a Whole Wheel Steering Vehicle with Yaw Velocity Feedback Rear Wheel Steering, Proc. I.Mech. E., Road Vehicle Handling Conference, ISBN 0 85298 515 0, 147-156.

7. Segel, L., (1956/7), Theoretical Prediction and Experimental Substantiation of the Response of the Automobile to Steering Control, Proc. I.Mech.E., Automobile Division, 310-330.

8. Senger, K-H., (1988), A Velocity Dependent Steering Gear Ratio for Four-Wheel Steered Vehicles, Proc. I.Mech.E., ISBN 0 85298 661 0, 165-170.

9. Sharp, R.S., Crolla, D.A., (1988), Controlled Rear Steering for Cars-a-Review, Proc. I.Mech.E., ISBN 0 85298 661 0, 149-163.

10. Takiquchi, T., Yasuda, N., Furutani, S., Kanazawa, H., Inoue, H., (1986), Improvement of Vehicle Dynamics by Vehicle - Speed - Sensing Four Wheel - Steering System, S.A.E. 860624.

7 Symbols

a	Sideslip Velocity Weighting Element
$\underline{A}$	Observer Plant Matrix
A_1	Frontal Cross Sectional Area of Car 2.0 m^2
b	Yaw Rate Weighting Element
$\underline{B}$	Observer Input Matrix
B_1	Steering Wheel Input Matrix
$\underline{C}$	State Measurement Matrix
C_{af}	Cornering Stiffness of Front Tyres 50. kN/rad
C_{ar}	Cornering Stiffness of Rear Tyres 73. kN/rad
C_y	Aerodynamic Side Force Coefficient 1.5
C_Ψ	Aerodynamic Yaw Torque Coefficient 1.5
$\underline{D}$	Input Measurement Matrix
$\mathbf{E}$	Expectation Value
$\underline{G}$	Process Noise Input Matrix
G_r	Rear Steering Gain
G_y	Lateral Gain 1.5 rad/m/s
G_Ψ	Yaw Gain 1.5 rad/rad/s
I	Yaw Inertia of Car 3150 Kg m^2
I_m	Idealised Yaw Inertia of Car 3150 Kg m^2
J	Performance Index

$\underline{K}$	Feedback Matrix
$\underline{L}$	Observer Feedback Matrix
1_a	Front Axle to C of G 1.21 m
1_b	Rear Axle to C of G 1.38m
M	Mass to Car 1800 Kg
M_m	Idealised Mass of Car 1800 Kg
$\underline{P}$	Solution of the Observer Riccati Equation
$\underline{Q}$	State Weighting Matrix
$\underline{R}$	Input Weighting Matrix
r	Yaw Rate (rad/s)
r_m	Yaw Rate from Vehicle Model (rad/s)
$\underline{S}$	Solution to the Regulator Riccati Equation
t	Time (s)
$\underline{u}$	Control Input Vector
$\underline{V}$	Measurement Noise Covariance
v	Side Slip Velocity (m/s)
v_m	Sideslip Velocity from Vehicle Model (m/s)
$\underline{\mathbf{v}}$	Measurement Noise Vector
V_1	Sidewind Velocity (m/s)
V_x	Forward Speed (m/s)
$\underline{W}$	Process Noise Covariance
$\underline{w}$	Process Noise Vector
$\underline{x}$	State Vector
$\underline{\hat{x}}$	Estimate of the State Vector
$\underline{y}$	Output Vector
δ_f	Front Steering Angle (rad)
δ_r	Rear Steering Angle (rad)
δ_w	Steering Wheel Input (rad)
ε_y	Lateral Control Error (m/s)
ε_Ψ	Yaw Control Error (rad/s)

The Generation of Control Laws for Active and Semi-Active Vehicle Suspension Systems via Linear Optimal Control Theory

R.S. Sharp[1] and D.A. Wilson[2]

Cranfield Institute of Technology[1] and University of Leeds[2]

Abstract An outline of the main background to linear optimal control theory is given and the application of the theory to the design of control laws for automotive active and semi-active suspension systems is discussed. The linear regulator under random disturbance is central to the development. It requires that the regulator is disturbed by white noise and that the cost function to be minimised is of quadratic form. The relatively simple single input case is considered first, the manner in which the suspension problem can be formulated to fit the background being described. Multiple input cases, involving correlations between the various inputs, are treated subsequently. These cases are typical of real vehicles and the correlations are influential in respect of suspension control and performance. References to literature covering mathematical details are made.

1 Introduction

An active suspension is one involving actuators with energy supply capabilities over and above what is implied by a conventional self-levelling arrangement. The term semi-active is used here to indicate systems which are essentially energy dissipative but which allow sophisticated control of damping forces, so that their control laws can be modelled on those appropriate to active devices. Ideally, when a semi-active element cannot, by its nature, mimic its active prototype, it does nothing. In both active and semi-active contexts, deriving control schemes from some form of linear optimal control theory has become popular.

The first applications of linear optimal control theory to suspension design were possibly those of Paul and Bender [19] and Bender [3]. A single mass system was optimised for combinations of ride comfort and suspension stroke, determined by a weighting parameter, with a free system structure. As far as is known, such complete optimisations cannot be calculated for systems of greater complexity. With such free structure optimisation, it may also transpire that the optimal system is physically unrealisable, as in Bender's case where further work on sub-optimal but practically feasible arrangements was carried out. Paul and Bender also treated a two mass system and obtained optimal control laws through reference to the theory of the stochastically disturbed linear output regulator with quadratic cost function, the linear quadratic Gaussian problem.

Thompson [25] has done much to popularise this approach, using the quarter car model to study performance and design issues. Many variations and elaborations of Thompson's basic study have followed and it is with explaining the range of these studies in physical terms, with an outline of the relevant mathematics, that this paper is concerned.

Notation

$A, A_c, A_1, B, B_1, C, G, G_1, M, P, P_1, Q, R, V, W, W_o, W_c$; matrices defined in text

D; time delay between front and rear inputs

E; expected value of

J; cost index

K, K_l; gain matrices with state estimator and without for limited measurements

S; mean square spectral density function

T_1, T_2, T_3, T_4; time constants

U; vehicle speed (constant)

b; vehicle half track width

c; road roughness constant

e; road displacement input vector

f; frequency

g_1, g_2; disturbance input distribution vectors for front and rear wheels

k_t; tyre spring stiffness

m, n, p, q; number of measurements, states, control inputs and controlled variables, respectively

m_u, m_s; unsprung and sprung masses

q_1, q_2; weightings on tyre load variation and suspension working space

r; index of road elevation spectral density

s; Laplace variable

t; time

u; control

v, w; random noise signals associated with measurements and road surface

x, x_1; system states

$\hat{x}$; estimated states

x_r, x_u, x_s; vertical displacements of road, unsprung mass and sprung mass

x_{ud}, x_{sd}; vertical velocities of unsprung and sprung masses

x_{rm}, x_{rp}, x_{rs}; mid-track, port and starboard side road elevations

y; measured variables

z; controlled variables
z_u; $(x_u - x_r)$
z_s; $(x_s - x_r)$
z_{ud}, x_{ud}
z_{sd}; x_{sd}
δ; Dirac delta function
∂; partial derivative
ν; wavenumber
ν_1; particular wavenumber below which road spectrum levels off
ξ; general white noise process
ξ_1, ξ_2; white noise inputs for mid-track road elevation and cross-level, respectively
θ_r; transverse road slope signal at mid-track
τ; time (different from t)
τ_1, τ_2; time constants for generation of two track road

2 The Theory of the Linear Quadratic Gaussian Regulator

The problem concerns an output regulator with linear elements subject to disturbance by a white noise excitation or by a number of uncorrelated white noise excitations. The function of the regulator is to minimise some response or responses to the disturbance, the optimal control being that feedback control law which achieves such a minimisation. With full state feedback available, the system state and output equations are [1,2,4,13,28]:

$$\dot{x} = A_x + Bu + Gw \quad \text{and} \quad z = C_x,$$

in which there are n state variables x, p control inputs u, q controlled variables z and the white noise disturbances w satisfying

$$E[w(t)w^T(\tau)] = W\delta(t - \tau),$$

in which E means "expected value of", δ is the Dirac delta function and W is a non-negative definite symmetric matrix.

A performance index or cost function of quadratic form is used to gauge the performance of the regulator. The necessary form of the cost function implies that squares of states or controls, products of states or controls and/or products of states with controls can appear in it with weighting constants chosen to represent the relative importance of the separate terms. Mathematically, the cost function is:

$$J = Lim_{t\to\infty} E[z^T(t)Qz(t) + u(t)^T Ru(t)],$$

in which Q and R are positive definite symmetric matrices.

The optimal control is a set of constant gains which relates states to controls in such a way that the cost function is minimised and it can be found via the non-negative definite solution for P of the steady state matrix-Riccati equation,

$$A^T P + PA + C^T QC - PBR^{-1}B^T P = 0. \tag{2.1}$$

This is a set of non-linear simultaneous algebraic equations for which standard methods of solution exist [1,2,4,12,13], provided that; (a) all system states are measurable; (b) the variances of all states can be reduced by the control available; and (c) those system modes which need stabilising via feedback contribute to the performance index. Condition (b) is associated with "controllability" and (c) with "observability". A mode is unobservable if it makes no contribution to the cost function, a consequence of which is that the process for the minimisation of the cost function contains no mechanism for ensuring the stability of the mode. The optimal control is given by:

$$U = -R^{-1}B^T Px,$$

with P from the Riccati equation.

Full state feedback optimal systems are not only stable in an absolute sense but have useful stability margins [14]. Small departures from the assumptions contained in the theory can therefore be expected not to spoil the conclusions reached. By choosing different forms for the cost function and different weighting constants, many "optimal" laws can be found. They should be viewed in this context and not thought of as being the best obtainable in any absolute sense.

Should all the states of the regulator not be measurable, either of two approaches is possible. The first is to employ a state estimator or Kalman filter to reconstruct states not actually measured, while the second is to use a different form of linear stochastic optimal control theory which does not demand full state information.

With the Kalman filter approach, measurement noise on each channel has to be "white" and its intensity has to be specified. Then, based on the idea that the system structure and parameters are known, an optimal reconstruction of all the states is made and the estimated states are utilised for feedback. Complete controllability and observability are necessary system attributes for the technique to be operable, observability in this case requiring that the measured variables, when fed to the system model, are such that responses in all the states to be estimated occur. The optimisation technique is substantially the same as in the full state feedback case [1,4,12,13].

In this case, the process is described by:

$$\dot{x} = Ax + Bu + Gw \text{ as before,}$$

the measurements are described by:

$$Y = Mx + v$$

with y and v of dimension m and

$$E[w(t)] = 0; E[v(t)] = 0; E[w(t)w^T(\tau)] = W\delta(t-\tau); E[v(t)v^T(\tau)]$$
$$= V\delta(t-\tau) \text{ and } E[v(t)w^T(\tau)] = 0.$$

The optimal control is:

$$u = -R^{-1}B^T P\hat{x},$$

P being derived from the steady state Riccati equation (2.1). This implies, of course, that the control gains are the same as in the full state feedback case but the gains act on the state estimates rather than on the states themselves.

The optimal state estimator is:

$$\dot{\hat{x}} = A\hat{x} + Bu + K(y - M\hat{x}), \text{ with } K = P_1 M^T V^{-1}$$

and with P_1 coming from the steady state Riccati equation;

$$AP_1 + P_1A^T - P_1M^TV^{-1}MP_1 + GWG^T = 0.$$

The continuous operation of the state estimator is illustrated in Fig.1.

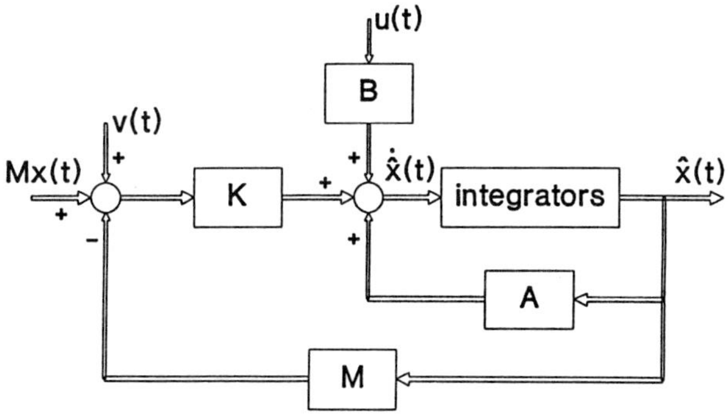

Figure 1. Illustration of the State Estimation Process

The optimal control depends on the levels of measurement noise assumed and if these are simply estimated, good practice is to repeat the calculations for different estimates so that results obtained are not unreasonably specific and restricted [30]. For very low measurement noise, the Kalman filter modified system model acts in such a way that the effects of initial state estimation errors decay very rapidly. On the other hand, as the noise levels increase, the Kalman filter gains, K, decrease towards zero, corresponding to the idea that the measurements contain less and less proper information and that, in the limit, no heed should be paid to them. Optimal solutions guarantee stability of the closed loop system [30].

In the second limited state feedback case, let the control law be:

$$u = -K_l y,$$

where K_l is a constant p by m matrix and the m measurements y are related to the states x by

$$y = Mx.$$

In this case, only controllable and observable modes will be stabilizable by the control. The other modes must be stable for a sensible optimisation. In contrast to the previous cases, the stability of the closed loop system, described by

$$\dot{x} = A_c x + Gw, \text{ where } A_c = A - BK_l M,$$

cannot be guaranteed and it is advisable to check it on a regular basis as the iterative optimisation process is carried out [28].

The cost function is given by:

$$J = trace[W_c(C^T QC + M^T K_l^T RK_l M)] = trace[W_0 W],$$

where

$$A_c W_c + W_c A_c^T + GWG^T = 0 \text{ and}$$

$$A_c^T W_0 + W_o A_c + C^T QC + M^T K_l^T RK_l M = 0$$

are Liapunov equations, giving W_c and W_0 [8].

From an initially assumed control gain matrix K_l, the cost can, in general, be reduced by a gradient descent method [20], since the cost function sensitivity matrix is given by [28]:

$$\partial J/\partial K_l = 2[RK_l MW_c M^T - B^T W_0 W_c M^T].$$

It can be argued that the Kalman filter approach to the availability of only limited state information represents the better option of the two, making the more comprehensive use of the measurements made but it may be case dependent, since full system designs and performance evaluations are likely to be needed before sure judgements can be made.

3 Application to Vehicle Suspension Problems

3.1 Systems having a single input

The classical single input case is the quarter car problem. The model is of one corner of a four wheeled vehicle or one end of a motorcycle with vertical degrees of freedom only, one for the wheel and one for the body. The vertical performance of a rolling tyre is similar to that of a spring, which is linear so long as the tyre load is greater than a small value of the order of 10% of the static load [5]. The disturbance input to the model arises from road roughness and consists of a prescribed vertical displacement of the base of the tyre spring. Measurements of undamaged road surfaces indicate that, on average, they have displacement spectral density functions of the form:

$$S(v) = c\nu^{-r}, 0.01 < \nu < 1,$$

in which ν is the wavenumber or spatial frequency (in cycle/m), c is a roughness constant and r has a value between 2 and 3 [22]. For a vehicle travelling over such a surface at constant speed U, the disturbance spectrum transforms to;

$$S(f) = cU^{(r-1)}f^{-r},$$

in which f is frequency [23]. A realistic aim for the suspension system is to minimise a cost function which includes weighted contributions from body acceleration, indicating discomfort, wheel to body relative displacement, indicating the demand for working space and wheel to road relative displacement, indicating variations from the static value of the tyre load. Depending on the road roughness, in a practical case there may be constraints on the working space, deriving from the limited wheel travel implied by mechanical design details. In its simplest arrangement, the suspension system consists of a perfect actuator applying equal and opposite forces to body and wheel without frequency response limitations but, in principle, no extra difficulty is introduced if a more realistic actuator/inner loop closure linear model replaces the perfect actuator.

In order to make the quarter car suspension problem conform to the requirements of the linear quadratic regulator theory, the car model must be augmented by a linear filter, the output of which represents the displacement input to the car and the input to which is white noise. Of course also, all other elements of the model must be linear and the optimisation cannot be directly constrained. Indirect constraints can be applied by choice of weighting constants in the performance index [23]. The filter may be an integrator, a low pass filter or some other form. Each of these implies a particular spectral form in the displacement input to the car model, the simplest case being that of the integrator, the output from which has a spectral density inversely proportional to f^2.

If the displacement input to the car model is $e(t)$, the equations for the car and the excitation process are:

$$\dot{x}_1 = A_1 x_1 + B_1 u + G_1 e \text{ and}$$

$$\dot{e} = A_2 e + G_2 w$$

and when these are combined by forming the augmented state vector $x = [x_1, e]^T$, we obtain [30]:

$$\dot{x} = Ax + Bu + Gw,$$

as required. In augmenting the system, it may have been rendered not stabilisable and/or not detectable. It may be necessary subsequently to partition the full problem into segments or to transform into new variables in order to achieve a well conditioned optimisation [9, 28].

The cost function must be of quadratic form, the optimisation involving minimising a weighted sum of squares of body acceleration, wheel to body displacement and wheel to ground force. If desired, the body acceleration can be fed to another linear filter with characteristics chosen to represent the sensitivities of humans to vertical vibrations of different frequencies and the output of this filter can be included in the performance index in preference to the body acceleration [7]. Also, other terms of quadratic form representing quantities like actuator force or mechanical work can be included in the cost function. The system in this form is depicted in Fig. 2.

Omitting the optional vibration sensitivity filter, the discomfort contribution to the cost function is contained within that representing the control effort, since the body acceleration and the control force are in proportion. If the spectral shape filter is a first order lag with time constant T_1, the state equations can be written

$$\begin{bmatrix} \dot{x}_r \\ \dot{x}_u \\ \dot{x}_s \\ \dot{x}_{ud} \\ \dot{x}_{sd} \end{bmatrix} = \begin{bmatrix} -1/T_1 & 0 & 0 & 0 & 0 \\ 0 & 0 & 0 & 1 & 0 \\ 0 & 0 & 0 & 0 & 1 \\ k_t/m_u & -k_t/m_u & 0 & 0 & 0 \\ 0 & 0 & 0 & 0 & 0 \end{bmatrix} \begin{bmatrix} x_r \\ x_u \\ x_s \\ x_{ud} \\ x_{sd} \end{bmatrix} + \begin{bmatrix} 0 \\ 0 \\ 0 \\ -1/m_u \\ -1/m_s \end{bmatrix} u$$

$$+\begin{bmatrix} 1/T_1 \\ 0 \\ 0 \\ 0 \\ 0 \end{bmatrix} \xi \tag{3.1}$$

where $1/T_1$ is $2\pi U \nu_1$, with ν_1 the wavenumber in cycle/m below which the road elevation spectrum levels off and ξ is the white noise disturbance input. The A, B and G matrices are clear, while C, Q and R are given below for a cost index containing tyre load variation, working space utilisation and actuator force.

$$C = \begin{bmatrix} -K_t & K_t & 0 & 0 & 0 \\ 0 & -1 & 1 & 0 & 0 \end{bmatrix} Q = \begin{bmatrix} q_1 & 0 \\ 0 & q_2 \end{bmatrix} R = [1] \tag{3.2}$$

If $T_1 \rightarrow \infty, \nu_1 \rightarrow 0$ and the shape filter becomes an integrator. In neither case can the control influence the output of the filter. In the latter case, the system has a zero eigenvalue associated with the integrator and the corresponding mode is neutrally stable. The problem mode can be separated from the remainder of the system by a variable transformation, $z_u = (x_u - x_r), z_{ud} = x_{ud}, z_s = (x_s - x_r)$ and $z_{sd} = x_{sd}$, which causes the neutrally stable mode to become decoupled. Then the optimal control,

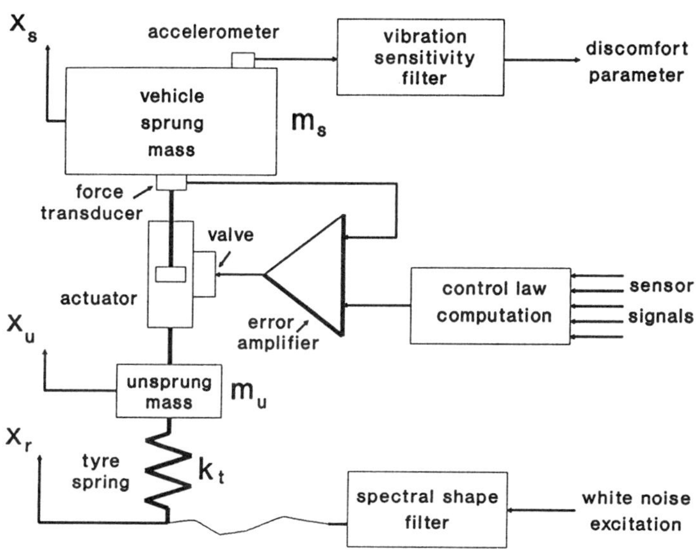

Figure 2. Quarter Car Model Augmented for Application of Linear Stochastic Optimal Regulator Theory

involving only four gains, can be found from the well behaved problem segment [28]. With $\nu_1 \neq 0$, the filter is stable, the optimal control contains five gains and no transformation of variables is needed [9, 29]. However, the uncontrollability of the system demands that special numerical procedures are employed to find the optimal control [9], although, curiously, the difficulty is not always referred to [29].

3.2 Systems with more than one input

If the different inputs to a multi-wheeled vehicle are uncorrelated, the procedures described in section 3.1 above are adequate for the generation of control laws. Normally, however, the inputs will be correlated and the correlation is likely to take either or both of two forms. Firstly, port and starboard inputs to a two track vehicle will be coherent for long wavelengths and incoherent for short wavelengths. Secondly, inputs to rear wheels will be time delayed versions of those at the front.

The dual track problem can be solved using a method due to Parkhilovskii [18], which has been developed by Rill [21]. Port and starboard disturbances are generated by combining signals which represent the vertical displacement and the transverse slope of the road at the vehicle centre line. The combining process involves the track width as a moment arm. The individual signals are obtained by linear filtering of independent white noise processes, thus allowing the regulator theory requirements to be satisfied. Rill gives the parameters of first order filters which will provide realistic vehicle inputs.

If x_{rm} represents the road elevation at mid-track, θ_r represents the cross-level (angular) excitation and b is the vehicle half-track width, port and starboard side road elevations are given by:

$$x_{rp} = x_{rm} + b\theta_r \text{ and } x_{rs} = x_{rm} - b\theta_r$$

and functions x_{rm} and θ_r are generated from:

$$\tau_1 \dot{x}_{rm} + x_{rm} = \xi_1 \text{ and } \tau_2 \dot{\theta}_r + \theta_r = \xi_2.$$

The process is illustrated by Fig. 3.

Any of three approaches to the time delay problem can be taken. The first is to ignore the front to rear correlation initially in establishing a sub-optimal state feedback control law and then to improve the law through an iterative procedure. Thompson and Pearce [26] gave an expression for the performance index in the half car problem, including the contribution of the time delay between front and rear;

$$J = cU.trace(G^T PG + 2g_1 g_2^T Pe^{[A_c]D}),$$

in which G is $[g_1, g_2]$, with g_1 and g_2 the input distribution vectors for the front and rear respectively and with time delay D between the two inputs.

This could be utilised to obtain successive improvements of the state feedback control or the improvement could be based on response calculations. Since a time delay is representable in the frequency domain by a phase lag which increases linearly with frequency, responses in that domain are easy to calculate. Corresponding time domain computations are also possible [10]. The quadratic form of the cost function and/or the road elevation spectrum could be varied, if desired, in these calculations, since regulator theory is not being referred to. Strictly, the resulting control would remain sub-optimal since only state feedback is allowed by virtue of the problem definition. Simplex [17] or Hooke and Jeeves [24] procedures are possible aids to updating the control gains. No new problems associated with the implementation in a vehicle of the control system design to account for input correlations are implied in this case.

The second approach is to represent the time delay by Pade approximations [7,11]. Again the laws derived are sub-optimal because of the imperfect representation of the time delays but they may be good for practical purposes. Corresponding to the Pade approximations used in the theory, Pade filters would be built into a practical system and the states of the filters would be utilised for control purposes. The nature and function of a fourth order Pade filter are illustrated in Fig. 4.

The third possibility is to refer to the theory of the linear quadratic regulator in discrete time form [12,13], making the sampling interval an integer submultiple of the time delay. The vehicle suspension problem can then be made to fit the regulator framework [15] and a true optimal

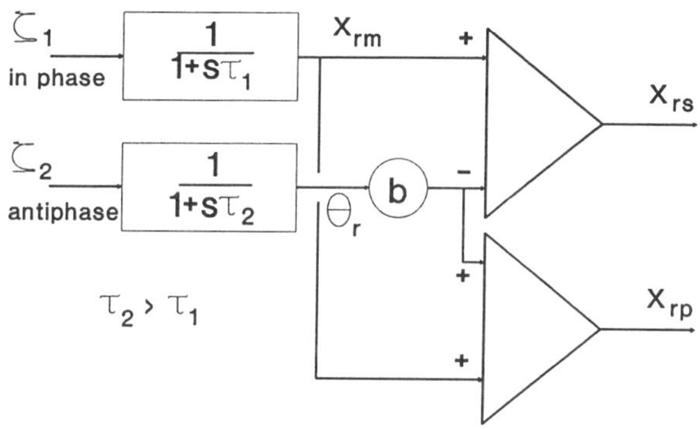

Figure 3. Generation of Port and Starboard Road Tracks from White Noise Signals

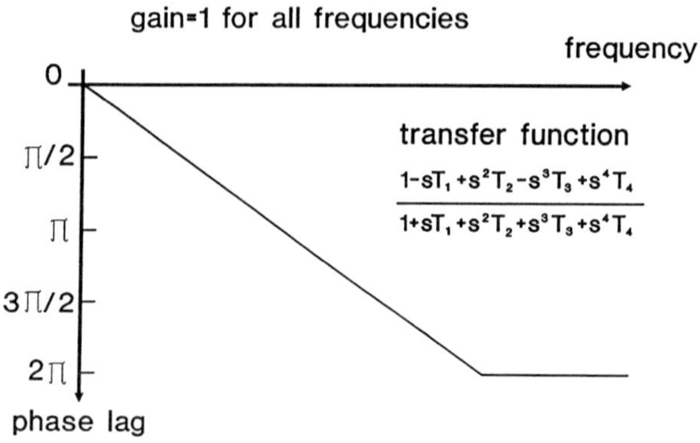

Figure 4. Illustration of a Fourth Order Pade Function

solution can be obtained. The full practical implications of this optimal strategy have not yet been worked out but the control involves the addition of feedforward elements to the same feedback as is derived if the front/rear correlation is ignored. Questions concern signal processing requirements and the consequences of vehicle speed variations but it has been established that the front state information is of considerable value in improving the rear suspension performance.

Dealing with the time delay between inputs to front and rear wheels naturally raises interest in using road surface preview information in some optimal fashion. Tomizuka [27] solved the optimal control problem in principle for a single mass model in discrete time, while Fruhauf et al [7] used the Pade approximation approach to represent the delays and a Kalman filter to estimate unmeasured states. Recently, Louam [16] has found a way of utilising optimal linear tracking theory to solve the suspension optimisation problem when road surface preview is available. Foag [6] employed a multi-criterion optimisation scheme which is capable of accommodating pure time delays but this scheme is not based on linear optimal control theory.

State estimation by Kalman filtering depends notionally on possessing a perfect system model but it is clear that in a real vehicle application, the possession of an accurate model is non-trivial. Vehicle dynamic properties will vary with load carried, tyre pressures etc. and the need arises to establish the consequences of modelling errors or to devise a scheme for regular and accurate updating of the system model used in the state estimation, in order to establish confidence in the practical viability of this strategy.

4 Conclusion

Linear optimal control theory provides a convenient and effective base for the computation of automotive suspension control laws which allow the assessment of performance capabilities and the design of real systems. Application of the theory to the suspension case requires that the vehicle problem be massaged to one of a number of standard formats, following which standard procedures can be utilised. Port/starboard and front/rear correlations in the vehicle situation pose special difficulties in terms of optimisation and the ways in which the difficulties can be overcome have been described. Recent work on the optimal use of road surface preview information has been mentioned. Sometimes of necessity and sometimes to economise on measurements, state estimation is involved. Performance losses due to imperfections in state measurements and in the system models used in the estimation processes are of interest in the context.

References

1. Anderson, B.D.O. and Moore, J.B., (1971), "Linear optimal control", Prentice-Hall, Englewood Cliffs NJ.

2. Athans, M. and Falb, P.L., (1966), "Optimal control: An introduction to the theory and its applications", McGraw-Hill, New York.

3. Bender, E.K., (1968), Optimal linear preview control with application to vehicle suspension, Trans, A.S.M.E. Journal of Basic Engineering, 90(2), 213-221.

4. Bryson, A.E. and Ho, Y.C., (1975), "Applied optimal control, optimisation, estimation and control", Wiley, New York.

5. Clark, S.K., (1981), "Mechanics of pneumatic tires (rev. ed.)", N.B.S. Monograph 122, Washington D.C.

6. Foag, W., (1988), A practical control concept for passenger car active suspensions with preview, Advanced Suspensions, Mechanical Engineering Publications, Bury St Edmunds, 43-50.

7. Fruhauf, F., Kasper, R. and Luckel, J., (1986), Design of an active suspension for a passenger vehicle model using input processes with time delays, Proc. 9th I.A.V.S.D. Symposium on the Dynamics of Vehicles on Roads and Tracks (ed. O. Nordstrom), Swets and Zeitlinger, Lisse, 126-138.

8. Golub, G.H., Nash, S. and Van Loan, C., (1979), A Hessenburg-Schur method for the problem $AX + XB = C$, I.E.E.E. Transactions on Automatic Control, AC-24(6), 909-913.

9. Hac, A., (1985), Suspension optimisation of a 2-DOF vehicle model using a stochastic optimal control technique, Journal of Sound and Vibration, 100(3), 343-357.

10. Hedrick, J.K. and Firouztash, H., (1974), The covariance propagation equation including time delayed inputs, I.E.E.E. Transactions on Automatic Control, AC-19(5), 587-589.

11. Houpis, C.H. and Lamont, G.B., (1985), "Digital control systems", McGraw-Hill, New York.

12. Kortum, W., (1979), Computational techniques in optimal state estimation-a tutorial review, Trans. A.S.M.E. Journal of Dynamic Systems, Measurement and Control, 101(2), 99-107.

13. Kwakernaak, H. and Sivan, R., (1972), "Linear optimal control systems", Wiley Interscience, New York.

14. Lehtomaki, N.A., Sandell, N.R. and Athans, M., (1981), Robustness results in linear quadratic Gaussian based multivariable control systems, I.E.E.E. Transactions on Automatic Control, AC-26(1), 75-93.

15. Louam, N., Wilson, D.A. and Sharp, R.S., (1988), Optimal control of a vehicle suspension incorporating the time delay between front and rear wheel inputs, Vehicle System Dynamics, 17, 317-336.

16. Louam, N., (1989), Private communication.

17. Nelder, J.A. and Mead, R., (1965), A simplex method for function minimisation, Computer Journal, 7, 308-313.

18. Parkhilovskii, I.G., (1968), Investigations of the probability characteristics of the surfaces of distributed types of roads, Avtom. Prom., 8, 18-22.

19. Paul, I.L. and Bender, E.K., (1966), Active vibration isolation and active vehicle suspension, MIT report PB-173-648.

20. Press, W.H., Flannery, B.P., Teukolsky, S.A. and Vetterling, W.T., (1986), "Numerical recipes; the art of scientific computing", Cambridge University Press, Cambridge.

21. Rill, G., (1984), The influence of correlated random road excitation processes on vehicle vibration, Proc. 8th I.A.V.S.D. Symposium on The Dynamics of Vehicles on Roads and Tracks (ed. J.K. Hedrick), Swets and Zeitlinger, Lisse, 449-459.

22. Robson, J.D., (1979), Road surface description and vehicle response, International Journal of Vehicle Design, 1, 25-35.

23. Sharp, R.S. and Hassan, S.A., (1986), The relative performance capabilities of passive, active and semi-active car suspension systems, Proc. I. Mech. E., 200(D3), 219-228.

24. Siddall, J.N., (1982), "Optimal engineering design", Marcel Dekker, New York.

25. Thompson, A.G., (1976), An active suspension with optimal linear state feedback, Vehicle System Dynamics, 5, 187-203.

26. Thompson, A.G. and Pearce, C.E.M., (1979), An optimal suspension for an automobile on a random road, Society of Automotive Engineers paper 790478.

27. Tomizuka, M., (1976), Optimal linear preview control with application to vehicle suspension-revisited, Trans. A.S.M.E. Journal of Dynamic Systems, Measurement and Control, 98(3), 309-315.

28. Wilson, D.A., Sharp, R.S. and Hassan, S.A., (1986), The application of linear optimal control theory to the design of active automotive suspensions, Vehicle System Dynamics, 15, 105-118.

29. Yoshimura, T., Ananthanarayana, N. and Deepak, D., (1986), An active vertical suspension for track/vehicle systems, Journal of Sound and Vibration, 106(2), 217-225.

30. Yue, C., Butsuen, T. and Hedrick, J.K., (1989), Alternative control laws for automotive active suspensions, Trans. A.S.M.E. Journal of Dynamic Systems, Measurement and Control, 111(2), 286-291.

Powertrain Dynamics and Control

R.P. Jones

University of Warwick

Abstract This paper discusses the role of dynamic simulation and nonlinear optimisation in relation to the dynamics and control of energy flow within the powertrain of an automotive vehicle. These concepts are presented with reference to a mathematical model of an automotive powertrain incorporating a CVT. The mathematical model represents the energy flow between the principal rotational elements of the powertrain, together with other interaction effects associated with the engine mounts and driveline compliance. Dynamic simulation techniques are illustrated with this model through consideration of a simulation experiment involving the response of the CVT powertrain to a step reduction in demanded ratio. Nonlinear optimisation techniques are illustrated with a reduced order model of the powertrain through the computation of a minimum fuel schedule for engine torque and gear ratio. The latter involves a nonlinear optimal control problem with state and control inequality constraints.

1 Introduction

An important aspect of control engineering entails the use of mathematical modelling techniques in the selection and assessment of control strategies. These techniques can be conveniently divided into the three categories:

- dynamic simulation,
- feedback analysis,
- nonlinear optimisation.

Dynamic simulation relates to the use of nonlinear dynamic models to predict the effect of changes in system parameters or control strategy on transient performance. Feedback analysis is primarily concerned with control system stability under closed loop feedback and involves the use of linear dynamic models and methods of linear systems theory. Nonlinear optimisation relates to the computation of open loop control schedules or parameters of closed loop control laws which maximise or minimise some measure of the performance of a dynamic system.

This paper discusses the role of dynamic simulation and nonlinear optimisation in relation to the dynamics and control of energy flow within the powertrain of an automotive vehicle. The dynamic effects associated with powertrain control are highly nonlinear and embrace the generation of power within the engine; the delivery of power to the drive wheels via the transmission; and the resulting effect on the longitudinal motion of the vehicle. This paper discusses these effects relative to the central role played by the transmission.

The role of the transmission is to enable the power produced by the engine to be matched to the varying power demands of the vehicle as its speed and acceleration requirements change. This is achieved within the gearbox by varying the value of gear ratio r, which effectively results in a conversion between torque and speed. Note that, in this paper, r is defined as the ratio of output shaft speed over input shaft speed.

Conventional automatic and manually operated gearboxes provide a discrete variation in gear ratio, i.e.

$$r \epsilon \{r_1, r_2, \dots, r_n\} \tag{1.1}$$

where, typically, $n = 4$ in a car, and $n = 15$ in a heavy goods vehicle (HGV). An alternative approach is to employ an advanced form of gearbox, viz. a continuously variable transmission (CVT), capable of providing a continuous variation in gear ratio within an upper and lower limit. With a CVT

$$r \epsilon [r^{min}, r^{max}] \tag{1.2}$$

where, typically, $[r^{min}, r^{max}] = [0, 2]$. Note that it is possible for the minimum gear ratio of a CVT to be zero, an effect which is called geared neutral.

The dynamic simulation and nonlinear optimisation concepts presented in this paper are illustrated with reference to the dynamics and control of an automotive powertrain incorporating a CVT [1,2]. However, it should be emphasised that this CVT study has been chosen for illustrative purposes only, and that the mathematical modelling concepts presented are equally applicable to other automotive powertrain configurations [3].

2 Powertrain Model

2.1 Aim and Scope

The dynamic powertrain model considered in this paper in centred on a CVT and is described conceptually in figure 1. The model represents the dynamics of energy flow between the principal rotational elements of the powertrain, together with other interaction effects with the engine mounts

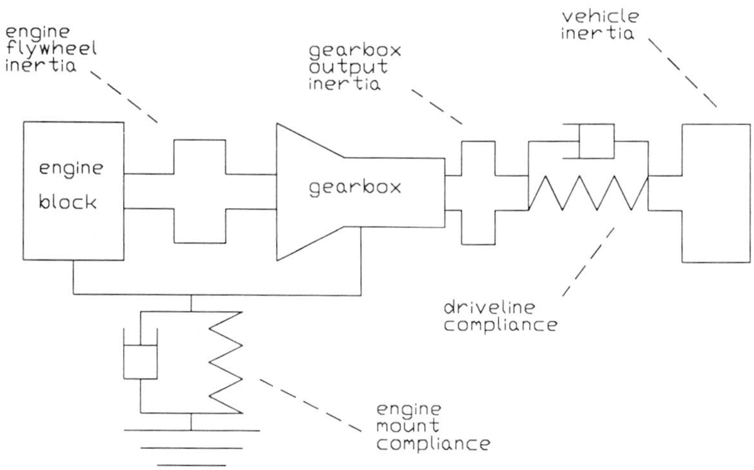

Figure 1. Conceptual model of powertrain

and driveline compliance, and is restricted to an upper frequency limit of 10 Hz. The dynamics associated with the movement of the engine and CVT on the engine mounts are included as a single degree of freedom rotation about the crankshaft. Dynamic engine effects associated with the combustion process and the intake of air and fuel, together with the cyclic nature of torque production, are ignored in this model. Also, the engine torque is assumed to be directly available as a control variable, together with the vehicle braking force and CVT ratio demand. The model is based on actual data corresponding to a medium size rear drive car, as presented in table 1.

2.2 Engine Power Constraints

The available engine power is represented by the inequality constraints

$$u_1^{min}(x_1) \leq u_1 \leq u_1^{max}(x_1) \tag{2.1}$$

and

$$x_1^{min} \leq x_1 \leq x_1^{max} \tag{2.2}$$

where x_1 represents the engine (flywheel) speed ($rad\ s^{-1}$), and u_1 the gross engine torque (Nm). Constraint 2.2 defines the maximum and minimum allowable engine speeds, and constraint 2.1 bounds the engine torque by a peak torque curve u_1^{max} and peak overrun torque u_1^{min}, which are both

Table 1. Parameters used in computational study

	a_1	-0.486×10^{-3}		c_1	12.4×10^{-9}
	a_2	0.308		c_2	0.359×10^{-6}
engine	a_3	86.6	fuel	c_3	78.7×10^{-9}
power	b_1	-38.1×10^{-3}	flow	c_4	-11.3×10^{-6}
constraints	b_2	-9.55	model	c_5	17.3×10^{-6}
	x_1^{min}	78.5		c_6	5.15×10^{-3}
	x_1^{max}	0.576×10^3		c_7	-0.207
	J_1	0.159	CVT	u_2^{min}	0.0
CVT	$\hat{J}_1$	0.05	ratio	u_2^{max}	2.0
power	k_{31}	0.237×10^3	actuation	T_2	0.05 or 1.0
transmission	d_{31}	17.3			
dynamics	r_d	0.275		M_3	1.34×10^3
	r_w	0.28	vehicle	a	0.206×10^3
engine	J_5	3.78	longitudinal	b	0.0
block	$\hat{J}_5$	1.2	dynamics	c	0.37
dynamics	k_{60}	9.0×10^3		μ	0.7
	d_{60}	47.4			

dependent on engine speed. The peak torque and peak overrun torque curves are derived as regression models from static torque and speed data obtained on an engine test bed. Specifically,

$$u_1^{max}(x_1) = a_1 x_1^2 + a_2 x_1 + a_3 \tag{2.3}$$

and

$$u_1^{min}(x_1) = b_1 x_1 + b_2 \tag{2.4}$$

as described in figure 2.

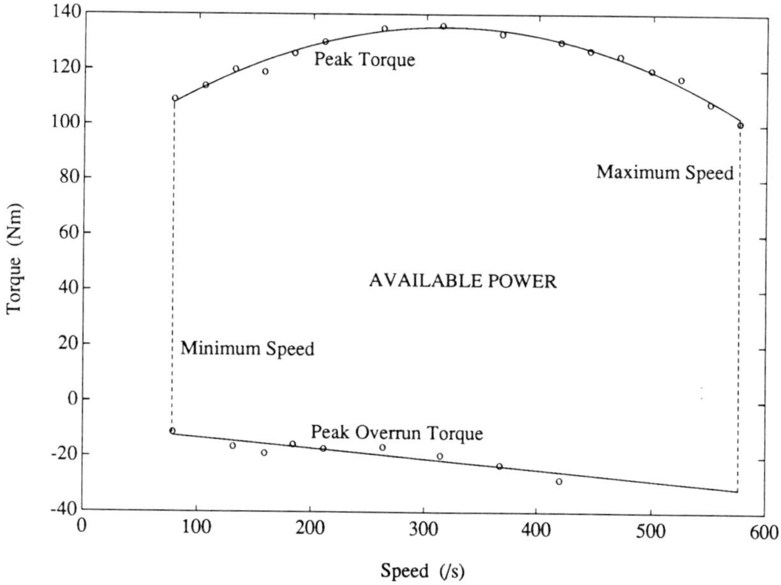

Figure 2. Engine power constraints

2.3 CVT Ratio Actuation

The CVT ratio actuation dynamics are represented by the 1st order lag

$$T_2 \dot{x}_2(t) = -x_2(t) + u_2(t) \tag{2.5}$$

where x_2 represents the CVT (gear) ratio, u_2 the demanded ratio, and T_2 is the CVT ratio change time constant. The CVT ratio is constrained by

$$0 \leq u_2^{min} \leq u_2 \leq u_2^{max} \tag{2.6}$$

where u_2^{max} and u_2^{min} are the maximum and minimum CVT ratios, respectively.

2.4 CVT Power Transmission Dynamics

The dynamics of the CVT power transmission elements are represented by

$$\begin{aligned}[J_1 + \hat{J}_1 x_2^2(t)]\dot{x}_1(t) &= u_1(t) - \frac{\hat{J}_1}{T_2} x_2(t)x_1(t)[-x_2(t) + u_2(t)] \\ &-x_2(t)\{k_{31}x_4(t) + d_{31}[x_2(t)x_1(t) - \frac{1}{r_d r_w} x_3(t)]\}\end{aligned} \tag{2.7}$$

and

$$\dot{x}_4(t) = x_2(t)x_1(t) - \frac{1}{r_d r_w} x_3(t) \tag{2.8}$$

where x_3 represents the vehicle speed (ms^{-1}), x_4 the wind-up in the driveline (rad), J_1 the engine flywheel inertia, $\hat{J}_1$ the CVT output shaft inertia, d_{31} the damping rate and k_{31} the stiffness rate of the driveline, r_d the differential ratio and r_w the wheel radius. The term

$$\frac{\hat{J}_1}{T_2} x_2(t)x_1(t)[-x_2(t) + u_2(t)] \tag{2.9}$$

in equation 2.7 represents the momentum change within the rotational elements of the CVT due to (continuous) ratio variation, whereas

$$k_{31}x_4(t) + d_{31}[x_2(t)x_1(t) - \frac{1}{r_d r_w} x_3(t)] \tag{2.10}$$

represents the aggregate driveline torque at the CVT output shaft.

2.5 Vehicle Longitudinal Dynamics

The longitudinal motion of the vehicle is represented as

$$\begin{aligned}M_3\dot{x}_3(t) &= -u_3(t) - [a + bx_3(t) + cx_3^2(t)] \\ &+\frac{1}{r_d r_w}\{k_{31}x_4(t) + d_{31}[x_2(t)x_1(t) - \frac{1}{r_d r_2} x_3(t)]\}\end{aligned} \tag{2.11}$$

where u_3 represents the (wheel) brake force (N), M_3 the vehicle mass, and the coefficients a, b and c define the road load.

The wheel torque is constrained, from wheel spin considerations, by

$$0 \leq u_3 \leq \mu M_3 g \tag{2.12}$$

where μ is the braking force (adhesion) coefficient, and g represents gravitational acceleration.

2.6 Engine Block Dynamics

The dynamics of the engine block and CVT casing are represented by

$$\begin{aligned}
[J_5 + \hat{J}_5]\dot{x}_5(t) &= -u_1(t) - [k_{60}x_6(t) + d_{60}x_5(t)] \\
&+ \frac{x_2(t)-1}{J_1 + \hat{J}_1 x_2^2(t)}\Big(\frac{J_1\hat{J}_1}{T_2}x_1(t)[-x_2(t) + u_2(t)] + \hat{J}_1 x_2(t) u_1(t) \\
&+ J_1\{k_{31}x_4(t) + d_{31}[x_2(t)x_1(t) - \frac{1}{r_d r_w}x_3(t)]\}\Big)
\end{aligned} \tag{2.13}$$

$$\dot{x}_6(t) = x_5(t) \tag{2.14}$$

where x_5 and x_6 represent the rate ($rads^{-1}$) and magnitude (rad) of the engine rotation about the crankshaft, J_5 and $\hat{J}_5$ the engine block and CVT housing inertias, and d_{60} and k_{60} the damping and stiffness rates of the engine mounts, respectively.

The term

$$k_{60}x_6(t) + d_{60}x_5(t)$$

in equation 2.13 represents the engine mount effects whereas

$$\begin{aligned}
&\frac{x_2(t)-1}{J_1 + \hat{J}_1 x_2^2(t)}\Big(\frac{J_1\hat{J}_1}{T_2}x_1(t)[-x_2(t) + u_2(t)] + \hat{J}_1 x_2(t) u_1(t) \\
&\quad + J_1\{k_{31}x_4(t) + d_{31}[x_2(t)x_1(t) - \frac{1}{r_d r_w}x_3(t)]\}\Big)
\end{aligned}$$

represents the torque reaction on the CVT casing, resulting from changes in CVT ratio, and torque at the input and output shafts. The corresponding torque reaction at the engine is provided by $-u_1(t)$.

3 Dynamic Simulation

The dynamic powertrain model presented in section 2 consists of 6 coupled nonlinear differential equations in the states $x_1, \ldots, x_6$ and control inputs $u_1, \ldots, u_3$. Dynamic simulation involves the use of such models to predict the effect of changes in system parameters or control strategy on transient performance. This is achieved through solution of the nonlinear dynamic equations using numerical integration techniques. In practice, this can be carried out using a general purpose high level simulation language, e.g. ACSL [4].

The dynamic powertrain model presented in section 2 has a wide dynamic range with a frequency spectrum which includes a (approximate) component of 0.04 Hz for the vehicle mass, 0.87 Hz for the driveline, and 6.24 Hz for the engine mounts. This has implications for the choice integration algorithm since, for computational efficiency and/or accuracy, variable step or stiff algorithms may be required [5].

This section illustrates the use of dynamic simulation techniques through consideration of a simulation experiment carried out on the model of section 2. The simulation experiment was concerned with investigating the response of the CVT powertrain to a step reduction in demanded ratio. For this experiment, the CVT ratio actuation time constant was chosen to be 0.05 s, and the vehicle was initially in a steady state travelling at a speed of 17.8 ms^{-1} ($40mph$) with a CVT ratio of 2 and corresponding engine speed of 115.45 $rads^{-1}$ ($1103rpm$). A step reduction in the CVT ratio demand from 2 to 0.8 was introduced at time $0.2s$.

Figure 3 displays the CVT ratio reponse which achieves its new value af-

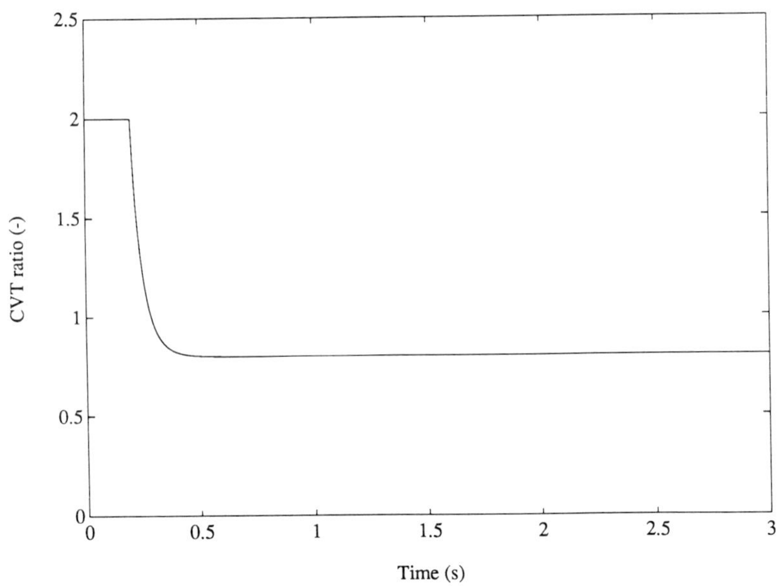

Figure 3. CVT ratio response

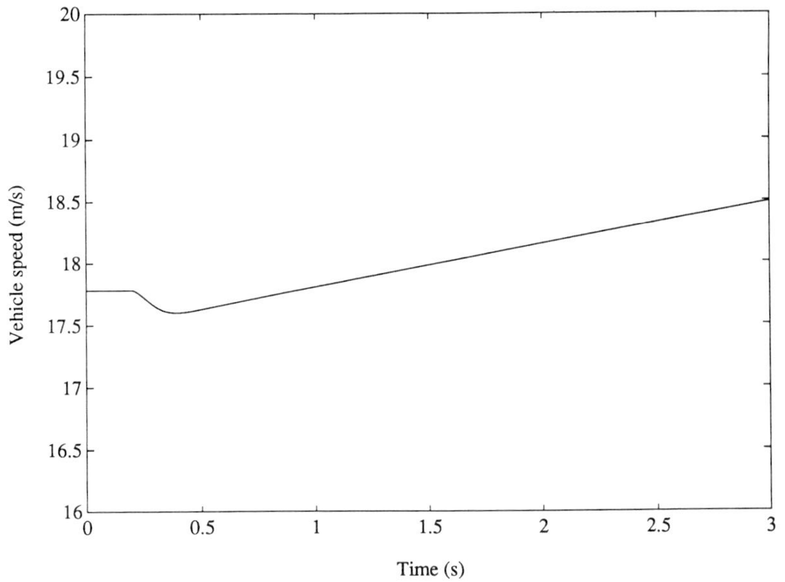

Figure 4. Response in vehicle speed

ter approximately $0.2s$. The resulting response in vehicle speed is displayed in figure 4, indicating that vehicle speed initially falls by $0.18ms^{-1}(0.4mph)$ after $0.15s$ before climbing to $18.49ms^{-1}(41.4mph)$ after $2.8s$.

Figure 5 displays the resulting response in driveline torque. This indicates that the driveline torque initially reverses sign and falls rapidly from its initial value of 24.8 Nm to a value of -131.6 Nm, before climbing to a peak of 62.6 Nm and settling at a new steady state value of 60.8 Nm after approximately $0.5s$.

The engine speed response is displayed in figure 6. This shows the engine speed rapidly rises to a value of $287.1rads^{-1}(2741rpm)$ after $0.3s$ before climbing slowly to 300.19 $rads^{-1}(2867rpm)$ after $2.8s$.

Finally, the engine mount excitation is displayed in figure 7. This shows that the engine block rotates about the crankshaft at a frequency of approximately 6 Hz with peak displacements of $0.006rad$ and $-0.014rad$, and that the initial steady displacement of $-0.003rad$ settles down to a new value of $-0.007rad$ after approximately $1.3s$.

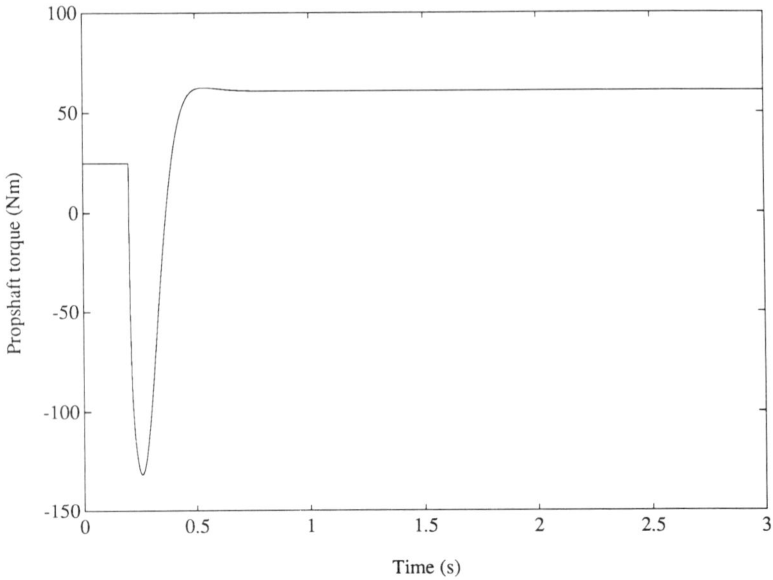

Figure 5. Response in driveline torque

4 Nonlinear Optimisation

Nonlinear optimisation relates to the computation of open loop control schedules or parameters of closed loop control laws which maximise or minimise some measure of the performance of a dynamic system. This section illustrates the use of nonlinear optimisation techniques in the scheduling of engine and transmission controllers. This is illustrated through the computation of a schedule for engine torque $u_1(t)$ and gear ratio $x_2(t)$ which minimises the fuel consumption when a CVT vehicle experiences a load demand, expressed in terms of a desired speed profile $\bar{x}_3(t)$ defined on a time interval [0,T]. The technique involves a nonlinear (fuel) optimal control problem with state and control inequality constraints [6].

Inspection of equations 2.5, 2.7, 2.8, 2.11, 2.13 and 2.14 indicates that the powertrain and engine mount dynamic subsystems are loosely coupled, in the sense that $x_1, \ldots, x_4$ effect x_5 and x_6, whereas x_5 and x_6 do not influence $x_1, \ldots, x_4$. Restricting attention to consideration of the powertrain alone, then the engine mount dynamics can clearly be ignored. If, furthermore, driveline compliance effects are neglected, i.e. setting $\dot{x}_4 \approx 0$, then the dynamic equations reduce to

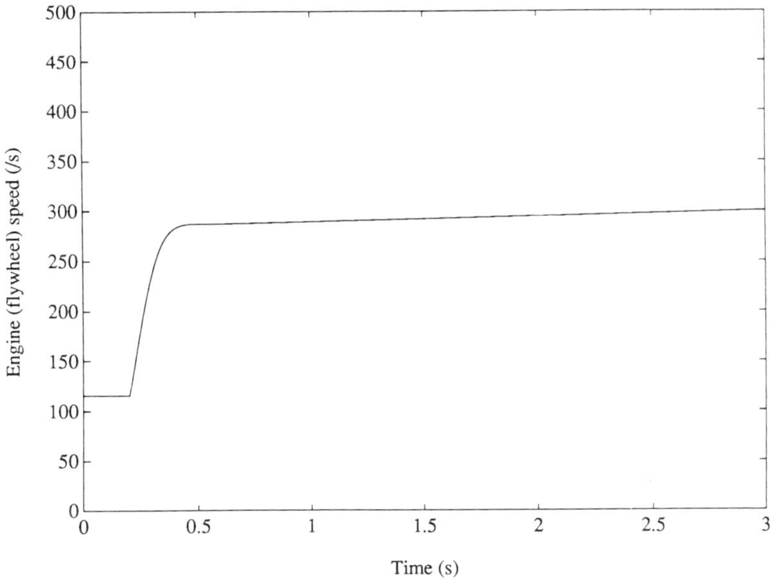

Figure 6. Response in engine speed

$$[\alpha + \beta x_2^2(t)]\dot{x}_1(t) = u_1(t) - \frac{\beta}{T_2}x_2(t)x_1(t)[-x_2(t) + u_2(t)]$$
$$-\mu x_2(t)u_3(t) - x_2(t)[\gamma + \delta x_2(t)x_1(t) + \epsilon x_2^2(t)x_1^2(t)] \quad (4.1)$$

$$T_2\dot{x}_2(t) = -x_2(t) + u_2(t) \quad (4.2)$$

$$x_3(t) = \mu x_2(t)x_1(t) \quad (4.3)$$

where $\alpha = J_1, \beta = \hat{J}_1 + \mu^2 M_3, \mu = r_d r_w, \gamma = a\mu, \delta = b\mu^2$, and $\epsilon = c\mu^3$. This approximation is valid for studies over a time scale associated with the vehicle acceleration, e.g. computation of fuel usage over a duty cycle.

The CVT powertrain model of section 2 has been reduced to the non-linear dynamic model given in equations 4.1 and 4.2, involving 2 state variables and 3 control variables. Associated with this model are the state and control inequality constraints 2.1 - 2.4, 2.6 and 2.12. Note that the constraints on u_1 are dependent on x_1, whereas the constraints on u_2, u_3 are independent of the states.

The optimal control problem considered here involves the minimisation of a cost functional in the form

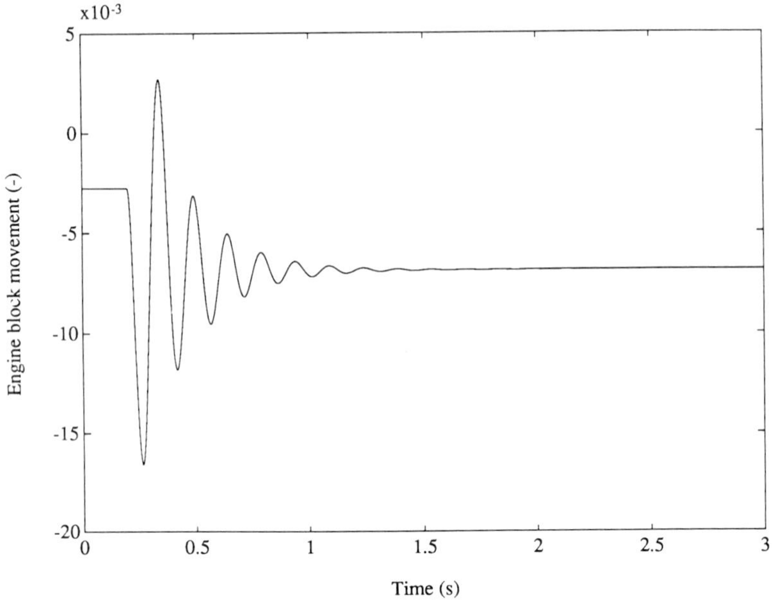

Figure 7. Response of engine block

$$J = v[\bar{x}_3(T) - x_3(T)]^2 + \frac{1}{T}\int_0^T \{\lambda[\bar{x}_3(t) - x_3(t)]^2 + f(x_1(t), u_1(t))\}dt \quad (4.4)$$

with respect to the triple $(u_1(.), u_2(.), u_3(.))$, and subject to the nonlinear state variable model 4.1 - 4.3 and state and control inequality constraints 2.1 - 2.4, 2.6 and 2.12. In this cost functional, $f(x_1, u_1)$ represents the fuel flow rate which is dependent on engine torque and speed, and the term $[\bar{x}_3 - x_3]^2$ represents the error between the desired and actual vehicle speed. The parameters λ and v provide a means of trading off the requirement to minimise fuel usage against the requirement for the vehicle speed x_3 to track the desired speed $\bar{x}_3$.

The fuel flow rate characteristics are derived as a regression model based on fuel flow, torque and speed measurements obtained on an engine test bed. Specifically, the fuel flow rate is represented as a function of engine torque and speed by the expression

$$f(x_1, u_1) = c_1x_1^3 + c_2u_1^3 + c_3x_1^2u_1 + c_4x_1^2 + c_5x_1u_1 + c_6x_1 + c_7 \quad (4.5)$$

which is described graphically in the contour plot presented in figure 8.

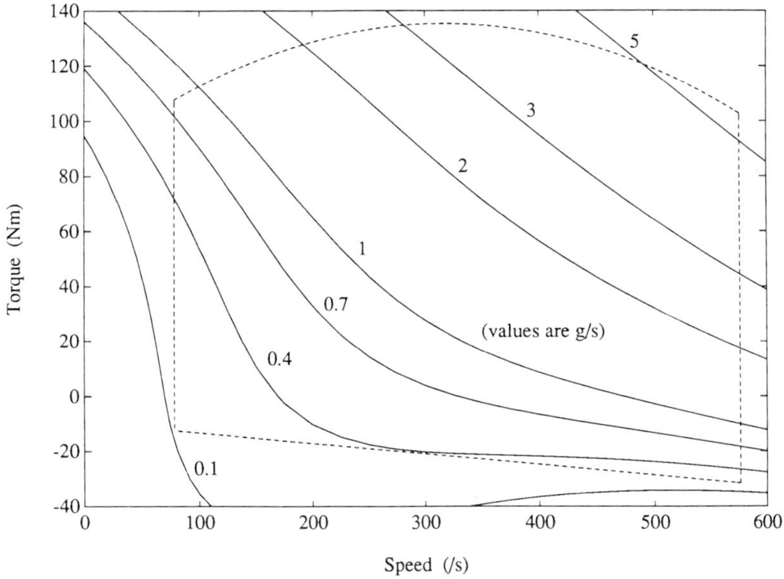

Figure 8. Fuel flow contour plot

The problem of scheduling engine torque and transmission ratio has been formulated as a nonlinear optimal control problem with state and control inequality constraints. A computational approach to this problem based on numerical solution by an iterative gradient technique is detailed elsewhere [2]. This paper illustrates the approach by discussing the results of a computational investigation, based on this method, which was carried out for the typical vehicle speed time history demand presented in figure 9. This speed profile is made up from a phase consisting of a constant acceleration from rest to $13.33ms^{-1}(48kmh^{-1})$ over $15s$ followed by a steady $20s$ period; a second phase consisting of a further constant acceleration to $22.22ms^{-1}(80kmh^{-1})$ over $15s$ followed by a steady $15s$ period; and a final phase consisting of a constant deceleration to $17.778ms^{-1}(64kmh^{-1})$ over $20s$ followed by a steady $15s$ period.

Figures 9 - 12 present the outcome of applying this optimisation algorithm for the particular choice of weighting parameters $\lambda = 30$ and $v = 10$ and with a CVT time constant of $T_2 = 1.0$. A comparison of the desired and actual vehicle speed time histories is shown in figure 9, where it can be seen that the tracking accuracy has been traded off against the (conflicting) requirement to minimise the fuel usage. The tracking error is bounded by $+1.98ms^{-1}$ and $1.57ms^{-1}$ and has a rms value of $0.64ms^{-1}$. The total quantity of fuel used in executing this optimal manoeuvre was found to be

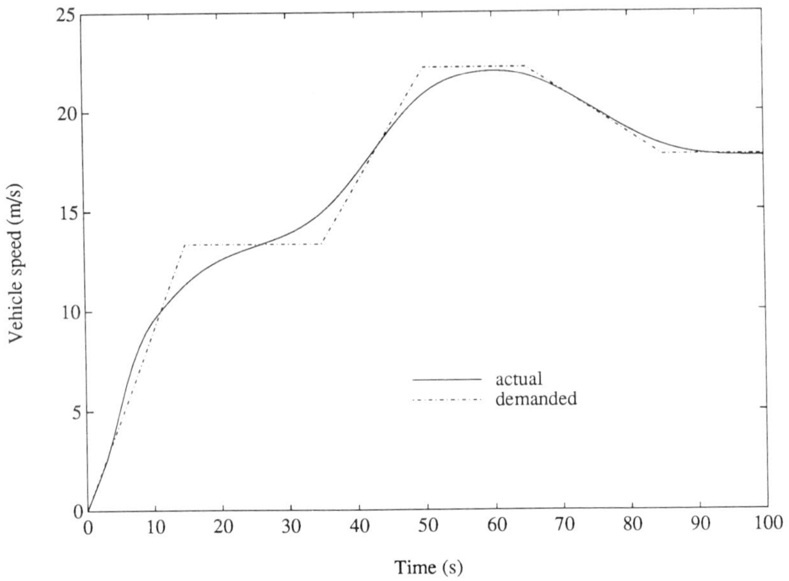

Figure 9. Tracking of vehicle speed

80.97 g.

The resulting optimal engine torque is displayed in figure 10. This shows that the engine torque climbs to values of 51.3 Nm and 61.7 Nm during periods of demanded acceleration, and falls to values of 31.7Nm, 13.4 Nm and 25.0 Nm during periods of demanded deceleration or cruise.

Figure 11 displays the optimal engine speed. This shows a rapid climb in engine speed to a value of 237.0$rads^{-1}$(2263rpm) and fall to 141.0$rads^{-1}$ (1347rpm), followed by a slow climb to 292.3$rads^{-1}$(2791rpm) and fall to 223.7$rads^{-1}$(2136rpm). The increase in engine speed corresponds to periods of demanded acceleration, and decrease to periods of demanded deceleration or cruise.

The optimal CVT ratio is given in figure 12. This shows an initial rapid climb in ratio to a value of 0.92 and slow additional climb to 1.01, followed by a fall and climb to values of 0.86 and 1.05.

5 Conclusion

This paper discusses the role of dynamic simulation and nonlinear optimisation in relation to the dynamics and control of energy flow within the powertrain of an automotive vehicle. Dynamic simulation relates to the

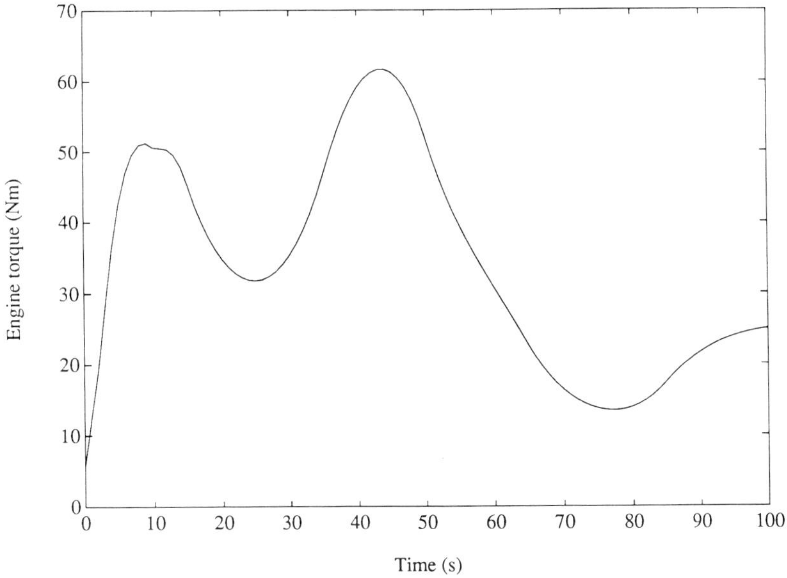

Figure 10. Optimal engine torque

use of nonlinear dynamic models to predict the effect of changes in system parameters or control strategy on transient performance. Nonlinear optimisation relates to the computation of open loop control schedules or parameters of closed loop control laws which maximise or minimise some measure of the performance of a dynamic system. These concepts are presented in this paper through consideration of a mathematical model of a powertrain incorporating a CVT.

The dynamic powertrain model considered in this paper represents the dynamics of energy flow between the principal rotational elements of the powertrain, together with other interaction effects associated with the engine mounts and driveline compliance, and is restricted to an upper frequency limit of 10 Hz. The model consists of 6 coupled nonlinear differential equations together with associated state and control inequality constraints.

Dynamic simulation techniques are illustrated with this model through consideration of a simulation experiment involving the response of the CVT powertrain to a step reduction in demanded ratio.

Nonlinear optimisation techniques are illustrated with a reduced order model of the powertrain through the computation of a minimum fuel schedule for engine torque and gear ratio. The latter involves a nonlinear optimal control problem with state and control inequality constraints.

The dynamic simulation and nonlinear optimisation concepts presented

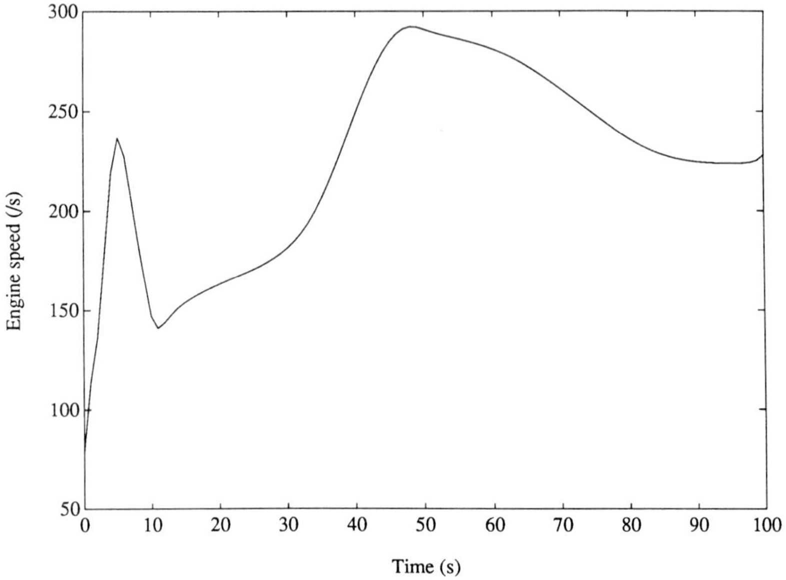

Figure 11. Optimal engine speed

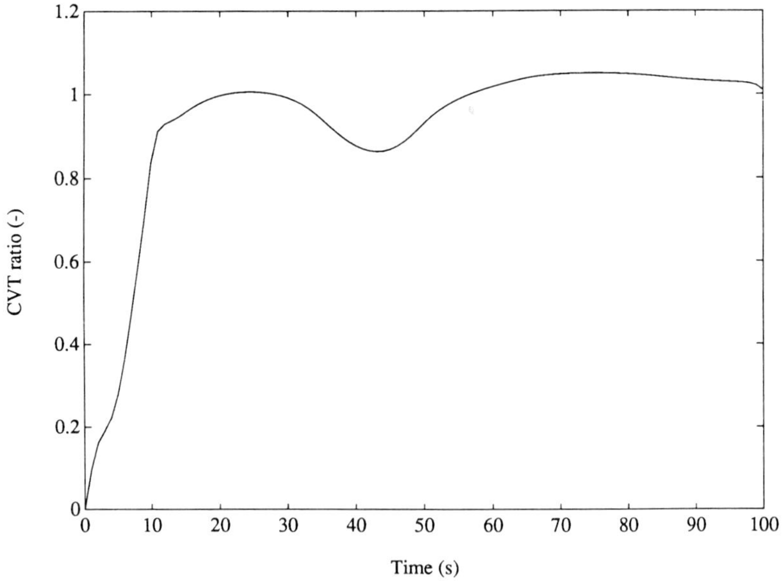

Figure 12. Optimal CVT ratio

in this paper are illustrated with reference to the dynamics and control of an automotive powertrain incorporating a CVT. However, it should be emphasised that this CVT study has been chosen for illustrative purposes only, and that the mathematical modelling concepts presented are equally applicable to other automotive powertrain configurations.

References

1. Jones, R.P. et al., (1986), 'Modelling and simulation of an automotive powertrain incorporating a Perbury continuously variable transmission', *in* 'Integrated Engine/Transmission Systems', 41-49, IMechE.

2. Jones, R.P., 'Fuel optimal control of an automotive CVT powertrain', Proc. 5th IMA Conference on Control Theory, (Eds. Nichols, N. and Owens, D.H.), Oxford University Press, To appear.

3. Crossley, P.R., Jones, R.P. and Howarth, S.I., (1985), 'Modelling and simulation of a HGV powertrain for transmission control studies', ISATA '85 Proceedings, Vol. 2, 131-150.

4. Spriet, J.A. and Vansteenkiste, G.C., (1982), 'Computer-aided Modelling and Simulation', Academic Press.

Modelling, Simulation and Control of Dynamometer Torque Loop in an Engine Test Cell

P.J. King[1], K.J. Burnham[1], D.J.G. James[1], A.J. Martyr[2], S.R. Sharpe[2]

Coventry Polytechnic[1], Froude Consine Limited[2].

Abstract In recent years industry has become more aware of the potential advantages of implementing adaptive control. Such schemes are already well established in the chemical and heating industries, where system time constants are relatively slow and extensive use is made of commercially available auto loop tuning and user interactive adaptive closed-loop schemes. Due to developments in computer technology, the self-tuning controller now offers a realistic option for many industrial systems which exhibit fast dynamics and non-linear behaviour; the IC engine testing dynamometer torque loop being a typical example of such systems.

The paper considers the development of a self-tuning control algorithm with switched controller structure appropriate for dynamometer torque control within an engine test cell and its performance is compared to that obtained using a conventional fixed PID approach. General issues arising in the implementation of self-tuning control are discussed including potential problems and methods of alleviating them.

1 Introduction

Traditionally manually tuned PI and PID controllers have found much favour in industry; with tuning following simple empirical design rules to give 'good' performance for a specific set of operating conditions or 'satisfactory' optimised performance for all foreseen normal operating conditions. It is recognised that the basic problem associated with such controllers lies in their inability to adapt to variations in plant behaviour. Self-tuning controllers, however, are able to adapt to such variations by making use of an updated mathematical model of the plant, thus providing the potential for improved plant performance and utilisation. Self-tuning schemes are believed to be appropriate due to the fact that model characteristics vary

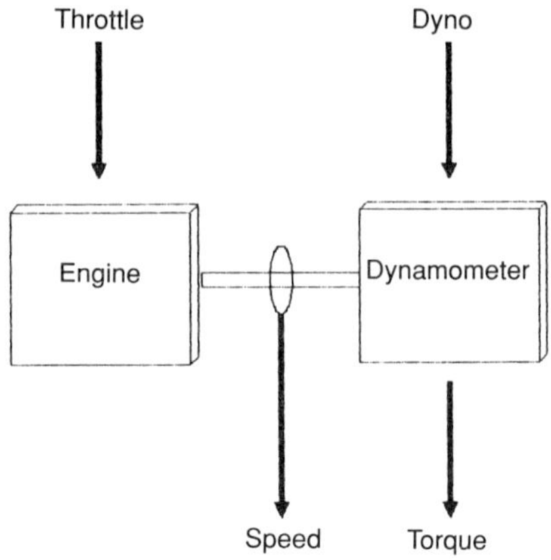

Figure 1. Schematic of an engine test cell

significantly from one engine to the next and that the useful life expectancies of the models generated are relatively short.

The dynamometer torque loop within an engine test cell is typical of many industrial systems which exhibit fast dynamics and non-linear behaviour; having speed dependent dynamics and non-linear gain characteristics. Essentially a dynamometer, of whatever generic type, is a torque absorbing device which is coupled directly to the engine via a shaft producing an interactively coupled system, which may be illustrated schematically as in Figure 1. Its function is to subject an engine to conditions which simulate, as closely as possible, those which could be realistically experienced in practice.

The engine testing dynamometer in the form of a water brake was originally designed by William Froude, and was described in his paper 'On a New Dynamometer for Measuring the Power Delivered to the Screws of Large Ships' [10], which was read before the Institution of Mechanical Engineers 25th July 1877. Froude Consine is the successor to the company he helped to found and has evolved to be world leader in the testing of automobile, aerospace and marine engines. Nowadays, demands of the market on the designers of dynamometers and their control systems are currently directed towards simulation of real-life conditions which could realistically be experienced at the engine flywheel; a feature which is only possible with high response active (four quadrant) dynamometers and their associated digital controllers.

In parallel with the obvious requirement to keep abreast of the new technology, government legislation is now demanding more stringent tests regarding fuel emissions etc., and the requirement for tighter control of the engine test is considered to be of paramount importance to give both reliable and repeatable test sequences. Consequently the need to evaluate the potential of self-tuning control schemes appropriate for the dynamometer torque loop is seen to be highly relevant.

For the purpose of this paper, the dynamometer under study is a Froude Consine EC type single disk (absorbing only) eddy-current dynamometer and a good summary of its main features, including design considerations is given in [6]. Before undertaking investigative studies to evaluate the potential of appropriate self-tuning schemes, an understanding of the mathematical model of an engine and its test cell is seen to be a clear prerequisite, and it is against this background that the present study is being conducted. The suitability of two existing self-tuning control schemes for the dynamometer torque loop are discussed and the performance of a self-tuning controller with switched controller structure is compared by simulation studies to that achieved using a conventional fixed PID approach.

2 Model Formulation

Self-tuning control schemes are characterised by two coupled sub-algorithms; one for updating an assumed discrete linear model of the plant, the other for generating the appropriate control action. It is assumed that variations in system characteristics, including effects due to non-linearities, are reflected in the time varying nature of the parameters in the assumed linear model, which are recursively estimated at each time step. The effectiveness of the control action generated will be dependent on the designers ability to match an appropriate discrete-time model to a real system, noting that some model mismatch will inevitably occur. In principle, once an appropriate model structure has been identified any analytical control law design procedure can be applied within the second sub-algorithm.

It is assumed that the system to be controlled may be modelled by a discrete linear single-input single-output difference equation or ARMAX model

$$A(q^{-1})y(t) = q^{-k}B(q^{-1})u(t) + C(q^{-1})e(t) \tag{2.1}$$

where the polynomials $A(q^{-1}), B(q^{-1})$ and $C(q^{-1})$ are defined as

$$A(q^{-1}) = 1 + a_1q^{-1} + a_2q^{-2} + \ldots + a_{n_a}q^{-n_a}$$

$$B(q^{-1}) = b_0 + b_1q^{-1} + b_2q^{-2} + \ldots + b_{n_b}q^{-n_b},\ b_0 \neq 0$$

$$C(q^{-1}) = 1 + c_1q^{-1} + c_2q^{-2} + \ldots + c_{n_c}q^{-n_c}$$

$u(t), y(t)$ and $e(t)$ are the input, output and white noise sequences respectively, $k \geq 1$ is the system time delay expressed as an integer multiple of the sampling interval, q^{-1} is the backward shift operator defined as $q^{-i}y(t) \equiv y(t-i)$ and the zeros of the polynomial $C(q^{-1})$ are taken to lie inside the unit circle.

An equivalent state-space formulation of (2.1), which is commonly used in self-tuning control [16], is represented by

$$\underline{x}(t+1) = P\underline{x}(t) + Qu(t) + Re(t) \tag{2.2}$$

$$y(t) = H\underline{x}(t) + e(t) \tag{2.3}$$

where $\underline{x}(t)\ \varepsilon\ \Re^N, N = n_j + k, n_j = \max\{n_a, n_b, n_c\}$, is the vector of state variables and the matrices P, Q, R and H are initially given in the implicit delay observable canonical form

$$P = \begin{bmatrix} 0 & \cdot & \cdot & \cdot & 0 \\ 1 & & & 0 & \cdot \\ & \cdot & & & 0 \\ & & \cdot & & -a_{n_j} \\ & & & \cdot & \cdot \\ & & & \cdot & \cdot \\ 0 & & & 1 & -a_1 \end{bmatrix} \quad Q = \begin{bmatrix} b_{n_j} \\ \cdot \\ b_0 \\ 0 \\ \cdot \\ 0 \end{bmatrix} \quad R = \begin{bmatrix} 0 \\ \cdot \\ 0 \\ c_{n_j} - a_{n_j} \\ \cdot \\ \cdot \\ c_1 - a_1 \end{bmatrix} \quad H^T = \begin{bmatrix} 0 \\ \cdot \\ \cdot \\ \cdot \\ 0 \\ 1 \end{bmatrix}$$

Remarks:

- selection of sampling interval is possibly the most crucial step, and ideally is taken to be one tenth of the fastest time constant to be identified.

- order of $B(q^{-1})$ polynomial must be chosen such that foreseen variation in system dead time can be accommodated.

- the model structure adopted is that which presents least computational burden yet adequately models the system.

- model mismatch will inevitably occur at times and *a priori* engineering knowledge of system is desirable to help minimise this mismatch. In the absence of such knowledge extensive off-line investigations may be necessary.

- it is normal to adopt a white noise model by setting $C(q^{-1}) = 1$.

- state-space framework provides a greater insight into properties such as stability, controllability, observability and closed loop dynamic behaviour. Also, in terms of controllability, potential problems arising from overparameterisation may be readily detected.

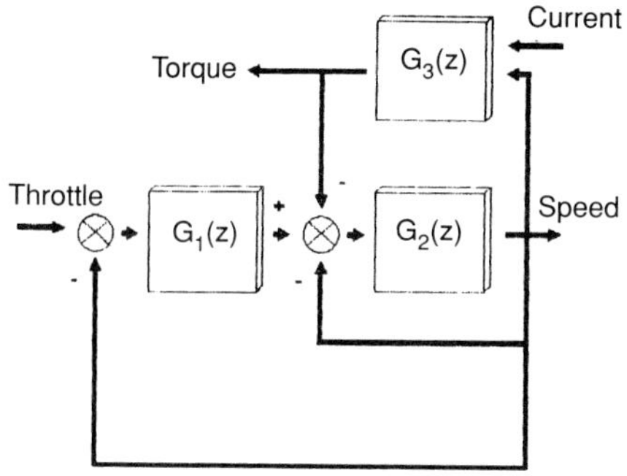

Figure 2. Adopted model structure

The model structure adopted for the dynamometer torque loop is shown in Figure 2, and is based on that adopted in linear idle-speed control [7,8]. By combining dynamic data from off-line tests with steady-state data appropriate for the eddy-current dynamometer, discrete transfer functions $G_1(z), G_2(z)$ and $G_3(z)$ have been identified. In order to accommodate varying characteristics and non-linear effects, the parameters of these transfer functions are then allowed to be time-varying in the overall ARMAX representation.

3 Parameter estimation for self-tuning control

An approach widely used in self-tuning control is that of recursive least squares (RLS). Setting the noise colouring polynomial $C(q^{-1})$ in (2.1) to unity and re-arranging leads to

$$y(t) = X^T(t)\Theta(t) + \zeta(t) \tag{3.1}$$

where $\Theta^T(t) = [-a_1 \dots - a_{n_a}; b_0 \dots b_{n_b}], X^T(t) = [y(t-1) \dots y(t - n_a); u(t-k) \dots u(t - n_b - k)]$ are the parameter and observation vectors respectively, and $\zeta(t)$ is a sequence of fitting errors which becomes equal to the noise sequence $e(t)$ upon convergence.

The standard form of the RLS algorithm for generating the estimated parameter vector $\hat{\Theta}t$ is given by

$$\hat{\Theta}(t) = \hat{\Theta}(t-1) + \Phi_1(t)[y(t) - X^T(t)\hat{\Theta}(t-1)] \tag{3.2}$$

where

$\Phi_1(t) = \Phi(t-1)X(t)/[1+X^T(t)\Phi(t-1)X(t)]$ is the gain vector, $\Phi(t) = [I - \Phi_1(t)X^T(t)]\Phi(t-1)/\lambda$ is the error covariance matrix, and λ (typically $0.8 < \lambda \leq 1$) is the forgetting factor.

Remarks:

- use of a forgetting factor allows the tracking of slowly varying parameters by placing emphasis on more recent observations; giving rise to an effective memory length of $(1-\lambda)^{-1}$. Use of a fixed forgetting factor may, during prolonged periods of steady-state operation, lead to covariance 'blow-up' and possible instability [1]. This problem may be overcome by using a variable forgetting factor (VFF) [9,18] which regulates the information content of the estimator. The value of the VFF is a function of the estimation prediction error and a lower bound must be chosen to ensure insensitivity to effects of measurement noise [2].

- covariance matrix reset provides an effective means of accommodating sudden parameter variation [3]. It has the effect of removing all *a priori* information from within the algorithm, thus facilitating fast adaptation to the new values. Reset action can be made to coincide with fault detection [12] (based on modelling errors) or coincident with set-point changes, termed logical reset [11].

- the diagonal elements of the covariance matrix, by definition, are positive and problems may occur with the basic form of RLS when parameter estimates converge and estimation errors become small. Due to the finite word storage capabilities of computers, these elements may become zero or even negative leading to numerical instability. An approach to alleviating this problem is to use square root filter techniques [14] which allows elements of Φ to be stored in square root form.

- In the case of the dynamometer, variation in non-linear gain causes rapid variations in certain parameters, and partial covariance reset is triggered whenever the value of the VFF falls below some pre-specified threshold value.

An alternative approach for generating estimates of the parameter vector $\hat{\Theta}(t \mid t)$ is to make use of the Kalman filter (KF) [19]. Essentially this is a two stage prediction/correction process,

Prediction:

$$\hat{\Theta}(t \mid t-1) = \hat{\Theta}(t-1 \mid t-1) \tag{3.3}$$

$$\Theta(t \mid t-1) = \Theta(t-1 \mid t-1) + R_w$$

Correction:

$$\hat{\Theta}(t \mid t) = \hat{\Theta}(t \mid t-1) + \Theta_2(t)[y(t) - X^T(t)\hat{\Theta}(t \mid t-1)] \qquad (3.4)$$

where

$$\Phi_2(t) = \Phi(t \mid t-1)X^T(t)/[r_v + X^T(t)\Phi(t \mid t-1)X(t)]$$

$$\Phi(t \mid t) = [I - \Phi_2(t)X(t)]\Phi(t \mid t-1)$$

R_w is the process noise covariance matrix, r_v is the output noise variance with $\Phi(t \mid t)$ and $\Phi_2(t)$ being the error covariance matrix and Kalman gain vector respectively.

Remarks:

- Note the similarity of the two approaches both in form and computational requirements: the difference being the covariance matrix update stage. In the case of RLS the covariance matrix is scaled by the forgetting factor whereas in the KF the process noise covariance matrix R_w is simply added.

- r_v, the noise variance, controls the filtering action. The larger r_v the more emphasis the algorithm places on previous parameter estimates, thus providing an element of filtering by regulating the importance of the possibly noisy measurements.

- R_w effectively allows the incorporation of an individual forgetting factor in the estimation of each parameter. This prevents the problem of sympathetic drift that occurs when weighting the parameter variation equally, as is the case when using RLS with forgetting factors [15].

In the case of the dynamometer, the selectivity offered by the KF is particularly advantageous since speed dependent variation is found to be relatively much slower than non-linear gain variation [13]. Making use of engineering knowledge of typical dynamometer characteristics, the algorithm may be tuned to prevent problems of sympathetic drift which are possible when using RLS with forgetting factors.

4 Control Law Evaluation

Attention is focused on two control strategies appropriate for dynamometer torque control; these being self-tuning PID [5] and self-tuning pole-placement [17].

Adopting a model structure with $n_a = 2, n_b = 0$ and $k = 1$, the self-tuning PID controller continuously adjusts the proportional gain and the integral and derivative times according to

$$K_p = 1/b_0, \;\; T_i = a_1/b_0 \text{ and } T_d = a_2/b_0$$

respectively, where b_0, a_1 and a_2 are the estimated system parameters. The overall resulting control action is then given by

$$u(t) =$$

$$u(t-1)+K_p v(t-1)+[r(t)-y(t)]+T_i[r(t-1)-y(t-1)]+T_d[r(t-2)-y(t-2)]$$

where $r(t)$ denotes the set-point reference.

Adopting the same model structure for the self-tuning pole placement control law, use is made of state variable feedback to provide an overall system input of the form

$$u(t) = F\hat{\underline{x}}(t) + mr(t)$$

where F is the feedback vector chosen such that the eigenvalues of the closed-loop matrix $(P + QF)$ are as specified, $\hat{\underline{x}}(t)$ is the estimated state vector obtained from a steady-state Kalman filter [4] and m is a scalar multiplier which ensures unity steady-state gain.

Remarks:

- the principles of pole-placement and PID are well understood by practising engineers and, as such, self-tuning versions are more readily acceptable.

- although parameter a_2 is included, it is found that for the dynamometer being investigated its value approximates to zero at times (model mismatch). As a result T_d tends to zero resulting in PI control and the pole-placement scheme produces a first order type 0 system.

- PID, with its inherent integral action, is able to handle ramp inputs. However, due to its error driven nature, its performance deteriorates when subject to step changes and is required to be de-tuned. The pole-placement controller, however, is ideally suited to step changes but in the case of the dynamometer, which is predominantly a first order type 0 system, its response to ramp inputs deteriorates due to problems of steady-state error.

Noting the above observations on the suitability of PID and pole-placement controllers for the eddy-current dynamometer being investigated, an alternative approach incorporating the desirable features of both techniques is considered. Essentially, the resulting scheme, whilst adopting the same estimation procedures, incorporates a switched controller structure involving both the PID and pole-placement algorithms.

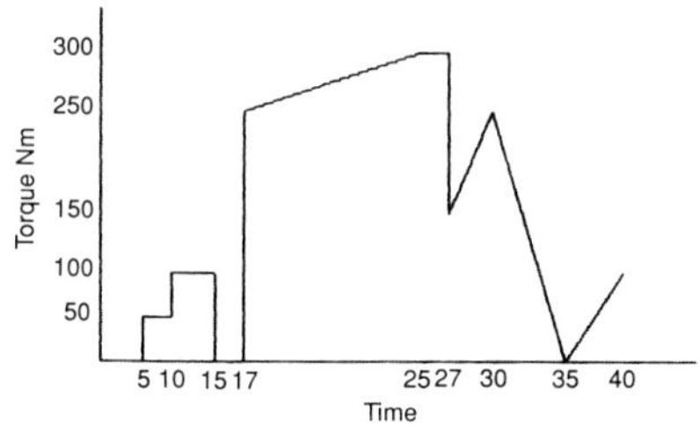

Figure 3. Desired torque profile

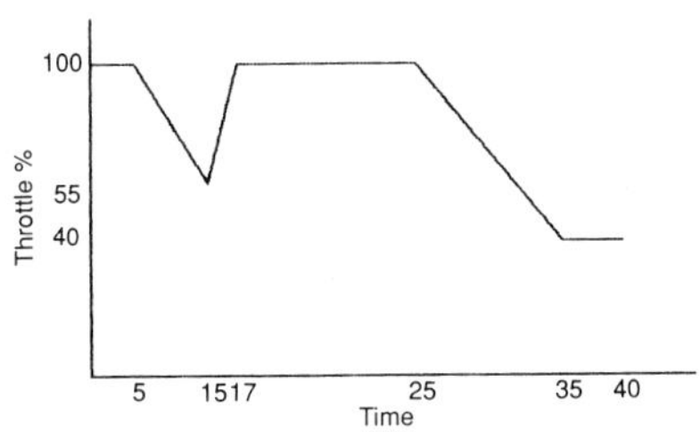

Figure 4. Throttle profile

5 Simulation Studies

The dynamometer torque loop was subjected to the test cycle shown in Figure 3. It involves ramps, steps and periods of constant torque demand, and is preceded by an initial five second identification period. In addition, the throttle position, expressed as a percentage of full scale, was forced (simultaneously) to follow the profile shown in Figure 4 and engine speed allowed to vary as a function of resulting dynamometer load.

A manually tuned fixed PID controller is used in order to provide the basis for realistic comparison purposes. The fixed PID controller is required to be tuned/de-tuned as a best compromise for all foreseen conditions such

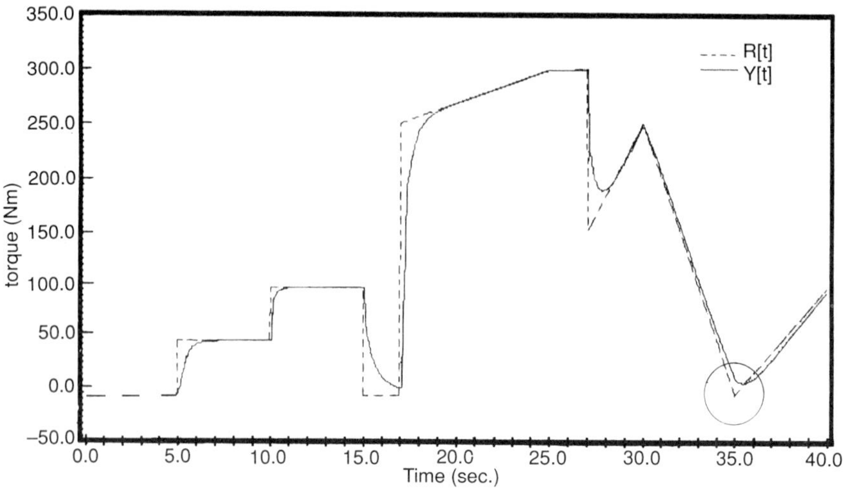

Figure 5. Fixed PID response

that 'satisfactory' performance is obtained. This results in a PI controller with $K_p = 0.031$ and $T_i = 0.39$.

The response shown in Figure 5 highlights the shortfalls of the fixed PID approach when applied to the time varying model of the dynamometer torque loop; there being inconsistencies in both step and ramp responses. The self-tuning controller, however, with its on-line parameter estimation procedure (Kalman filter with partial covariance reset action) is able to adapt to the time varying nature of the dynamometer model. The response shown in Figure 6 highlights the potential advantages to be gained from the use of the self-tuning scheme with switched controller action. (In each case, the dashed line indicates the five second indentification period and the encircled sections of Figures 5 and 6 are illustrated respectively in enlarged form in Figures 7 and 8 together with the additional trace corresponding to the control input response which is scaled by a factor of ten.) Results suggest that, to a large extent, use of the self-tuning scheme overcomes the major shortfalls associated with the de-tuned PID approach, giving rise to a much tighter control.

6 Conclusions

The paper has highlighted the potential advantages of adopting self-tuning control for systems exhibiting fast dynamics and non-linear characteristics; with particular emphasis being directed towards dynamometer torque

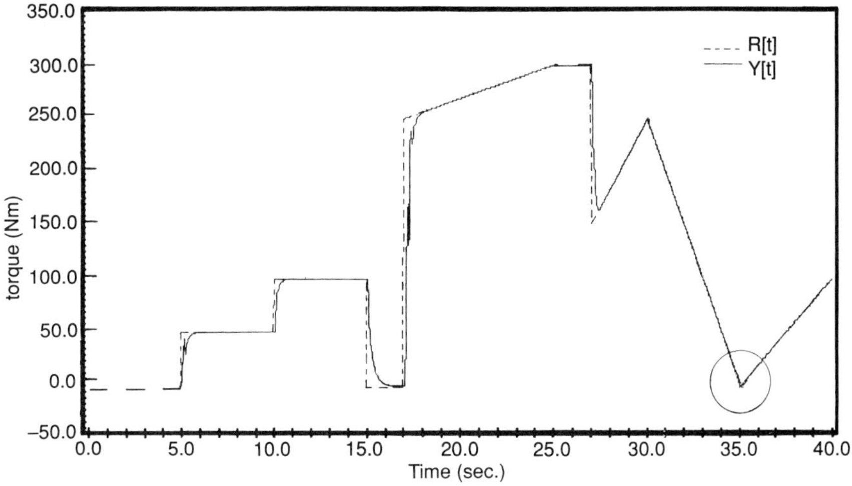

Figure 6. Proposed self-tuning response

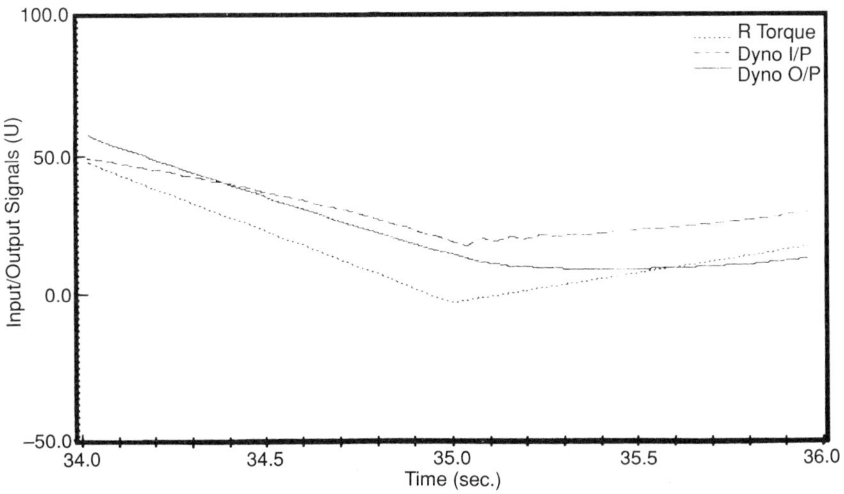

Figure 7. Section of fixed PID response

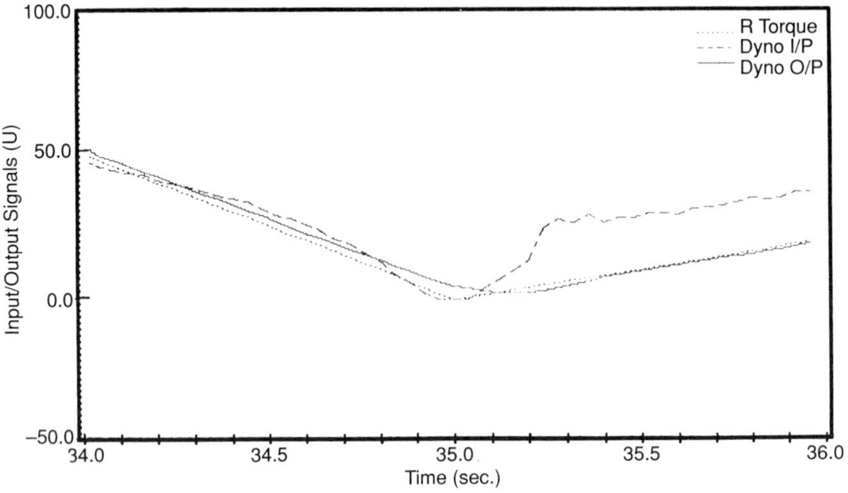

Figure 8. Section of self-tuning response

control within an engine test cell. Problems specific to the implementation of self-tuning PID and pole-placement techniques appropriate for an eddy-current dynamometer have been discussed and the performance of a self-tuning scheme having switched controller structure has been compared by simulation studies to that of a conventional fixed PID approach. Preliminary results are encouraging, indicating that tighter control of the dynamometer torque loop is possible by adopting self-tuning techniques.

References

1. Åström, K.J., (1980), 'Applications of Adaptive Control', Ed. Narendra K.S. and Monopoli R.V., Academic Press.

2. Burnham, K.J., James, D.J.G. and Shields, D.N., (1985), Self-tuning control algorithm for bilinear systems - role of variable forgetting factor, Proc. IASTED, 4th Int. Symp. MIC Grindelwald, Switzerland, ACTA Press, Zurich, 101-105.

3. Burnham, K.J., James, D.J.G. and Shields, D.N., (1987), Self-tuning control of bilinear systems, J. Optimal Control Applications and Methods, Vol. 8, 147-157.

4. Caines, P.E., (1972), Relationship between Box-Jenkins-Astrom control and Kalman linear regulator, Proc. IEE, Vol. 119, No. 5, 601-670.

5. Cameron, F. and Seborg, D.E., (1983), A self-tuning controller with PID structure, Int. J. Control, Vol. 38, No. 2, 401-417.

6. Chalmers, B.J. and Dukes, B.J., (January 1980), High-performance eddy-current dynamometers, IEE PROC., Vol. 127, Pt. B, No. 1, 20-28.

7. Coats, F.E. and Fruechte, R.D., (1983), Dynamic engine models for control development - Part II: Application to idle speed control, Int. J. of Vehicle Design, Special SP4, 75-88.

8. Cook, J.A. and Powell, B.K., (August 1988), Modelling of an internal combustion engine for control analysis, IEE Control Magazine, Vol. 8, No. 4, 20-25.

9. Fortescue, T.R., Kershenbaum, L.S. and Ydstie, B.E., (1981), Implementation of self-tuning regulators with variable forgetting factors, Automatica, Vol. 17, No. 6. 831-835.

10. Froude, W., (25th July 1877), On a New Dynamometer for Measuring the Power Delivered to the Screws of Large Ships, read before the Institution of Mechanical Engineers.

11. Goodhart, S.G., Burnham, K.J. and James, D.J.G., (September 1989), Self-tuning control of a non-linear industrial test rig, Proc. Systems Science X, Poland.

12. Hägglund, T., (December 1983), 'New Estimation Techniques for Adaptive Control', Ph.D Thesis, Lund Univ., Sweden.

13. King, P.J., Burnham, K.J., James, D.J.G., Martyr, A.J. and Sharpe, S.R., (September 1989), Modelling and simulation of a combined engine and dynamometer system, Proc. Systems Science X, Poland.

14. Peterka, V., (1975), A square root filter for real time multivariate regression, Kybernetica, Vol. 11, 53-67.

15. Randall, A., Burnham, K.J. and James, D.J.G., (September 1989), A study of Kalman filtering techniques for joint state and parameter estimation in self-tuning control, Proc. Systems Science X, Poland.

16. Warwick, K., (1985), Further Developments in Self-Tuning Control, in Harris C.J. and Billings S.A., Ed. 'Self-Tuning and Adaptive Control', IEE Control Series 15, Peter Peregrinus Ltd.

17. Warwick, K., (1981), Self-tuning regulators - a state-space approach, Int. J. Control, Vol. 33, No. 5, 839-858.

18. Wellstead, P.E. and Sanoff, S.P., (1981), Extended self-tuning algorithm, Int.J.C., Vol. 34, No. 3, 433-455.

19. Young, P.C., (1974), Recursive approaches to time series analysis, IMA Bulletin, May/June, 209-224.

Automotive Noise

A.P. Dowling

Cambridge University Engineering Department

Abstract The main sources of vehicle noise are reviewed and the physical principles involved in the prediction of the sound field are highlighted. The noise generated by an internal combustion engine is considered in some detail. The various approximations commonly made in the literature are discussed with reference to idealized examples with such simple geometries that analytical solutions are possible. The sound radiated by tyres and the effects of the unsteady flow over the vehicle on the internal noise level are mentioned briefly.

1 Introduction

Environmental pressures to reduce noise levels near roads have ensured that the maximum permitted sound radiation by vehicles is constantly being decreased. The European Commission introduced a new standard in October 1989 and it is likely that the maximum legal level will be reduced still further in the 1990s. The second reason why automotive noise needs to be considered by the vehicle designer is that low interior noise levels increase passenger comfort and are an important selling point. The onus is therefore on the designer to reduce both the radiated and internal sound fields. This is not an easy task. Internal combustion engines rely on inherently unsteady combustion to generate their power and such unsteady combustion is a very effective sound source. In this paper we review some of the mathematical and computational techniques available to illuminate the physical processes involved in noise generation.

Section 2 begins with a discussion of the sound field generated by unsteady combustion. This topic is introduced by considering a problem with such a simple geometry that it can be solved analytically: combustion at the centre of a heavy spherical shell. Comparison with this exact problem enables us to discuss the approach commonly adopted in the prediction of engine noise and to assess its limitations.

At high forward speeds tyre noise exceeds engine noise and tyres become the most significant acoustic source. The mechanism of sound generation by tyres is not clearly understood, but we discuss some of the current theories in Section 3.

Unsteadiness in the external flow over a vehicle can lead to interior noise by inducing panel or, more significantly, window vibration. Wind-induced noise is briefly described in Section 4.

2 Engine Noise

Unsteadiness in the rate of heat release in an internal combustion engine is a major source of vehicle noise. Before discussing the complications introduced by the engine geometry, let us investigate the sound field generated by combustion in unbounded space. Suppose that burning leads to a rate of heat release $q(\underline{x}, t)/$ unit volume. For simplicity let us assume that a linear amount of heat is added to fluid which is initially at rest with density ρ_1 and sound speed c_1, although the analysis can be extended to nonlinear combustion rates and particle velocities [1].

The linearized equations of conservation of mass and momentum have the usual form.

$$\frac{\partial \rho}{\partial t} + \rho_1 \nabla \bullet \underline{u} = 0 \tag{2.1}$$

and

$$\rho_1 \frac{\partial \underline{u}}{\partial t} + \nabla p = 0, \tag{2.2}$$

where viscous forces have been neglected. The particle velocity $\underline{u}(\underline{x}, t)$ may be eliminated by combining equations (2.1) and (2.2) to give

$$\frac{\partial^2 \rho}{\partial t^2} - \nabla^2 p = 0. \tag{2.3}$$

In a reacting gas the density, $\rho(\underline{x}, t)$, varies due to changes in both the pressure, $p(\underline{x}, t)$, and the specific entropy, $s(\underline{x}, t)$.

$$\frac{D\rho}{Dt} = \frac{\partial \rho}{\partial p}\bigg|_s \frac{Dp}{Dt + \frac{\partial \rho}{\partial s}}\bigg|_p \frac{Ds}{Dt}. \tag{2.4}$$

Now $\partial \rho / \partial p \,|_s$ is just the inverse square of the speed of sound, while for a perfect gas $\partial s / \partial \rho \,|_p = -c_p/\rho$, where c_p denotes the specific heat capacity at constant pressure. It follows from the second law of thermodynamics that

$$\rho T \frac{Ds}{Dt} = q. \tag{2.5}$$

Hence, after linearization, equation (2.4) reduces to

$$\frac{\partial \rho}{\partial t} = \frac{1}{c_1^2}\frac{\partial p}{\partial t} - \frac{\gamma - 1}{c_1^2} q, \tag{2.6}$$

where γ is the ratio of specific heat capacities.

Elimination of ρ between equations (2.3) and (2.6) leads to an inhomogeneous wave equation for the pressure perturbation

$$\frac{\partial^2 p'}{\partial t^2} - \nabla^2 p' = \frac{\gamma - 1}{c_1^2}\frac{\partial q}{\partial t}. \tag{2.7}$$

The solution of this wave equation can be written down immediately (see for example reference [2] Chapter 7) and is

$$p'(\underline{x}, t) = \frac{\gamma - 1}{4\pi c_1^2}\int \frac{\partial q}{\partial t}(\underline{y}, t - \mid \underline{x} - \underline{y} \mid / c_1)\frac{d^3 \underline{y}}{\mid \underline{x} - \underline{y} \mid}. \tag{2.8}$$

Evidently, in an unbounded region sound pressure is generated by the rate-of-change in the rate of heat release. Steady burning is silent. However in an internal combustion engine the burning takes place within a confined region. We will investigate the new effects this introduces by first considering such a simple geometry that an analytical solution is possible.

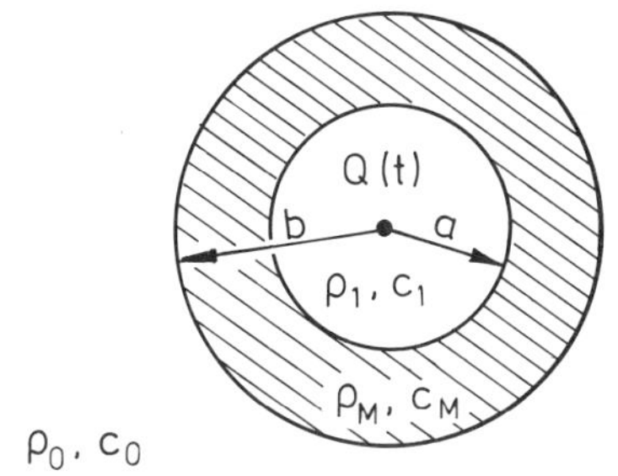

Figure 1. Combustion at the centre of a spherical shell

As a first example consider the geometry illustrated in Figure 1. Combustion occurs with a total rate of heat release $Q(t)$ concentrated at the centre of a spherical cavity of radius a. The gas within the cavity is hot and has a mean density ρ_1 and sound speed c_1. The cavity is surrounded by a heavy spherical shell of thickness $b - a$. We will suppose the shell to be composed of fluid with density ρ_M and sound speed c_M. We could equally well (and perhaps more realistically) consider the shell to be made of an elastic material, but the only consequences of a non zero shear wave speed are to make the algebra more complicated without adding significantly to an understanding of the phenomenon. Outside the shell the surrounding gas has mean density ρ_o and sound speed c_o.

The pressure perturbations therefore satisfy the equations

$$\frac{1}{c_1^2}\frac{\partial^2 p'}{\partial t^2} - \nabla^2 p' = \frac{\gamma-1}{c_1^2}\dot{Q}(t)\delta(\underline{x}) \ \text{in}\, r < a \tag{2.9a}$$

$$\frac{1}{c_M^2}\frac{\partial^2 p'}{\partial t^2} - \nabla^2 p' = 0 \ \text{in}\, a < r < b \tag{2.9b}$$

$$\frac{1}{c_o^2}\frac{\partial^2 p'}{\partial t^2} - \nabla^2 p' = 0 \ \text{in}\, b < r. \tag{2.9c}$$

It is convenient to introduce $\hat{p}(\underline{x},\omega)$ the Fourier transform of $p'(\underline{x},t)$

$$\hat{p}(\underline{x},\omega) = \int p'(\underline{x},t)e^{-i\omega t}dt. \tag{2.10}$$

The general solution of (2.9a) has the form

$$\hat{p}(r,\omega) = (Ie^{-i\omega r/c_1} + Re^{i\omega r/c_1})/r \ \text{in}\, r \leq a, \tag{2.11}$$

and integrating (2.9a) over a small sphere centred on the origin shows that

$$I + R = \frac{i\omega(\gamma-1)\hat{Q}}{4\pi c_1^2}. \tag{2.12}$$

The momentum equation (2.2) enables the Fourier transform of the radial particle velocity, $\hat{u}(\underline{x},\omega)$, to be expressed in terms of the unknown coefficients I and R.

In a similar way we can write

$$\hat{p}(r,\omega) = (Ae^{-i\omega r/c_M} + Be^{i\omega r/c_M})/r \ \text{in} \ a \leq r \leq b \tag{2.13}$$

and

$$\hat{p}(r,\omega) = Te^{-i\omega r/c_o} \ \text{in} \ r \geq b. \tag{2.14}$$

It is then a matter of straightforward algebra to use equation (2.12) and continuity of pressure and particle velocity across $r = a$ and $r = b$ to determine the coefficients I, R, A, B and T in terms of $\hat{Q}$. These results simplify when the shell is heavy.

For finite shell thickness, we find that, when ρ_M is very large in comparison with ρ_o and ρ_1 and the frequency is well away from the resonant frequencies (which are roots of $\tan(\omega a/c_1) = \omega a/c_1$ or $\sin\omega(b-a)/c_M = 0$),

$$\hat{p}(r,\omega) = \frac{i\omega(\gamma-1)\hat{Q}}{4\pi c_1^2} \times \left(\frac{e}{r}^{-i\omega r/c_1} - e^{-i\omega a/c_1}\frac{\sin(\omega r/c_1)}{r}\frac{1+i\omega a/c_1}{\sin(\omega a/c_1)-(\omega a/c_1)\cos(\omega a/c_1)}\right) \tag{2.15}$$

This expression for the pressure field is identical to that obtained when the combustion takes place in a spherical cavity of radius a with rigid walls. The first term on the right-hand side of (2.15) is the direct field. It describes the pressure field that the heat release rate $Q(t)$ would produce in an unbounded region of fluid with sound speed c_1. The second term is the reverberant field resulting from multiple reflections from the walls. It is interesting to note that for a compact cavity, $\omega a/c_1 << 1$, this reverberant field becomes independent of the radius r and leads to the same pressure perturbation throughout the cavity.

The radiated pressure outside the shell has a rather complicated form

$$\hat{p}(r,\omega) = \frac{e^{-i\omega(r-b)/c_o}}{r}\frac{\hat{p}(a,\omega)}{1+i\omega b/c_o}\frac{\rho_o\omega a}{\rho_M c_M \sin\omega(b-a)/c_M}. \tag{2.16}$$

for $\tan(\omega a/c_1) \neq \omega a/c_1$ and $\sin\,\omega(b-a)/c_M \neq 0$. The radiated sound field is evidently a function of ρ_o, c_o, ρ_M, c_M and the geometry. We can interpret this expression if we evaluate the particle velocity on $r = b$.

$$\hat{u}(b,\omega) = \hat{p}(a,\omega)\frac{a}{ib\rho_M c_M \sin\,\omega(b-a)/c_M}. \tag{2.17}$$

The particle velocity on the outer edge of the shell is evidently a function of the shell properties and the applied force only. Simple algebra shows $\hat{u}(b,\omega)$ to be equal to the velocity that would be induced on $r = b$ due to radial forcing/unit area of $\hat{p}(a,\omega)$ on the same 'shell' in vacuo. Finally a comparison of (2.16) and (2.17) gives

$$\hat{p}(r,\omega) = \frac{\hat{u}(b,\omega)}{1+i\omega b/c_o}\frac{\rho_o i\omega b^2}{r}e^{-i\omega(r-b)/c_o} \tag{2.18}$$

Hence, away from the resonant frequencies, we have found that for a heavy shell the problem of determining the radiated sound field due to the heat release rate $Q(t)$ decouples into (i) determining the cavity pressure that would be generated by $Q(t)$ if the cavity were bounded by rigid walls (ii) investigating the vibration of the outer surface of the shell in vacuo induced by a specified force on its inner surface (iii) calculating the sound field radiated by the vibration of the outer surface of the structure.

At the resonant frequencies the problem does not decouple in this way. These frequencies correspond to resonances of the cavity, $\tan(\omega a/c_1) = \omega a/c_1$, and to resonances of the shell, $\sin\,\omega(b-a)/c_M = 0$. At these frequencies the energy radiates away as sound controls the cavity pressure, which therefore depends crucially on the properties of the external fluid. The cavity resonance can be avoided if we restrict our attention to frequencies ω which are less than $\pi c_1/a$. For $a \simeq 5cm, c_1 \simeq 800$ m/s, this implies frequencies less than 8 kHz. For larger diameter cavities the maximum permitted frequency is lower. Damping in the shell may be modelled by giving the sound speed c_M a small imaginary part. Then sin $\omega(b-a)/c_M$ cannot vanish for real ω.

These results can be generalized to an engine where unsteady combustion occurs within a cylinder head which is enclosed by the complicated geometry of the engine block. For frequencies less than the lowest acoustic resonant frequency of the cylinder and provided the engine block has sufficient mechanical damping for the level of its response to resonant forcing to be determined by its mechanical damping and not the energy radiated away as sound, the analysis can again be decoupled into three elements. These are (i) the cylinder pressure generated by the unsteady heat release in the absence of wall vibration (ii) the vibration of the engine block that would be produced in vacuo by an unsteady applied force/unit area equal to the cylinder pressure (iii) calculation of the sound field radiated by the vibration of the outer surface of the engine block. We will discuss the problem broken down in this way, but bearing in mind that such decoupling will not be valid at resonant frequencies of the cylinder or if the engine block has any very lightly damped modes of vibration.

The cavity pressure we calculated in our exact problem simplifies considerably if the cavity is compact, $\omega a/c_1 << 1$. Then, away from the origin, the reverberant field is larger than the direct field and (2.15) reduces to

$$p(\underline{x},\omega) \simeq \frac{3(\gamma-1)\hat{Q}}{i\omega\, 4\pi a^3}, \quad |\,\underline{x}\,| \neq 0. \tag{2.19}$$

In free space we found the pressure perturbation to be proportional to the rate-of-change in the rate of heat release (see equation 2.8)). In contrast, when the combustion occurs in a compact, confined region, equation (2.19) shows that $p'(\underline{x},t)$ is proportional to the total heat that has been released. This has a straightforward physical interpretation which extends the result to an arbitrarily shaped, compact volume and large pressure changes.

Suppose that the combustion is distributed and occurs at a rate $q(\underline{x},t)$ /unit volume within a volume V. V has rigid walls and is compact. The pressure p is almost uniform throughout V. The perfect gas relationship may be written in the form

$$pV = mR\bar{T}, \tag{2.20}$$

where $\bar{T}$ is the mass-averaged temperature, $\bar{T} = \int \rho T dV/m$. Changes in $\bar{T}$ are directly related to the heat release rate,

$$mc_v \frac{d\bar{T}}{dt} = \int q(\underline{x}, t)dV. \tag{2.21}$$

A combination of equations (2.20) and (2.21) leads to

$$\frac{dp}{dt} = \frac{\gamma - 1}{V} \int q(\underline{x}, t)dV. \tag{2.22}$$

For the particular case of concentrated combustion within a spherical cavity (2.22) reduces to (2.19). In general it provides a means of calculating the time history of the cylinder pressure if the rate of heat release is known (see for example references [3] and [4]). The rise in cavity pressure is proportional to the overall heat input.

In practice it is easier to measure the cylinder pressure directly than to predict the heat release rate, and measured cylinder pressures are frequently used as a starting point in the determination of the sound field. Figure 2 shows a plot of cylinder pressure, $p_{cyl}(t)$ as a function of crankangle for a typical diesel engine. In a diesel engine, the combustion occurs spontaneously when the fuel is injected, resulting in rapid heat release and a discontinuity in the gradient of the cylinder pressure. This rapid change means that the high frequency components of $\hat{p}_{cyl}(\omega)$ have appreciable amplitudes. Indeed the cylinder pressure spectrum only decays with frequency at the rate of 30 dB/decade at high frequencies [5]. The high frequency components of p_{cyl} are particularly important in the generation of the external sound field since they are transmitted fairly effectively through the engine structure and radiate sound efficiently. A reduction of 30 dB/decade means that

$$\hat{p}_{cyl}(\omega) \quad \propto \quad \omega^{-3/2} \text{ as } \omega \to \infty. \tag{2.23}$$

The curve of cylinder pressure against crankangle is found to have virtually the same form for all engine speeds, N. Hence, in (2.23), ω should be nondimensionalized with respect to N, which shows that at a particular frequency

$$\hat{p}_{cyl} \quad \propto \quad N^{3/2}. \tag{2.24}$$

Experimental results do indeed confirm that the sound energy radiated from a diesel engine increases in proportion to N^3 [5].

Attempts have been made in the literature [6] to fit an exponential curve of the form $1-e^{-kt}$ (where k is a constant) to the rapid rise in the cylinder pressure in a diesel engine. However it is apparent from an application of Tauber's theorem [7] that the condition (2.23) implies

$$P_{cy\ell}(t) \to t^{1/2} \text{ as } t \to 0^+ \tag{2.25}$$

where t denotes time measured from the initiation of the combustion. One can easily imagine such a form with an infinite gradient at $t = 0^+$ fitting the diesel cylinder pressure curve in Figure 2.

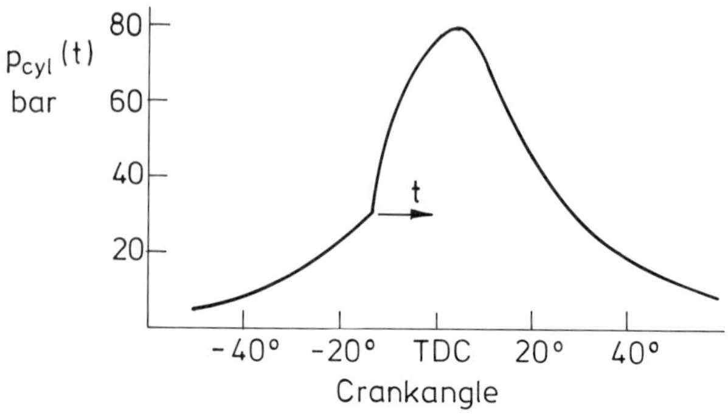

Figure 2. Cylinder pressure in a typical diesel engine

The rapid rise in cylinder pressure leads to significant high frequency components which in turn generate much of the radiated sound. Modest changes to the cylinder-pressure time-history near $t = 0$ can therefore lead to reductions in the radiated sound field. Priede [8] reports that changes to the combustion that make the gradient of $p_{cy\ell}(t)$ continuous ensure that $\hat{p}_{cy\ell}(\omega)$ decays like $\omega^{-5/2}$ at high frequencies rather than $\omega^{-3/2}$ as for standard diesel engines. This observation is entirely consistent with Abel's theorem on the asymptotics of transforms [7]. The decrease in the high frequency components of $p_{cy\ell}(t)$ leads to a reduction of 20 dB in the radiated sound, without changing the engine power. Split-fuel injection is a convenient way of achieving such smooth combustion in a direct diesel engine [9]. A small volume of fuel is injected to initiate the combustion, causing a gradual initial rise in $p_{cy\ell}(t)$. The main fuel injection then follows.

A typical variation of cylinder pressure with crankangle for a petrol engine is shown in Figure 3. In a petrol engine the combustion, after initiation by the spark, increases gradually as the flame kernal grows. The gradient of $P_{cy\ell}(t)$ is therefore continuous and the cylinder pressure spectrum decays more rapidly at high frequencies. A decay rate of 50 dB/decade is typical. This means that

$$\hat{p}_{cy\ell}(\omega) \;\propto\; \omega^{-5/2} \text{ as } \omega \;\to \infty \tag{2.26}$$

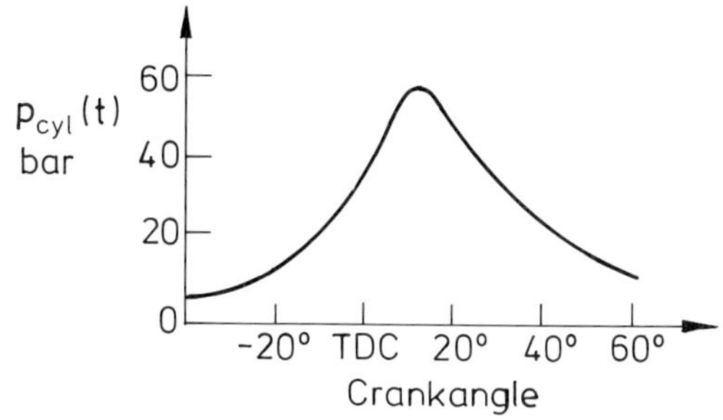

Figure 3. Cylinder pressure in a typical petrol engine

Again the curve of cylinder pressure against crankangle is virtually independent of the engine speed N. It therefore follows from (2.26) that at a fixed frequency (c.f. steps between (2.23) and (2.24))

$$\hat{p}_{cy\ell} \ \alpha \ N^{5/2} \tag{2.27}$$

The sound energy radiated from a petrol engine increases in proportion to N^5[5].

Attempts have been made in the literature to fit an exponential of the form e^{kt} to the cylinder pressure in a petrol engine [6]. However standard transform theory [7] and the high frequency asymptotic form (2.26) suggest that

$$p_{cy\ell}(t) \to t^{3/2} \quad \text{as} \quad t \to 0^+ \tag{2.28}$$

where t denotes time measured from the initiation of the combustion.

It is interesting to note that current attempts to maximize the combustion burn rate in petrol engines, in order to extend the lean burn limit, increase the gradient of $p_{cy\ell}(t)$ and can have an adverse effect on the radiated sound [10].

The unsteady cylinder pressures in both diesel and petrol engines provide forces which excite vibrations of the engine block. Mechanical impacts provide another form of excitation of the engine structure. One example of this is the so-called 'piston slap', where a moving piston impacts on the cylinder wall exerting lateral forces on the engine block [11]. These forces can be determined from an investigation of the dynamics of the piston.

Once the sources of excitation, the unsteady cylinder pressure and the forces due to mechanical impacts, have been described, the next stage is to determine the vibration of the engine structure. The velocities induced on the outer surface of the engine are of particular interest, since they are coupled to the sound field. This is really a vibration problem and it is not appropriate to give many details here.

Finite element analysis provides the most common means of investigating such forced vibration problems. Crocker [12] shows a typical mesh for a six-cylinder engine. It has more than 10^4 degrees of freedom and the solution for 90 modes took 40 hours on a 32-bit "Super Mini". Finite element analysis is most useful at low and modest frequencies. At higher frequencies, there are many modes and a fine mesh with many degrees of freedom is required. Moreover, these high order modes are sensitive to small details in the structure and may be affected by the differences between manufactured engines and the design drawing [12] or indeed between different but nominally identical engines.

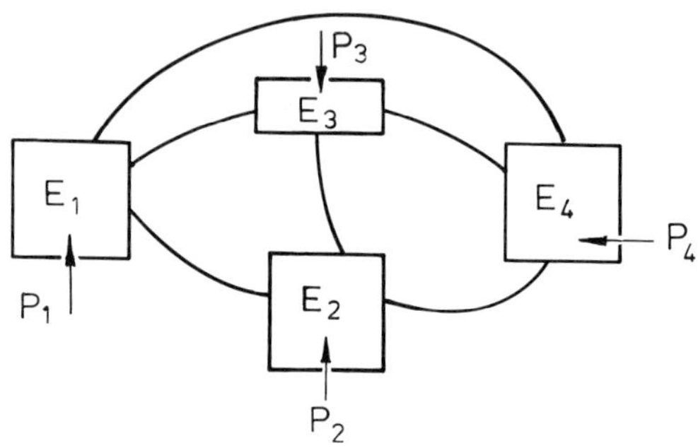

Figure 4. A vibrating structure divided into substructures for the application of SEA

At high frequencies when the modes are dense an energy balance for the structure averaged over many modes is often more useful. That is provided by Statistical Energy Analysis. The literature on SEA is extensive. An introduction to the approach is given, for example, in references [13]-[15]. In order to apply SEA a structure is divided into n substructures as shown in Figure 4. The time-averaged energy in each substructure is determined from a consideration of the energy input into the structure, the exchange of energy between substructures and the dissipation in each substructure. Let us denote the time-averaged energy/mode in subsystem i by E_i and the energy input by external forcing into subsystem i by P_i. The basic assumptions of SEA are that (i) the rate of energy dissipation in subsystem i is proportional of E_i. We will call that constant of proportionality η_i. (ii) The rate of power flow from subsystem i to subsystem j is proportional to the differences in their energy densities, equal to $\eta_{ij}(E_i - E_j)$, say. (iii) The driving forces are statistically independent so that their energy inputs add linearly [15]. Assumption (ii) is satisfied exactly for the case of two subsystems with conservative coupling. In other cases it may be approximately true [16].

An energy balance for subsystem i relates the rate of dissipation of energy in the subsystem to the rate of power input into the subsystem:

$$\eta_i E_i = P_i + \sum_{j=1}^{n} \eta_{ji}(E_j - E_i) \quad i = 1, \ldots, n \tag{2.29}$$

In principle, if P_i, η_i and η_{ij} are known, these equations can be solved to determine the energy in the modes. In practice, η_i and η_{ij} need to be determined experimentally, but once this has been done the effects of, say, altering a coupling coefficient can be investigated analytically.

As an introductory example, let us consider a system composed of two substructures, with only substructure 1 driven by external forces. Then

$$\eta_1 E_1 = \eta_{12}(E_2 - E_1) + P_1 \tag{2.30a}$$

$$\eta_2 E_2 = \eta_{12}(E_1 - E_2). \tag{2.30b}$$

These two equations can be solved immediately to give

$$E_2 = \frac{P_1 \eta_{12}}{\eta_1 \eta_2 + \eta_{12}(\eta_1 + \eta_2)}. \tag{2.31}$$

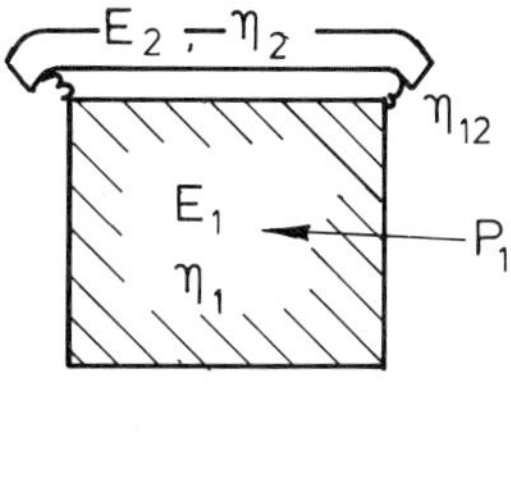

Figure 5. An engine with a cover

System 1 might, for example, represent an engine and 2 a cover as shown in Figure 5. Since the vibration of the surface in contact with the surrounding air generates sound, some attempts at reducing the radiated sound involve installing a cover over the engine. The idea is that if the cover can be mechanically isolated from the engine (η_{12} large), the cover would vibrate less than the surface of the engine, and hopefully radiate less sound. The performance of such a cover can be investigated via equation (2.31). For example, this equation shows that when the internal damping in the engine is very small, $\eta_1 << 1, E_2$ is independent of η_{12}. In this case reductions in the coupling coefficient η_{12} increase E_1, and have no effect on the radiated sound!

Limited use of SEA has been made in the analysis of engine vibrations. Challen & Croker [17] investigated the coupling between an engine block and the intake manifold, while Carr & Lalor [18] used SEA to analyse cylinder head/valve cover and crankcase/sump interactions. More extensive use of the method has been made in other areas and it is likely that its application to automotive vibration problems will increase.

Let us now assume that the outer surface vibration of an engine has been determined. This could have been calculated from a finite element analysis or measured, maybe by an array of accelerometers or alternatively from holograms of the surface motion [19]. The next stage is to determine the sound field generated by such surface vibration. We will begin by discussing the results of our simple problem with spherical symmetry.

When damping in the shell is modelled by giving c_M a small positive imaginary part (which we will later let tend to zero), $(\sin\omega(b-a)/c_M)^{-1}$ in equation (2.16) can be expanded as a series to give

$$\hat{p}(r,\omega) = \frac{\rho_o 3b(\gamma-1)\hat{Q}(\omega)}{\rho_M c_M a^2 2\pi r(1+i\omega b/c_o)} \sum_{n=0}^{\infty} \exp i\omega \left[-\frac{r-b}{c_o} - \frac{b-a}{c_M}(1+2n)\right] \tag{2.32}$$

If $Q(t)$ only contains frequencies which are low in comparison with c_o/b, the Fourier transform of $p'(r,t)$ described by equation (2.32) can be inverted to give

$$p'(r,t) = \frac{\rho o 3b(\gamma-1)}{\rho_M c_M a^2 2\pi r} \sum_{n=0}^{\infty} Q(t_n) \tag{2.33}$$

where $t_n = t - (r-b)/c_o - (b-a)(1+2n)/c_M$. The first sound to arrive at r, travels a distance $r-b$ in the fluid with sound speed c_o and $b-a$ through the shell material with sound speed c_M. Later waves are multiply-reflected in the shell, and by travelling back and forth n times they suffer an additional time delay of $n2(b-a)/c_M$. The pressure perturbation produced by each wave is proportional to the heat release rate, whereas were the combustion to occur in free space the pressure perturbation would be proportional to the rate-of-change of the rate of heat release.

If, on the other hand, $Q(t)$ only contains frequencies which are large in comparison with c_o/b, (2.32) leads to

$$p'(r,t) = \frac{\rho_o c_o 3(\gamma-1)}{\rho_M c_M 2\pi r a^2} \sum_{n=0}^{\infty} \int_{-\infty}^{t_n} Q(\tau)d\tau. \tag{2.34}$$

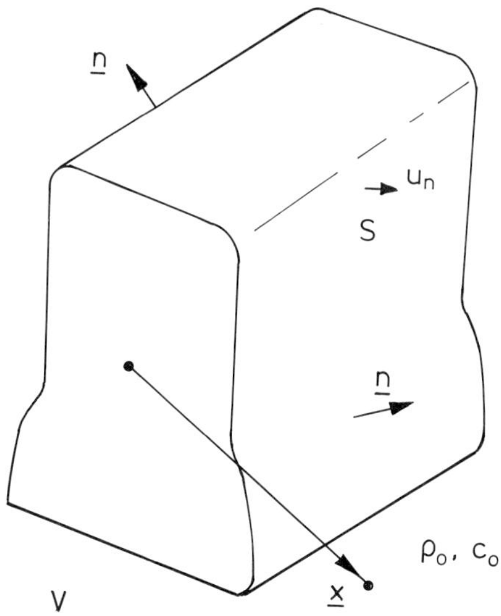

Figure 6. Sound generation by engine vibration

In this case the external pressure is proportional to the overall heat release, just like the internal pressures in (2.22).

For this problem with spherical symmetry it was possible to find an analytical expression for the radiated sound field. In general, the geometry is not that simple and a numerical solution is necessary. Each Fourier component of the pressure satisfies Helmholtz's equation

$$(\nabla^2 + \omega^2/c_o^2)\hat{p} = 0 \text{ in } V, \tag{2.35}$$

subject to the specified normal velocity on the surface S

$$\frac{\partial \hat{p}}{\partial n} = -i\omega\rho_o\hat{u}_n \text{ on } S, \tag{2.36}$$

and a radiation condition at infinity

$$r\left(\frac{\partial \hat{p}}{\partial r} + \frac{i\omega}{c_o}\hat{p}\right) \to \text{as } r \to \infty. \tag{2.37}$$

Equation (2.35) could be solved by finite difference methods, but then a three-dimensional problem needs to be solved for each frequency ω. If instead it is recast as surface integral equation the number of dimensions is reduced from three to two. That is the idea behind the boundary element method.

Let us introduce a Green function $G(\underline{y} \mid \underline{x})$ defined by

$$(\nabla^2 + \omega^2/c_o^2)G = \delta(\underline{y} - \underline{x}) \text{ in } V \tag{2.38}$$

An application of Green's theorem to equations (2.35) and (2.38) leads to

$$H(\underline{x})\hat{p}(\underline{x},\omega) = \int_S \left(G\frac{\partial \hat{p}}{\partial n} - \hat{p}\frac{\partial G}{\partial n} \right) dS, \tag{2.39}$$

which becomes

$$H(\underline{x})\hat{p}(\underline{x},\omega) = -\int_S \left(i\omega\rho_o\hat{u}_n G + \hat{p}\frac{\partial G}{\partial n} \right) dS, \tag{2.40}$$

after making use of the boundary condition (2.36). $H(\underline{x})$ is a Heaviside function defined by

$$\begin{aligned} H(\underline{x}) &= \text{within } V \\ &= \frac{1}{2} \text{ on } S \\ &= 0 \text{ outside } V \end{aligned} \tag{2.41}$$

For surfaces with simple geometrical shapes, it is possible to calculate analytically an exact Green function which satisfies $\partial G/\partial n = 0$ on S. Equation (2.40) then determines $\hat{p}(\underline{x},\omega)$ throughout V in terms of the known distribution of surface velocity $\hat{u}_n$. However, for general shaped surfaces, it is not possible to find an analytical expression for the exact Green function. Equation (2.40) is then an integral equation for the unknown function $\hat{p}(\underline{x},\omega)$. If we take $G(\underline{y} \mid \underline{x})$ to be the free-space Green function and consider a position $\underline{x}$ on the surface S, equation (2.40) becomes

$$\frac{1}{2}\hat{p}(\underline{x},\omega) = \frac{1}{4\pi}\int \left\{ \frac{i\omega\rho_o\hat{u}_n e^{-i\omega|\underline{x}-\underline{y}|/c_o}}{|\,\underline{x}-\underline{y}\,|} + \hat{p}\frac{\partial}{\partial n}\left(\frac{e^{-i\omega|\underline{x}-\underline{y}|/c_o}}{|\,\underline{x}-\underline{y}\,|} \right) \right\} dS(\underline{y}) \tag{2.42}$$

This equation for the surface pressure distribution can be solved numerically. First of all the surface S is divided into N elements, each sufficiently small for the pressure $\hat{p}_j(\omega)$ and the normal surface velocity $\hat{u}_{nj}(\omega), j = 1,\ldots,N$, not to vary appreciably over the element. Discretization of (2.42) then leads to N equations for the N unknowns $\hat{p}_j$. In general these equations can be solved to determine the surface pressure distribution. The pressure at a general position $\underline{x}$ in V may then be calculated from a discretized form of the integral in (2.40).

The system of N equations for $\hat{p}_j$ becomes singular at frequencies which correspond to the internal resonances of S. These difficulties are due to the method of solution and do not reflect physical effects. At these frequencies (2.42) has a non-unique solution. It is possible to have non-zero $\hat{p}_j$ even when $\hat{u}_n$ is identically zero all over S. This difficulty arises because (2.42) omits much of the information contained in (2.40). There are established ways of overcoming this difficulty, see for example [19] and [20]. Recent work by Kirkup et al. [21], in which the boundary element method is used to calculate the sound radiated by engine vibration, has produced promising results. Their method for overcoming the apparent singularities is presented in reference[22].

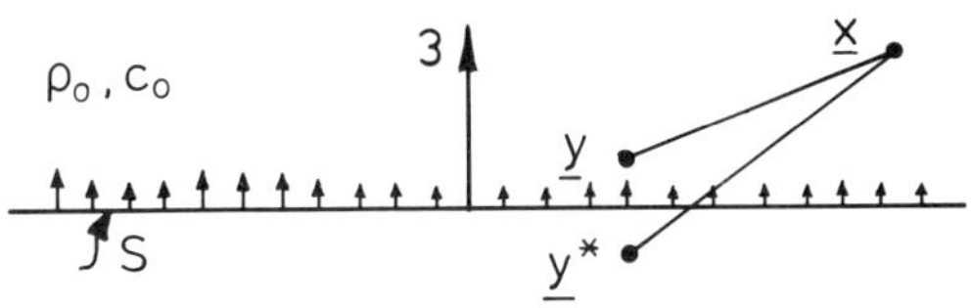

Figure 7. Sound radiation into a half-space

If the surface S is the infinite plane surface $x_3 = 0$ as shown in Figure 7, it is appropriate to choose G to be the half-space Green function:

$$G(\underline{y} \mid \underline{x}) = -\frac{e^{-i\omega|\underline{x}-\underline{y}|/c_o}}{4\pi \mid \underline{x} - \underline{y} \mid} - \frac{e^{-i\omega|\underline{x}-\underline{y}^*|/c_o}}{4\pi \mid \underline{x} - \underline{y}^* \mid}. \tag{2.43}$$

$\underline{y}^*$ is the image of $\underline{y}$ in the surface $y_3 = 0, \underline{y}^* = (y_1, y_2, -y_3)$. Then $\partial G/\partial n = 0$ on S and equation (2.40) simplifies to

$$H(\underline{x})\hat{p}(\underline{x}, \omega) = \frac{i\omega\rho_o}{2\pi} \int_S \frac{\hat{u}_n(\underline{y}, \omega)e^{-i\omega|\underline{x}-\underline{y}|/c_o}}{\mid \underline{x} - \underline{y} \mid} dS \tag{2.44}$$

This expression is known as Rayleigh's integral. It gives the pressure perturbation in a very convenient form, but is only valid for the particular case where S is an infinite plane surface.

Fahy [23] recommends using the integral in (2.44) to calculate the radiated sound "for surfaces which have dimensions and radii of curvature large compared with a wavelength". We can test such a procedure by applying it to our exact problem with spherical symmetry. It follows from equation (2.18) that a normal surface velocity

$$\hat{u}_n(b, \omega) = U_o \text{ on the sphere } r = b \tag{2.45}$$

produces a pressure perturbation

$$\hat{p}(r,\omega) = \frac{U_o}{1 + i\omega b/c_o} \frac{\rho_o i\omega b^2}{r} e^{-i\omega(r-b)/c_o} \tag{2.46}$$

in a fluid with mean density ρ_o and sound speed c_o. For large $\omega b/c_o$, the sphere's size and radius of curvature are large in comparison with the acoustic wavelength and (2.46) simplifies to

$$\hat{p}(r,\omega) = \frac{U_o \rho_o c_o b}{r} e^{i\omega b/c_o} e^{-i\omega r/c_o}. \tag{2.47}$$

After substitution for $\hat{u}_n$ from (2.45) into (2.44), the integral can be evaluated in a straighforward way, and for large $\omega b/c_o$ gives

$$\frac{U_o \rho_o c_o b}{r} 2i \sin(\omega b/c_o) e^{-i\omega r/c_o}. \tag{2.48}$$

We see from a comparison of (2.47) and (2.48) that this procedure does not lead to the correct form for the radiated pressure field.

When the surface size and its radii of curvature are large in comparison with the wavelength, the correct approximation is to apply ray theory. Ray theory is the leading term of an asymptotic series in inverse powers of the frequency, ω [2], [24]. It is obtained by substituting an expansion of the form

$$p'(\underline{x},t) = A(\underline{x},\omega) e^{i\omega(t-\tau(\underline{x}))} \quad \text{with} \quad A(\underline{x},\omega) = \sum_{n=0}^{\infty} (i\omega)^{-n} A_n(\underline{x}) \tag{2.49}$$

into the differential equation (2.35) and the boundary condition (2.36). The phase $\tau(\underline{x})$ and the amplitudes $A_n(\underline{x})$ are determined by requiring the coefficients of each power ω to vanish identically in the resulting equations.

If the surface S is convex and

$$u_n(\underline{x}) = U_o(\underline{x}) e^{i\omega t} \quad \text{on } S, \tag{2.50}$$

where $U_o(\underline{x})$ varies slowly over the surface. Then, in the limit $\omega \to \infty$, we need only retain the first term of the ray series expansion. This term relates the pressure at $\underline{x}$ to the surface velocity at $\underline{x}_s$, a position at the base of a surface normal passing through $\underline{x}$ as shown in Figure 8. It gives [25]

$$p'(\underline{x},t) \simeq \sqrt{\frac{A_s}{A_x}} \rho_o c_o U_o(\underline{x}_s) e^{i\omega(t-d/c_o)} \quad \text{as } \omega \to \infty. \tag{2.51}$$

The sound propagates as if travelling in straight lines perpendicular to the surface. On the surface

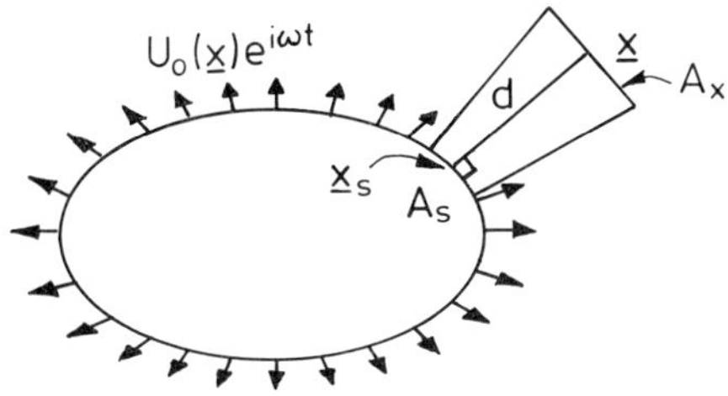

Figure 8. High frequency sound in rays due to normal surface vibration $U_0(\underline{x})e^{i\omega t}$, where $U_0(\underline{x})$ varies slowly

$$p'(\underline{x},t) = \rho_o c_o U_o(\underline{x})e^{i\omega t} \quad \text{on} \quad S, \tag{2.52}$$

which is the relationship between the pressure perturbation and particle velocity appropriate for a plane wave travelling in the direction of the normal. The factor $e^{-i\omega d/c_o}$ in (2.51) accounts for the time taken for sound to travel a distance d, while the area ratio A_x/A_s describes the geometrical spread of a ray tube. The square-root term in (2.51) ensures that the energy flow down each tube is invariant. In ray theory the pressure and velocity fields are related to each other as if the wave were locally plane.

However, often the surface velocities induced by engine vibration do not conform to the approximation we have made. Holograms of engine surface vibration [26] show that the displacement has the form of standing waves. As the frequency increases the wavelengths of these surface waves decrease. If the normal velocity associated with such modes is expressed in the form of equation (2.50), $U_o(\underline{x})$ varies increasingly rapidly with $\underline{x}$ as $\omega \to \infty$. The results in (2.51) and (2.52) do therefore not apply, since they were derived by assuming $U_o(\underline{x})$ to be slowly varying. The observed standing waves may be described by a linear combination of travelling waves. We will illustrate the features of the sound field generated by a high frequency wave travelling over a surface by investigating two problems that can be solved analytically.

As a first example, let us consider the case of a vibrating patch in an infinite plane baffle. Suppose that a patch of area Σ carries waves travelling in the x_1-direction with speed V and that the rest of the surface $x_3 = 0$ is at rest. Then the normal velocity is given by

$$\begin{aligned} u_n(x_1,x_2) &= U_1(x_1,x_2)e^{i\omega(t-x_1/V)} && \text{for } x_1,x_2 \text{ within } \Sigma \\ &= 0 && \text{for } x_1,x_2 \text{ outside } \Sigma. \end{aligned} \tag{2.53}$$

$U_1(x_1,x_2)$ varies slowly with position. Equation (2.44) may be used to write down an expression for the pressure field.

$$H(x_3)p'(\underline{x},t) = \frac{\rho_o i\omega}{2\pi}\int_\Sigma \frac{U_1(y_1,y_2)}{\mid \underline{x}-\underline{y}\mid} e^{i\omega(t-y_1/V-|\underline{x}-\underline{y}|/c_o)}dy_1dy_2 \qquad (2.54)$$

In the ray theory limit, $\omega \to \infty$, the integrals in (2.54) may be evaluated by standard asymptotic techniques.

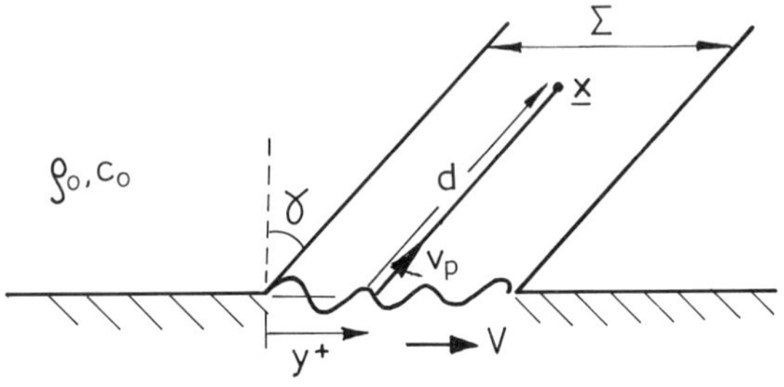

Figure 9. High frequency sound in rays due to normal surface vibration $U_1(x_1,x_2)e^{i\omega(t-x_1/c_0)}$, $V > c_0$, where $U_1(x_1,x_2)$ varies slowly

If the surface wave speed V is supersonic, $V = c_o/\sin\gamma$ say, the method of stationary phase shows that the pressure field may be expressed in the form

$$p'(x,t) = \frac{\rho_o c_o}{\cos\gamma}U_1(y_1^+,x_2)e^{i\omega(t-y_1^+/V-d/c_o)} \qquad (2.55)$$

y_1^+ is the root of $x_1 - y_1^+ = \sin\gamma d$, where $d = ((x_1-y_1^+)^2 + x_3^2)^{\frac{1}{2}}$ is the distance between $(y_1^+, x_2, 0)$ and $\underline{x}$. The point $(y_1^+, x_2, 0)$ is the base of the line passing through $\underline{x}$, which lies in the x_2-plane and makes an angle γ with the 3-axis as shown Figure 9. The expression in (2.55) has a simple physical interpretation. Sound propagates away from the surface in the form of a beam, which makes an angle γ with the 3-direction and whose area of intersection with any 1-2 plane is Σ. On the surface $x_3 = 0$, equation (2.55) reduces to

$$p'(\underline{x},t) = \frac{\rho_o c_o}{\cos\gamma}U_1(x_1,x_2)e^{i\omega(t-x_1/V)}. \qquad (2.56)$$

The component of the particle velocity in the normal direction is $U_1(x_1,x_2)e^{i\omega(t-x_1/V)}$ and $U_1(x_1,x_2)e^{i\omega(t-x_i/V)}/\cos\gamma$ is the resultant velocity in the direction of propagation of the beam. The relationship between pressure and particle velocity described in (2.56) is therefore just the plane wave result,

$$p' = \rho_o c_o v_p, \tag{2.57}$$

where v_p is the particle velocity in the direction of propagation of the wave.

Outside the beam or if V is subsonic, there is no stationary phase point. Then for large ω the main contribution to the integral in (2.54) comes from near the boundaries of the range of integration. This leads to a pressure field that decays as $\omega \to \infty$.

This example has shown us that it is misleading to interpret all surface vibration as a significant sound source. The distribution of surface vibration determines its effectiveness at generating sound. Surface vibrations composed of high frequency waves travelling supersonically over the surface produce beams which propagate away at an angle to the surface. High frequency subsonic surface waves are less efficient radiators. It is only their creation and destruction at the boundaries of the vibrating region that are noisy. We can illustrate that these conclusions are not an artefact of the plane geometry by considering a second example in which waves travel around the circumference of a circular cylinder.

Suppose that S is the surface of a circular cylinder of radius a and that a surface wave travels around the cylinder in the θ-direction with speed V.

$$u_n = U_o e^{i\omega(t-\theta a/V)} \quad \text{say} \tag{2.58}$$

where for a single-valued normal velocity $\omega a/V$ is equal to some integer, n say. The solution for the pressure field can be written directly

$$p'(\underline{x}, t) = -i\rho_o c_o U_o \frac{H_n^{(2)}(\omega\sigma/c_o)}{H_n^{(2)}(\omega a/c_o)} e^{i(\omega t - n\theta)} \tag{2.59}$$

where (σ, θ, x_3) are the polar co-ordinates of $\underline{x}$ and $H_n^{(2)}(z)$ denotes a Hankel function of the second kind. In the ray theory limit, $\omega \to \infty$, the large argument and large order form of the Hankel functions [27] may be used to simplify (2.59).

When V is supersonic, $V = c_o/\sin\gamma$, the resulting expression may be expressed in the form

$$p'(\underline{x}, t) = \frac{\rho_o c_o U_o}{\cos\gamma} \left(\frac{a \cos\gamma}{\sigma \cos(\gamma - \theta + \theta^+)} \right)^{\frac{1}{2}} e^{i\omega(t - \theta^+ a/V - d/c_o)}. \tag{2.60}$$

The position (a, θ^+, x_3) is at the base of a line that passes through $\underline{x}$ and meets the cylinder making an angle γ with the normal, as shown in Figure 10. The sound leaves the cylinder in the form of a ray, inclined at an angle γ, and travels a distance d to $\underline{x}$. The square-root in (2.60) accounts for the geometrical spread of the ray tube. On the cylinder

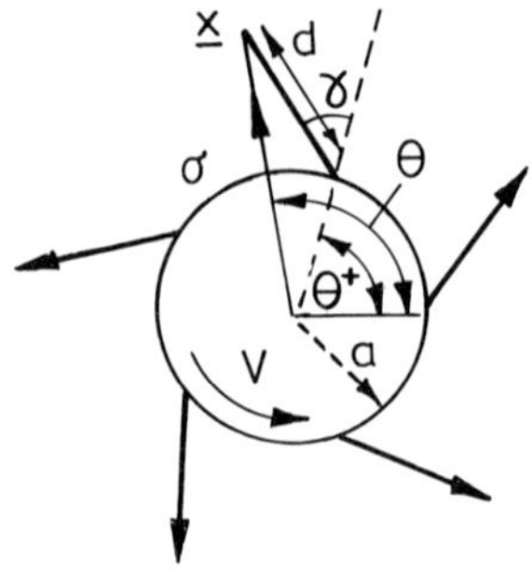

Figure 10. High frequency sound in rays due to normal surface vibration $U_0 e^{i\omega(t-\theta a/V)}$ on a cylinder radius a

$$p'(\underline{x}, t) = \frac{\rho_o c_o U_o}{\cos\gamma} e^{i\omega(t-\theta a/V)} \tag{2.61}$$

and we again recover the plane wave result that the pressure perturbation is $\rho_o c_o$ times the particle velocity in the direction of propagation of the wave.

If V is subsonic, $V = c_o/\cosh\alpha$ say, the pressure perturbation decays with increasing distance from the cylinder for $a < \sigma < ac_o/V$. At a radial distance of ac_o/V, waves with an angular velocity V/a appear to be travelling sonically. Beyond this radius sound waves propagate but are exponentially small. For large σ

$$p'(\underline{x}, t) \simeq 0(\sigma^{-\frac{1}{2}} e^{-(\alpha - \tanh\alpha)\omega a/V}). \tag{2.62}$$

In the ray theory limit of large ω, this pressure perturbation has negligible magnitude.

The sound power, P, emitted by the vibration of a surface S is given by

$$P = \int_S \overline{p' u_n} dS, \tag{2.63}$$

where the overbar denotes a time average. Richards [28] suggests writing

$$P = \sigma_{rad} \rho_o c_o \int_S \overline{u_n^2} dS. \tag{2.64}$$

For high frequency surface waves travelling around a cylinder, equation (2.61) shows that $\sigma_{rad} = (\cos\gamma)^{-1}$ when V is supersonic and $V = c_o/\sin\gamma$. It is apparent from (2.62) that σ_{rad} is exponentially small for high frequency disturbances when V is subsonic. In this simple example σ_{rad} varies from 0 to ∞, illustrating the difficulties inherent in estimating it accurately in practice.

To summarise, if the frequency is sufficiently high that the surface size and radii of curvature are large in comparison with wavelength, ray theory is appropriate. Ray theory predicts that a wave travelling supersonically over the surface with speed V generates sound which propagates in the form of beams making an angle $\sin^{-1}(c_o/V)$ with the surface normal. Standing surface waves composed of two waves travelling supersonically in opposite directions will produce two beams. Investigation of the beam directions can indicate where it is most appropriate to install sound absorbent material in an engine compartment. Alternatively ray theory could be used to estimate the pressure field radiated by an engine in an unconfined region. If the surface wave speeds are subsonic, they radiate high frequency sound less effectively. Most of the surface motion is then silent, with the far-field sound mainly coming from the creation and destruction of the surface waves.

It is often complicated to evaluate the geometrical spread of a ray tube which is required to calculate directly the pressure field radiated by supersonic surface motion. An alternative way of applying ray theory is just to use it to determine the surface pressure. It gives

$$p'(\underline{x}, t) = \frac{\rho_o c_o}{(1 - c_o^2/V^2)^{\frac{1}{2}}} u_n \text{ on } S \tag{2.65}$$

The radiated pressure field then follows once this has been substituted into (2.40)

$$H(\underline{x})\hat{p}(\underline{x}, \omega) =$$

$$\frac{\rho_o}{4\pi} \int_S \left\{ \frac{i\omega \hat{u}_n e^{-i\omega|\underline{x}-\underline{y}|/c_o}}{|\underline{x}-\underline{y}|} + \frac{c_o \hat{u}_n}{(1-c_o^2/V^2)^{\frac{1}{2}}} \frac{\partial}{\partial n}\left(\frac{e^{-i\omega|\underline{x}-\underline{y}|/c_o}}{|\underline{x}-\underline{y}|} \right) \right\} dS \tag{2.66}$$

The integrals can be evaluated in a straight-forward way.

We can check this procedure by applying it to our exact problem in which S is the surface of a sphere of radius b and $\hat{u}_n = U_o$ (see equation 2.45). In this case all the surface motion is in phase and V is infinitely large. For such a simple velocity distribution the integrals in (2.66) can be evaluated analytically, and we find that

$$\frac{1}{4\pi} \int_S \frac{i\omega \hat{u}_n}{|\underline{x}-\underline{y}|} e^{-i\omega|\underline{x}-\underline{y}|/c_o} dS = \frac{U_o c_o b}{r} i \sin(\omega b/c_o) e^{-i\omega r/c} \tag{2.67}$$

(c.f. equation 2.48) while

$$\frac{c_o}{4\pi}\int \frac{\hat{u}_n}{(1-c_o^2/V^2)^{\frac{1}{2}}}\frac{\partial}{\partial n}\left(\frac{e^{-i\omega|\underline{x}-\underline{y}|/c_o}}{|\,\underline{x}-\underline{y}\,|}\right)dS = \frac{U_o c_o b}{r}\cos(\omega b/c_o)e^{-i\omega r/c_o} \tag{2.68}$$

$r = |\,\underline{x}\,|$. When these two terms are substituted into (2.66), the predicted radiated pressure agrees with the high frequency form of the exact pressure field in (2.47).

We have concentrated in particular on engine noise because it is a dominant source of sound within the interior of vehicles. At low forward speeds, it is also the most significant component of the external radiated noise. However as the vehicle speed increases the noise generated by the tyres becomes increasingly important.

3 Tyre Noise

At vehicle speeds above about 40 mph for passenger cars and 55 mph for trucks, tyres are found to be a dominant source of radiated sound. While the annoyance due to engine noise can be reduced by enclosing the engine compartment and lining it with sound absorbent materials and by installing efficient silencers, there is no effective way of attenuating tyre noise once it has been generated. In spite of its importance, the mechanisms involved in the generation of sound by tyres are not clearly understood.

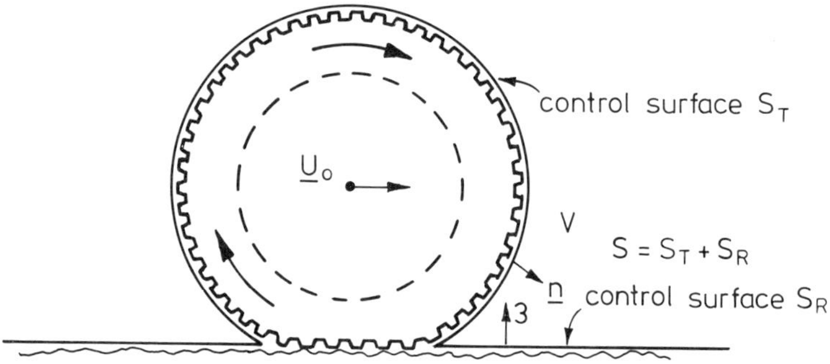

Figure 11. A tyre and road surrounded by a control surface S

Let us consider a rotating tyre in contact with a rough road and moving forward with a constant velocity $\underline{U}_o$. We will surround the tyre and surface roughness with a control surface S, as shown in Figure 11. S has two parts, S_R and S_T, both of fixed shape and moving with a velocity $\underline{U}_o$. S_R consists of part of a plane surface positioned just above the highest surface roughness and parallel to the road. Let us choose our co-ordinate system so that S_R lies in the plane $x_3 = 0$. The surface S_T is large enough to enclose the tyre at all times. Applying Lighthill's theory of aerodynamic sound [29] to the region V outside S gives

$$4\pi c_o^2 H\rho'(\underline{x},t) = \frac{\partial^2}{\partial x_i \partial x_j}\int_V \left[\frac{T_{ij}}{r\mid 1-M_r\mid}\right] d^3\underline{\eta}$$
$$-\frac{\partial}{\partial x_i}\int_S \left[\frac{p\delta_{ij}+\rho u_i(u_j-U_{oj})}{r\mid 1-M_r\mid}\right] n_j dS(\underline{\eta})$$
$$+\frac{\partial}{\partial t}\int_S \left[\frac{\rho(u_i-U_{oi})+\rho_o U_{oi}}{r\mid 1-M_r\mid}\right] n_i dS(\underline{\eta}) \quad (3.1)$$

(see [2], equation 9.41). ρ' is the density perturbation, T_{ij} denotes Lighthill's quadrupole, $T_{ij} = \rho u_i u_j + (p' - c_o^2\rho')\delta_{ij}$. $\underline{\eta}$ is a co-ordinate in a reference frame that moves with velocity $\underline{U}_o$. The square brackets denote that the functions they enclose are to be evaluated at a retarded time τ defined by

$$c_o(t-\tau) = \mid \underline{x} - \underline{\eta} - \underline{U}_o\tau \mid \quad (3.2)$$

r is the distance between the observer at $\underline{x}$ and the source at emission time:

$$r = \mid \underline{x} - \underline{\eta} - \underline{U}_o\tau \mid, \quad (3.3)$$

and M_r is the component of the Mach number, $\underline{U}_o/c_o$, in the direction of the observer:

$$M_r = \frac{\underline{U}_o\cdot(\underline{x} - \underline{\eta} - \underline{U}_o\tau)}{c_o r}. \quad (3.4)$$

When investigating the sound generation processes in tyres it is usual to linearize the fluid motion. If we also linearize in the height and slope of the roughness, equation (3.1) simplifies to

$$4\pi H p'(\underline{x},\underline{t}) = -\frac{\partial}{\partial x_i}\int_{S_T} \frac{p'(\underline{y}, t - \mid \underline{x}-\underline{y}\mid / c_o)}{\mid \underline{x}-\underline{y}\mid} n_i dS$$
$$+\frac{\partial}{\partial t}\int_{S_T} \frac{\rho_o u_n(\underline{y}, t - \mid \underline{x}-\underline{y}\mid / c_o)}{\mid \underline{x}-\underline{y}\mid} dS$$
$$-\frac{\partial}{\partial x_3}\int_{S_R} \frac{p'(\underline{y}, t - \mid \underline{x}-\underline{y}\mid / c_o)}{\mid \underline{x}-\underline{y}\mid} dy_1 dy_2 \quad (3.5)$$

Finally, the pressure perturbation on S_R can be eliminated by adding equation (3.5) for a point (x_1, x_2, x_2) to the corresponding expression for $\underline{x}^* = (x_1, x_2, -x_3)$. This gives

$$
\begin{aligned}
4\pi H p'(\underline{x},t) &= -\frac{\partial}{\partial x_i}\int_{S_T}\frac{p'(\underline{y},t-\mid\underline{x}-\underline{y}\mid/c_o)}{\mid\underline{x}-\underline{y}\mid}n_i dS \\
&- \frac{\partial}{\partial x_i^*}\int_{S_T}\frac{p'(\underline{y},t-\mid\underline{x}^*-\underline{y}\mid/c_O)}{\mid\underline{x}^*-\underline{y}\mid}n_i dS
\end{aligned}
$$

$$
+\rho_o\frac{\partial}{\partial t}\int_{S_T}\left(\frac{u_n(\underline{y},t-\mid\underline{x}-\underline{y}\mid/c_o)}{\mid\underline{x}-\underline{y}\mid}+\frac{u_n(\underline{y},t-\mid\underline{x}^*-\underline{y}\mid/c_o)}{\mid\underline{x}^*-\underline{y}\mid}\right)dS \quad (3.6)
$$

a result which could also have been obtained by using the half-space Green function in the derivation of (3.1). As discussed in section 2, equations of this form are integral equations for $p'(\underline{x},t)$. In principle $p'(\underline{x},t)$ is completely determined once u_n is known. In practice a numerical solution might be necessary.

The normal fluid velocity u_n may be decomposed into contributions due to the forward velocity and to perturbations caused by the tread and road roughness by writing

$$
u_n = \underline{U}_o \bullet \underline{n} + u_n' \quad (3.7)
$$

The term $\underline{U}_o \bullet \underline{n}$ leads to a 'source' that is constant in a subsonic, uniformly moving reference frame and so does not generate sound ([2], page 190). A smooth tyre on a flat road would be silent.

An early theory of tyre noise was proposed by Hayden [30]. He concentrated on the contribution to u_n' from air squeezed out of the grooves between tread blocks as the blocks were compressed on contact with the ground. This process is usually referred to as "air pumping".

Recent work has emphasized the sound generated by tyre vibrations induced by unsteady forces associated with the contact and release of the tread blocks with the road. These forces excite modes of vibration of the tyre and produce surface velocities u_n'. Heckl [31] models a tyre by an elastic circular ring and finds that its natural modes of vibration consist of waves running subsonically around the circumference. At high frequencies such subsonic surface waves do not radiate sound efficiently (see equation 2.62). However their radiation efficiency can be dramatically increased if their near-field pressure fluctuations excite acoustic resonances. Jennewein & Bergmann [32] find the high frequency spectrum of tyre noise to be dominated by resonances of the air columns in tread grooves. They claim that the near-field pressures due to tyre vibration excite large velocity perturbations in and out of the column or cavity near its resonant frequency. These lead to large localized values of u_n' which can generate significant far-field sound. On the other hand, Watanabe et al. [33] identify the vibration of the sidewall of cross-bar tyres as their major noise source by using intensity techniques.

We must conclude that the processes of sound generation by tyres are still not understood. The problem is complicated with many complementary effects. There is a role for model problems designed to illustrate the principles and to quantify these effects.

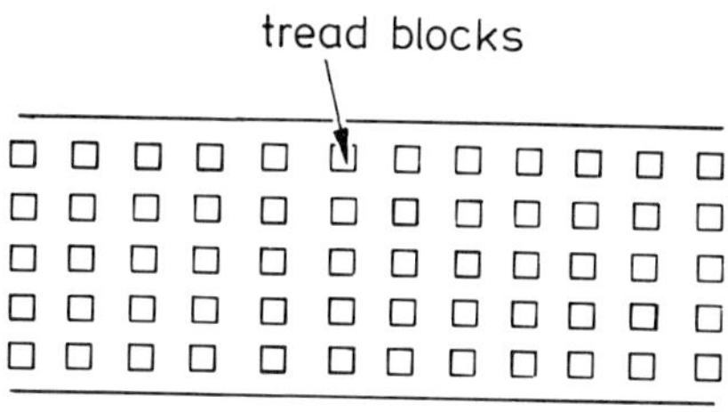

Figure 12. A tyre with a regular pattern of tread blocks

Even without a theoretical description of the source processes, designers have been able to reduce the annoyance of tyre noise by detuning the tyres. Consider a tyre with the tread pattern shown in Figure 12. The tread blocks all across the tyre impact with the road virtually simultaneously, leading to intense forces on the tyre. Moreover the uniform circumferential spacing between tread blocks will produce an annoying pure tone. By varying the length of the tread blocks and the spacing between them, both across and around the tyre, the frequency content of the noise can be broadened. Vorih [34] pioneered this technique. The current state of the art is described in [35]. The idea is to assume that a δ-function of sound is produced everytime a tread block impacts on the road. The spectrum is then evaluated for a particular arrangement of tread blocks. Finally the length and spacing between individual tread blocks is redesigned until a sufficiently 'broad-band' spectrum is obtained, with energy fairly uniformly distributed throughout the high harmonics of the tyre rotation frequency. Of course, since the description of the noise generated by the impact is so crude, such a procedure can only predict the position of the peaks in the spectrum, not their magnitude.

4 Flow Noise

Flow (or wind) noise refers to the interior noise due to the external flow of air over a vehicle. At the moment flow noise is only the dominant source of internal sound in passenger cars at speeds in excess of 100 mph. But it can be obtrusive at lower speeds. If techniques of engine containment and anti-sound [36], [37] continue to reduce the engine contribution to the internal sound, it is likely that flow noise will become relatively more important in the future. Measurements on a range of production cars [38], [39] have shown that flow noise does not correlate with the overall drag coefficient. There is a scope to minimize the flow noise without inducing a drag penalty.

The flow perturbations within the interior of a car's passenger compartment are linear and we will again use Green's theorem to discuss the sound generation. We choose S to be a control surface enclosing the passenger compartment as shown in Figure 13. Then (2.40) leads to

$$4\pi H(\underline{x})\hat{p}(\underline{x},\omega) = \int_S \left\{ \frac{i\omega\rho_o\hat{u}_n}{|\underline{x}-\underline{y}|} e^{-i\omega|\underline{x}-\underline{y}|/c_o} + \hat{p}\frac{\partial}{\partial n}\left(\frac{e^{-i\omega|\underline{x}-\underline{y}|/c_o}}{|\underline{x}-\underline{y}|}\right)\right\} dS \tag{4.1}$$

This is an integral equation for $\hat{p}(\underline{x},\omega)$, the Fourier transform of the pressure perturbation. In principle $\hat{p}(\underline{x},\omega)$ is completely specified once $\hat{u}_n$ is known, but in practice a numerical solution may be necessary.

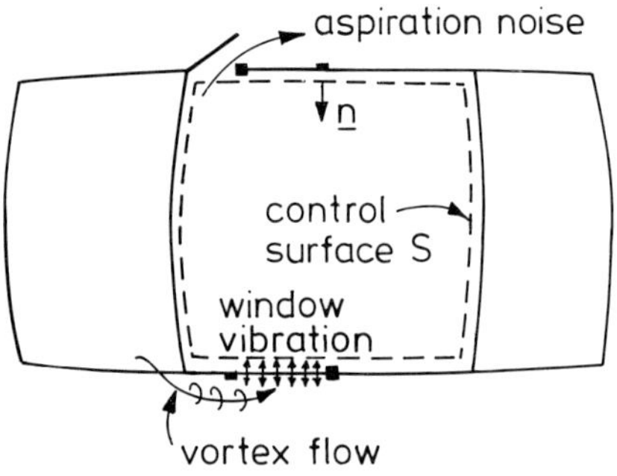

Figure 13. Sources of flow noise

Engine sound and vibration can vibrate panels surrounding the passenger compartment and so produce internal pressure fluctuations. Similarly, normal surface velocities $\hat{u}_n$ can also be induced by the unsteady pressure field of the air-flow over the vehicle. This is referred to as the flow noise and there are two main types.

Whenever there is an opening, which might be an open window or sun-roof, or gaps in door joints or poor window seals, there is a direct interaction between the passenger compartment and the external flow. Then the pressure fluctuations in the external air-flow drive an unsteady flow through the gap which couples to the internal pressure field. This is called 'aspiration noise' and the effective cure is to improve the sealing.

The second type of flow noise is usually referred to as 'wind rush noise'. This is due to the vibration of the car body by fluctuating pressures in the external flow. The vibration of side windows is a particularly important source. This is partly because a vortex flow from the windscreen corners impinges on the side windows resulting in strong unsteady pressures and partly because the windows are fairly flexible and unlike the metallic panels are not covered with damping material. Buchheim et. al. [38] present a highly simplified theory of wind rush noise, which nevertheless gives reasonable agreement with their experiment. Similar flow-induced noise occurs in both underwater and aeronautical applications. There more sophisticated analyses have been developed (see [40] for a recent review), which could be applied to vehicle geometries.

5 Conclusion

The main sources of vehicle noise have been discussed within a unified framework. Engine noise is often dominant and, in many ways, is well understood. Theories of tyre noise and for wind-induced noise are much less developed and there is considerable scope for mathematical modelling of these processes.

References

1. Dowling, A.P., (1986), Thermoacoustic sources, IUTAM Symposium on Aero- and Hydro-Acoustics, 183-209, Springer.

2. Dowling, A.P., and Ffowcs Williams, J.E., (1983), Sound and Sources of Sound, Ellis Horwood.

3. Lyn, W.T., (1963), Study of burning rate and nature of combustion in diesel engines, 9th Symposium (International) on Combustion, 1069-1082. The Combustion Institute, Academic Press.

4. Head, H.E., (1984), A model to predict combustion noise and ignition delay in diesel engines, I. Mech. E. Conf. Publication, Vehicle Noise and Vibration, C156/84.

5. Baxa, D.E., (1982), Noise Control in Internal Combustion Engines, Wiley.

6. Priede, T., (1980), In search of origins of engine noise, an historical review, SAE Trans. 800534.

7. Van der Pol, B., and Bremmer, H., (1950), Operational Calculus Based on the Two-sided Laplace Integral, Cambridge University Press.

8. Priede, T., (1982), Road vehicle noise, Noise and Vibration, ed R.G. White and J.G. Walker, 389-437, Ellis Horwood.

9. Pettitt, R.A., (1988), Noise reduction of a four litre direct injection diesel engine. I. Mech. E. Conf. Publication, Advances in the Control and Refinement of Vehicle Noise, C28/88.

10. Croker, M.D., (1988), The analysis and control of engine noise quality. I. Mech. E. Conf. Publication, Advances in the Control of Refinement of Vehicle Noise, C16/88.

11. Rohrle, M.D., (1975), Affecting diesel engine noise by the piston, SAE Paper 750799.

12. Croker, M.D., (1985), Engine structure analysis for low noise - the options, SAE Paper 850970.

13. Lyon, R.H., (1975), Statistical Energy Analysis of Dynamical Systems: Theory and Applications, M.I.T. Press.

14. Fahy, F.J., (1982), Statistical energy analysis, Noise and Vibration, ed. R.G. White and J.G. Walker, 165-186, Ellis Horwood.

15. Woodhouse, J., (1981), An introduction to statistical energy analysis of structural vibration, Applied Acoustics, 14, 455-469,

16. Hodges, C.H. and Woodhouse, J., Theories of noise and vibration transmission in complex systems, Rep. Prog. Phys., 49, 107-170.

17. Challen, B.J. and Croker, D.M., (1982), A review of recent progress in diesel engine noise reduction, SAE Paper 820517.

18. Carr, I.E. and Lalor, N., (1988), The prediction and reduction of noise from engine covers using energy methods, I. Mech. E. Conf. Publication, Advances in the Control and Refinement of Vehicle Noise, C20/88.

19. Koopmann, G.H. and Benner, H., (1982), Method for computing the sound power of machines based on the Helmholtz integral, J. Acoust. Soc. Am., 71, 78-89.

20. Hodson, D.C. and Sadek, M.M., (1983), A technique for the prediction of the noise field from an arbitrary vibrating machine, Proc. Instn. Mech. Engrs., 197*C*, 189-197.

21. Kirkup, S.M., Henwood, D.J. and Tyrrell, R.J., (1988), Engine noise-practicalities and prediction Part 3: noise prediction using the boundary element method, I. Mech. E. Conf. Publication, Advances in the Control and Refinement of Vehicle Noise, C33/88.

22. Kirkup, S.M., Henwood, D.J. and Tyrrell, R.J., (1989), The prediction of the effect of engine noise shields using a combination of boundary, shell and finite elements. IMA Conference on Mathematics in the Automotive Industry, Warwick.

23. Fahy, F.J., (1982), Structure-fluid interaction, Noise and Vibration, ed R.G. White and J.G. Walker, 247-266, Ellis Horwood.

24. Pierce, A.D., (1981), Acoustics: An Introduction to its Physical Principles and Applications, McGraw-Hill.

25. Crighton, D.G., Dowling, A.P., Ffowcs Williams, J.E. and Leppington, F.G., Chapter 16, Course notes for Short Course on Aero- and Hydro-Acoustics, Stanford.

26. Ochiai, K., Aisaka, M. and Sakata, S., (1975), Simple model technique for better understanding of diesel engine vibration and noise. SAE Paper 750834.

27. Abramowitz, M. and Stegun, I.A., (1965), Handbook of Mathematical Functions, Dover.

28. Richards, E.J., (1981), On the prediction of impact noise, III: energy accountancy in industrial machines, J. Sound Vib., 76, 187-232.

29. Lighthill, M.J., (1952), On sound generated aerodynamically, Part I. General theory, Proc. Roy. Soc. *A*211, 564-587.

30. Hayden, R.E., (1971), Roadside noise from the interaction of a rolling tyre with the road surface, in M.J. Crocker (ed), Noise and Vibration Control Eng., Purdue, Noise Control Foundation, Poughkeepsie, N.Y.

31. Heckl, M., (1986), Tyre noise generation, Wear, 113, 157-170.

32. Jennewein, M. and Bergmann, M., (1984), Investigations concerning tyre/road noise sources and the possibilities of noise reduction, I. Mech. E. Conf. Proc. on Vehicle Noise and Vibration, C150/84.

33. Watanabe, T., Tomita, N., Kishinami, S., Yamaguchi, H., Konishi, S. and Kawana, A., (1988), The noise generation mechanism of cross-bar tyres and countermeasures against it. Int. J. of Vehicle Design, 9, 641-653.

34. Vorih, W.J., (1971), Designing quiet tread spacings for tires, 99th Rubber Division meeting of the American Chemical Society, Miami Beach, Flo.

35. Mambretti, F.N., (1988), Tyre noise simulation at a computer, I. Mech. E. Conf. Proc. on Advances in the Control and Refinement of Vehicle Noise, C35/88.

36. McDonald, A.M., Quinn, D.C., Saunders, T.J., Elliott, S.J., Nelson, P.A. and Stothers, I.M., (1988), Adaptive noise control of automobile interior noise, I. Mech. E. Conf. Proc. on Advances in the Control and Refinement of Vehicle Noise.

37. Ross, C.F., (1989), Active cancellation of cabin noise, 20th International Symposium on Automotive Technology and Automation, Florence, Italy.

38. Buchheim, R., Dobrzynski, W., Mankau, H. and Schwabe, D., (1982), Vehicle interior noise related to external aerodynamics, Int. J. of Vehicle Design, 3, 398-410.

39. Sakoda, M., Nakagawa, K. and Koike, M., (1985), Reduction of aerodynamic noise, JSAE Review, 66-72.

40. Leehey, P., (1988), Structural excitation by a turbulent boundary layer: an overview, ASME J of Vibration, Acoustics, Stress and Reliability in Design, 110, 220-225.

Computational Techniques for Engine Noise Prediction

S.M. Kirkup[1], R.J. Tyrrell[2] and D.J. Henwood[3]

University of Salford[1], Ricardo Consulting Engineers Limited[2] and Brighton Polytechnic[3]

Abstract The prediction of noise from the design of an engine requires computational models for predicting both the vibratory and acoustic properties. In this paper recent developments in the use of the finite element method for predicting the vibratory properties and the application of the boundary element method for the prediction of the acoustic properties are considered. The methods are demonstrated on the Ricardo asymmetric crankcase simulation rig.

1 Introduction

Noise is the unwanted sound that accompanies all vehicles. Laws restricting the noise output of newly-designed vehicles, and the need to satisfy customers has meant that it has become an important factor for automotive engineers to consider. In most operating conditions the engine is the main contributor to the total vehicle noise. Hence the burden of designing low-noise vehicles has largely fallen to the engine designer.

The relationship between the noise produced by engines, the vibration of the engine, the design of the engine structure and the excitation forces has interested engine designers for several decades. Equipment exists for the measurement of each of these properties and their relationship is usually analysed after they are resolved into their frequency components. The measurements will then enable the engineer to understand how the noise arises and perhaps decide how the design may be modified to reduce the noise. However, development in this way is costly and hence, over recent years, a computer-aided approach to the design of low-noise engines has been pursued.

The steady improvement in both numerical techniques and computer equipment over recent decades, has meant that it is possible to make computational predictions of both the vibratory and acoustic properties of engines. The object of this paper is to describe the advanced computational techniques that are employed by Ricardo for the noise analysis of engines. The use of the finite element method (FEM) for the prediction of vibratory properties is included. However, the main concern is with the use of the boundary element method (BEM) for acoustic prediction.

2 Outline of the Process of Designing Low-Noise Engines

The function of the engine is to convert fuel into the mechanical forces which power the vehicle. The vibration of the engine structure is a consequence of the engine's internal forces and combustion of the fuel. The noise is the perceived acoustic response of the air to the surface vibration.

In the design of an engine, the noise is analysed after the basic design of the engine is complete. There is then scope for making alterations to the design to reduce the noise. The process of designing an engine for low noise may be broadly divided into two parts which will be termed the pre- and post-prototype stages.

2.1 The pre-prototype stage

In the process of designing an engine, methods for predicting the vibratory and acoustic properties from a draft of the design of the engine would be useful. Such information would indicate whether or not the engine would be too noisy when built and perhaps how the design may be improved. Once the design of the engine is such that the predicted noise is satisfactory, a prototype of the engine is then built.

Figure 1 illustrates this part of the design process. The efficiency and even the success of this process is dependent on the accuracy of the predictions, the ability to interpret the predicted data, and the ability to make successful alterations to the engine design based on this interpretation.

2.2 The post-prototype stage

Once the prototype is built, the vibratory and acoustic properties can be measured. However, it may be more feasible to predict the acoustic properties from the surface vibration, should such an approach be available. If the vibratory and acoustic properties of the prototype are satisfactory then the design is complete. If the prototype is unsatisfactory then, provided only minor modifications in the design are required to correct it, these alterations can be carried out and the design would then be complete. It is possible to find that the prototype engine is much too noisy and that any modifications to the prototype are insufficient to reduce the noise to an acceptable level. In this case the prototype would have to be abandoned and

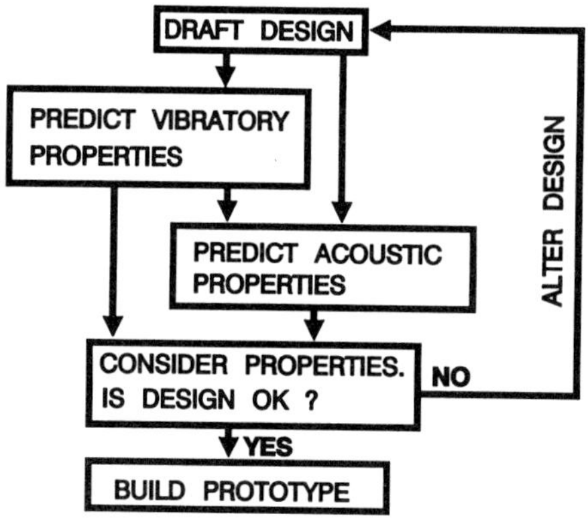

Figure 1. Part 1 of the Process of Low-Noise Design

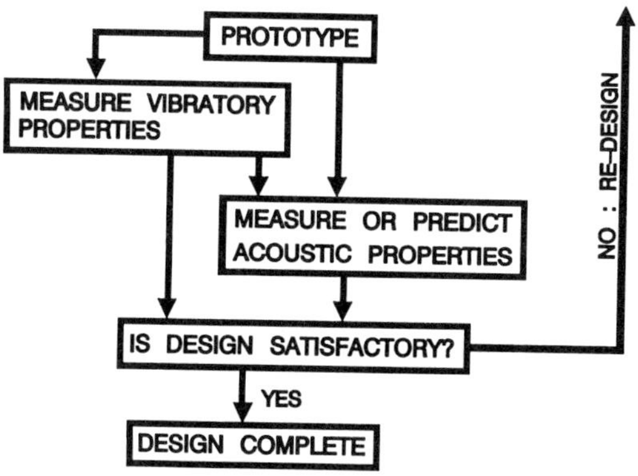

Figure 2. Part 2 of the Process of Low-Noise Design

the engine would have to be re-designed. This part of the design process is illustrated in figure 2.

3 The Design Process in Detail

Noise has been an important factor in the design of engines for several decades. In this period, engineers have become more experienced in designing low-noise engines, using simple relationships between design parameters and noise. However, this approach is severely limited since the relationship between the design and the noise produced by an engine is not trivial. A more valid approach is to base noise prediction methods on the principles that govern the properties of the engine structure and the air which surrounds it.

In the design process, illustrated in figures 1 and 2, the design engineer has to be able to interpret the vibratory and acoustic information and judge whether or not the engine design is satisfactory. Furthermore, if the design is unsatisfactory then he must decide how he should alter the design of the engine to reduce its noise. Clearly, the more reliable and comprehensive the vibratory and acoustic information presented to him, the better placed he is to make these judgements.

3.1 The evaluation of the vibratory properties

The response of the structure to the internal forces, the structural mode shapes and the resonant frequencies are some of the vibratory properties that are useful to the design engineer. Vibratory information about the structure can be measured using, for example, accelerometers under test operating conditions (see [4]). The required vibratory properties can be obtained through a Fourier analysis of this information. Alternatively, the vibratory properties of an engine structure can be predicted, usually by the FEM, which is based on the equations of motion. This approach is considered in section 5.

3.2 The evaluation of the acoustic properties

Sound power spectra, surface intensity patterns, the sound pressure at points in space, and radiation ratio curves for individual mode shapes are some of the acoustic properties that may be useful to the design engineer. Acoustic properties can generally be measured (see [4]). Methods for predicting the acoustic properties from the surface vibration can be based on the Rayleigh integral method (RIM) or the boundary element method (BEM). There is also a popular simple method for estimating the sound power which will be termed the sound power integral method (SPIM). These methods are considered within the context of engine noise in section 6.

4 The Simulation of Engine Structures at Ricardo

In order to gain an understanding of the effect of different designs of engine on the vibratory and acoustic properties, the noise refinement centre at Ricardo embarked on building and testing four engine-like structures which are known as crankcase simulation rigs (CSRs). The engine-like structures had all the main features of a heavy duty diesel engine crankcase/ cylinder block/ cylinder head assembly, but with the minimum of irregular details which characterise a real engine design. In this paper the computational methods are demonstrated on one of these structures known as the asymmetric CSR (CSRASYM).

5 Prediction of the Vibratory Properties

The finite element method is now regarded as a reliable tool for modelling the vibratory behaviour of elastic structures. Indeed, versatile commercial software packages are now available for carrying out this task. At Ricardo the PAFEC software [7] is implemented which, when applied by experienced users, is able to predict the vibratory properties of complex structures such as engine blocks. The vibratory properties that are generally of most interest to engine designers are the modeshapes with corresponding resonant frequencies and the forced response.

6 Prediction of the Acoustic Properties

In this section the suitability of the three acoustic prediction methods (the SPIM, RIM and BEM) for the problem of predicting the acoustic properties of air surrounding the engine block are considered. Each of these methods has been applied to the CSRs by Ricardo, see references [1,2,3,5,6,8]. Table 1 summarises the effectiveness of the methods. For more details see [6].

Table 1. A Comparison of the Methods for the Prediction of Engine Noise

	SPIM	RIM	BEM	Measuring
Sound Pressure	x	?	√	√
Surface Sound Intensity	x	?	√	√
Sound Power	?	? or √	√	? or √
Radiation	x	? or √	√	?

6.1 The 'no sound interaction' model (SPIM)

Assuming that the sum of the sound powers from each point on the surface is equal to the total sound power allows us to derive the sound power integral method (SPIM). The method is very crude and it is inappropriate for the prediction of the other acoustic properties. The method is equivalent to assuming the radiation ratio is always equal to one. Loosely speaking, for engine the radiation ratio is roughly unity for frequencies greater than approximately 1kHz and the model can therefore yield a rough prediction of sound powers at higher frequencies.

6.2 The baffled plate model (RIM):

The background to this method is that it is assumed that the engine is made up of a set of flat plates which radiate independently. The Rayleigh integral method (see [6]) is then applied to each face. In order to use this method for the prediction of sound power or radiation ratio the engine surface must be idealised into a set of flat plates. Hence this method is likely to be less satisfactory for V-shaped engines than for slab-sided in-line engines.

The sound pressure can perhaps only be found to satisfactory accuracy at points near to and on the parts of the engine surface where the response is strongest at any particular frequency. The same is true for the surface intensity. The sound power and hence the radiation ratio calculated via the RIM should usually be obtainable to reasonable accuracy.

6.3 The closed vibrating surface model (BEM)

From the theoretical point of view, this model is clearly most suited to the problem of predicting the acoustic properties of the air surrounding a vibrating engine. It is able to represent closely the physical situation of the engine in free-space or in an anechoic chamber. Basing the method on this model leads to the boundary element method (BEM); see reference [6] for more details. The main difference between the physical problem and the model is that in order to use the BEM the engine surface must be idealised as a simple closed surface so that many details on the engine's surface will need to be left out.

There have been a number of reports on the application of the BEM to engine and machine noise prediction problem, see [5,6] for references. However, previous research generally discusses the potential of the BEM and generally presents results at just one or two frequencies.

6.4 Summary

The potential for success or otherwise in the use of different methods for evaluating the acoustic properties of the air surrounding a vibrating engine for noise purposes is illustrated in table 1. It must be realised that for engine noise purpose the most important region of the frequency spectrum is usually approximately the 400Hz to 2400Hz region, and, for noise purposes, the acoustic properties need not be evaluated to high accuracy.

7 Experimentation with the CSRASYM

In this section the relevant experiments on the CSRASYM are described.

7.1 Measurements on the CSRASYM

In order to take measurements of the acoustic and vibratory properties, the CSRASYM was suspended on elastic cords in a semi-anechoic test cell. Surface vibration response to banger-excitation was measured at some 250 points using a miniature accelerometer. The sound pressure was measured at 10 microphone points on a hemisphere of radius 1.4m and at 8 microphone points on a cylinder of radius 1.4m which joins onto the hemisphere. The sound power was determined by calculating a numerical approximation based on an integal over the hemisphere and cylinder. The surface acceleration, the sound pressures, the sound powers, and the cylinder pressure were converted into 10Hz narrow band frequency spectra. For more details see references [1,2].

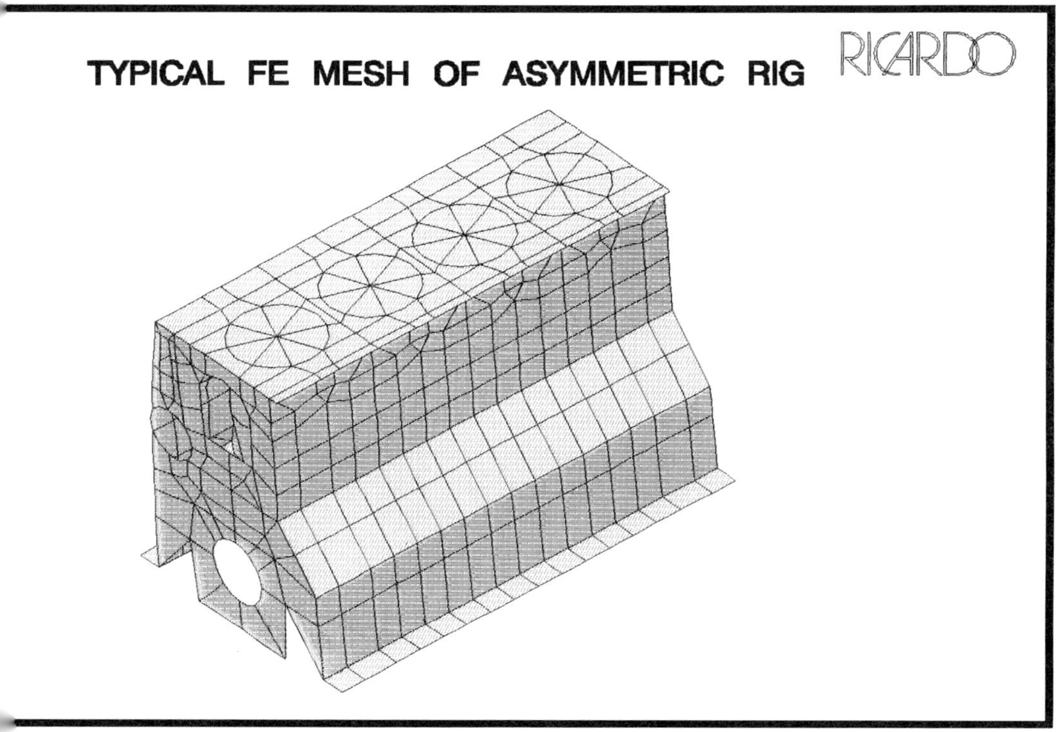

Figure 3.

7.2 Application of the finite element method

Ricardo applied the FEM to the CSRs as a preliminary test of the effectiveness of the method. For the CSRs large finite element models containing up to 4,000 high order shell and solid elements with up to 50,000 degrees of freedom were used. The finite element model of the CSRASYM is illustrated in figure 3. The comparison of computed and measured vibratory properties is a severe test on the FEM. In order to predict the forced response a knowledge of the acting internal forces is required. These forces may be specified by measurement or analysis, though usually it is not feasible to obtain a detailed description of them. See [7] for more details.

For the CSRs, calculations were based on the concept of a unit magnitude flat in-cylinder pressure spectrum. The use of this excitation resulted in forced vibration data in the form of transfer functions which are readily comparable with the exprimentally derived results. Pressure forces were applied directly to model cylinder flame faces. Forces on the main bearings are calculated by means of calculated transfer paths through the piston/

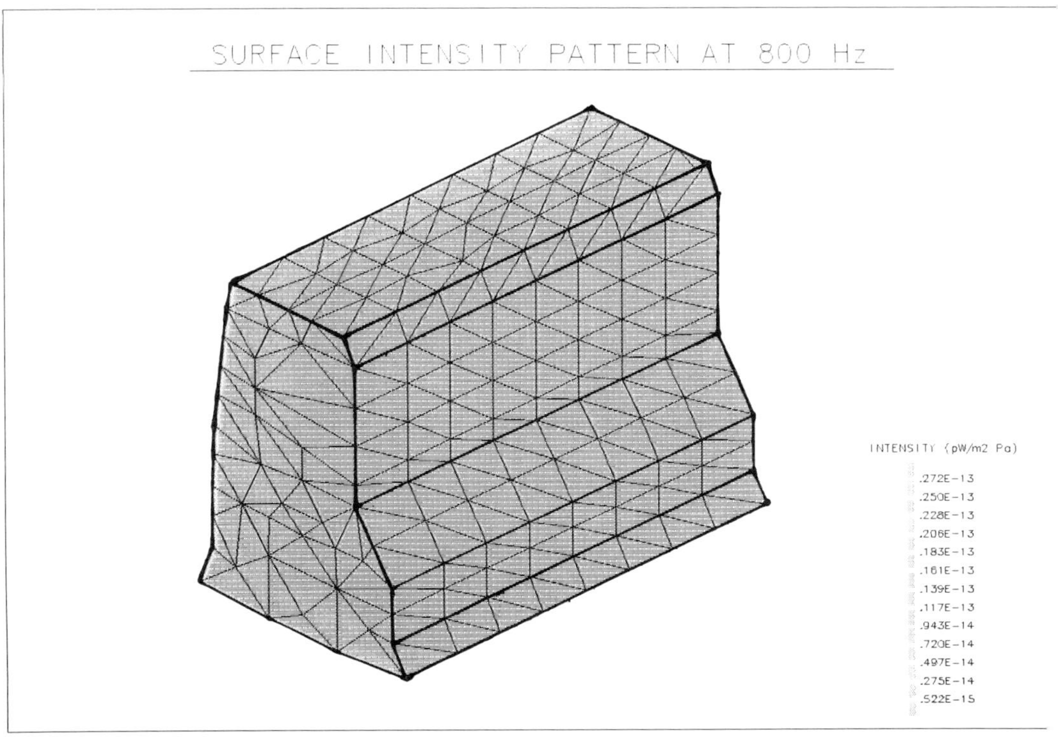

Figure 4.

connecting rod/ crankshaft assembly. As forced vibration analysis yields vast amounts of data, Ricardo have implemented the model synthesis calculations as a front end to the programs for calculation of acoustic properties. By this means the need to store large amounts of surface vibration data is avoided.

7.3 Application of the boundary element method

In order to apply the BEM to the CSRASYM the surface is idealised into a simple closed surface. The surface is then approximated by around 550 planar triangular elements. Generally, the positions of the elements are chosen so that their vertices are at the accelerometer points. The boundary element mesh of the CSRASYM is shown in figure 4.

On each boundary element the surface vibration is specified by interpolating the values at the vertices of the components of the velocity normal to the element. At vertices where there was no accelerometer reading the velocity is prescribed zero values.

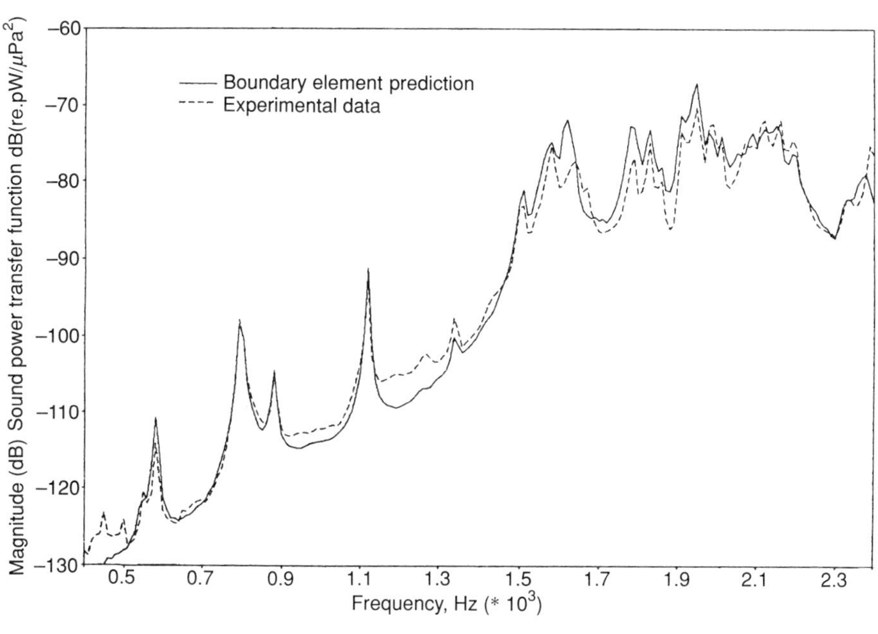

Figure 5. Comparison of the Measured and Computed Sound Power Transfer Function Spectra

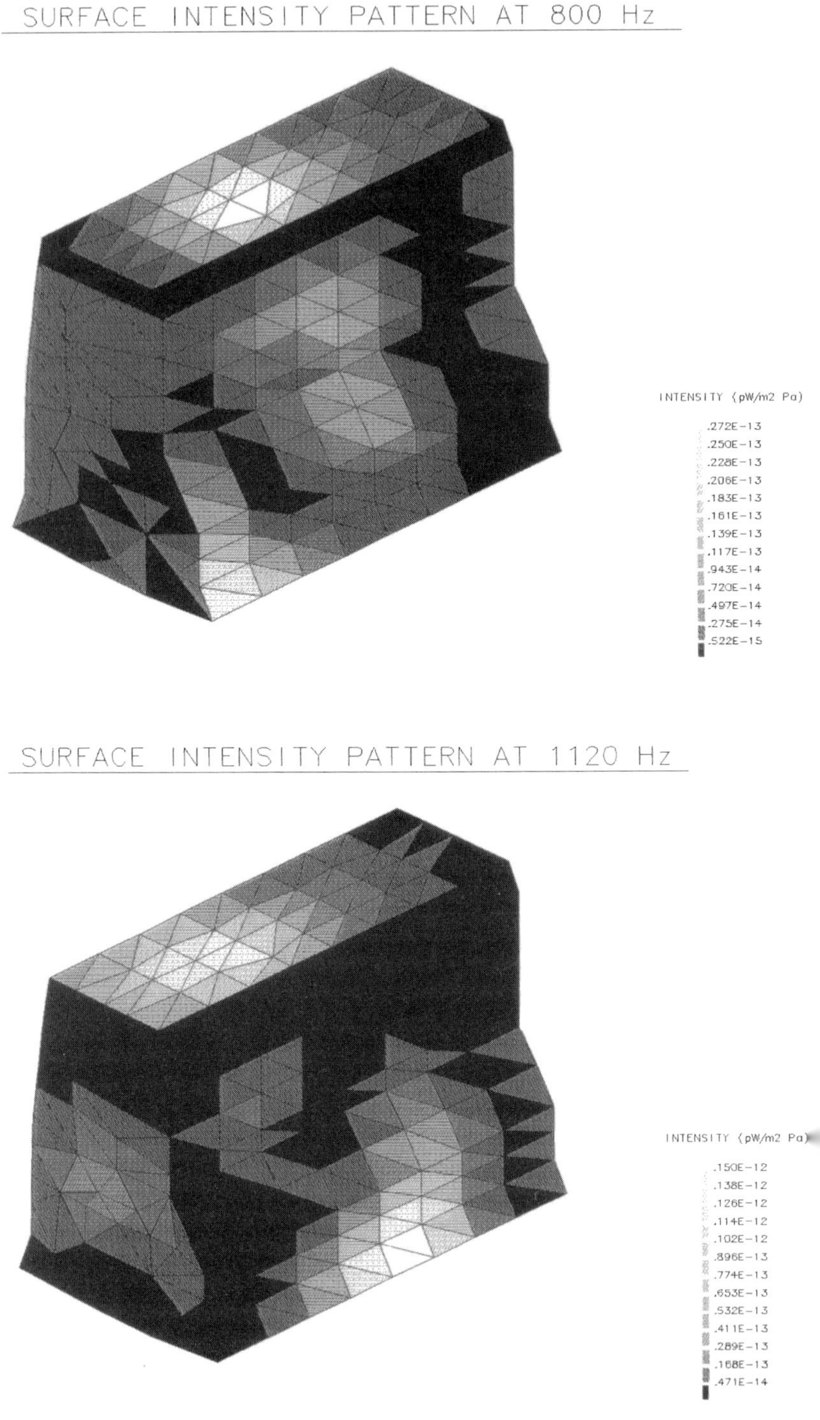
SURFACE INTENSITY PATTERN AT 800 Hz
INTENSITY (pW/m2 Pa)
.272E-13
.250E-13
.228E-13
.206E-13
.183E-13
.161E-13
.139E-13
.117E-13
.943E-14
.720E-14
.497E-14
.275E-14
.522E-15
SURFACE INTENSITY PATTERN AT 1120 Hz
INTENSITY (pW/m2 Pa)
.150E-12
.138E-12
.126E-12
.114E-12
.102E-12
.896E-13
.774E-13
.653E-13
.532E-13
.411E-13
.289E-13
.168E-13
.471E-14

Figure 6.

8 Computed Results

Modeshapes, resonant frequencies and forced response of the CSR for frequencies up to 2500Hz have computed via the model illustrated in figure 3. The resonant frequencies were generally predicted to within 5%. Modeshapes prediction and forced vibration analyses yielded results which compared well with those measured on the CSRs.

Figure 5 presents a comparison of the measured and computer sound power transfer function (with respect to cylinder pressure) over the range 400Hz to 2400Hz. The computation required for obtaining the results in figure 5 was approximately 100 hours of CPU time on the MV20000 computer at Ricardo.

In figure 5, the graphs of the sound power spectrum have peaks at 800 Hz and at 1120 Hz. The peaks occur at these frequencies because the rig has structural resonances at 795 Hz and 1121 Hz. Figures 6 illustrate the surface sound intensity with respect to cylinder pressure at 800 Hz and 1120 Hz; the shading scheme is meant to indicate the average intensity on each boundary element.

9 Concluding Discussion

In this paper it has been shown that computational methods can be used to calculate both the vibratory and acoustic behaviour of engine structures in a useful and cost effective manner. Attention in the vibratory calculation is now concentrated on improving the efficiency of model generation, improving computational efficiency and, not least, improving the methods of calculation of exciting forces.

In this paper the BEM has been applied to the engine noise prediction problem, and its practicability in comparison with other techniques has been assessed. Making the acoustic software, described in [6], versatile, easy-to-use, robust, and reasonably efficient has enabled us to make a considerable advance in the computer-aided low-noise engine design area.

References

1. Croker, M.D., (1987), Engine noise: practicalities are prediction, part 1 - Hardware evaluation, SAE Noise and Vibration Conference, SAE 870977.

2. Croker, M.D., (1987), Summary report of experimental work on the Ricardo crankcase simulation rigs - Part 1, Ricardo report 87/716.

3. Croker, M.D. and Tyrrell, R.J., (1989), Engine noise: practicalities and prediction, part 4 - Hardware optimisation, SAE Surface vehicle noise and vibration conference.

4. Hassall, J.R. and Zaveri, K., (1979), "Acoustic noise measurements", Bruë and Kjaer.

5. Kirkup, S.M., Henwood, D.J. and Tyrrell, R.J., (1988), Engine noise - practicalities and prediction Part 3: Noise prediction using the boundary element method, International conference on advances on the control and refinement of vehicle noise, The Institution of Mechanical Engineers, Paper no: c33/88, 139-145.

6. Kirkup, S.M., (1989), "Solution of exterior acoustic problems by the boundary element method", PhD thesis, Brighton Polytechnic.

7. PAFEC Ltd., Strelley Hall, Strelley, Nottingham, U.K.

8. Tyrrell, R.J., Croker, D.M., (1987), Engine noise: practicalities and prediction, part 2 - Finite element and Vibration Conference, SAE 870978.

10 Acknowledgement

The authors are grateful to their colleagues at Ricardo who have contributed to this work. The authors would also like to thank the directors of Ricardo Consulting Engineers Ltd and Brighton Polytechnic for funding this research.

Prediction of the Effect of Engine Noise Shields Using a Combination of Boundary, Shell and Finite Elements

S.M. Kirkup[1], D.J. Henwood[2] and R.J. Tyrrell[3]

University of Salford[1] Brighton Polytechnic[2] and Ricardo Consulting Engineers Limited[3]

Abstract A reduction in engine noise is sometimes attempted by fitting one or more noise shields around the engine block structure. The change in noise resulting from fitting the shield(s) will depend on the nature of the engine's vibration, the size, shape and position of the shield(s) and the material from which the shields are built. In general the shield itself will vibrate because of acoustic and/or mechanical coupling.
In this paper it is shown how the problem can be solved using boundary and shell elements to model the acoustic field exterior to boundaries and shields and finite elements to model the vibratory properties of the shields.

1 Introduction

Once the design of the engine block itself has been finalised, shielding may be used to effect a further reduction in noise. Shielding may take the form of complete or substantial enclosure of the engine block or it may consist of one or more shields which together only partially enclose the engine. Both methods add bulk to the engine block and hence neither method is an easy option for design engineers. However, whereas complete or substantial enclosure is generally obstructive to routine maintenance operations, partial shielding need not be. For a discussion on the practicalities and success of enclosure and shielding see [7]. In this paper only partial enclosure of the engine using a set of close fitting shields is considered. An illustration is given in figure 1.

On fitting one or more noise shields around the engine block, the change in noise will depend on the nature of the engine's vibration, the size, shape and position of each shield and the material from which each shield is made.

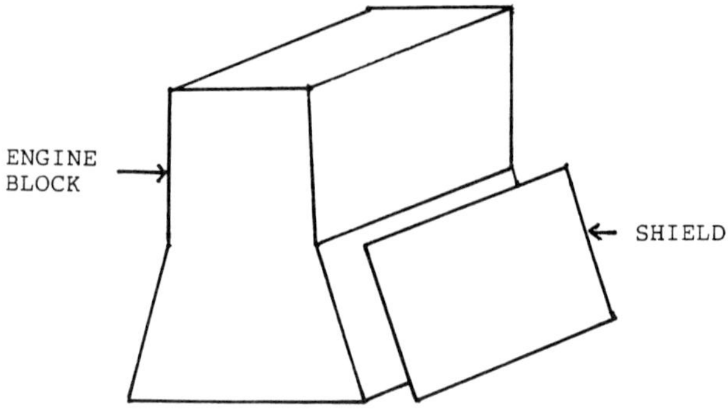

Figure 1. An Illustration of Engine Block and Shield

Diagnostic information about the vibratory and acoustic properties of the engine block (see [2] or [5]) will help the design engineer decide where the shields should be placed. If a computational tool for predicting the effect of the shield on the noise is available then a computer-aided approach to developing the shield design could be applied. However, if no such tool is available, then the shield may only be developed by building a physical model of the shield, testing and taking measurements. The overall process of designing the engine-shield system is illustrated in figure 2.

Clearly a computer-aided approach to noise shield design development is to be preferred to the method of carrying out tests and measurements. An attempt at predicting the effect of engine noise shields is given in [6]. However, the mathematical models on which the methods in [6] are based are not suitable for the frequency range of interest and acoustic-structure coupling is not modelled. Moreover, the comparisons of experimental and computed results given in [6] are generally poor. A subroutine ABSEMGEN (written by the first author) is described in [4]. It is able to predict the acoustic field around a set of vibrating bodes and coupled, forced shells. The subroutine implements the boundary and shell element method (BSEM), as described in [3], and also includes acoustic structure coupling at the shield. Subroutine ABSEMGEN is based on an elegant computational model and provides us with an easy-to-use and versatile software tool for the prediction of the effect of engine noise shields.

In this paper the subroutine ABSEMGEN is outlined. The subroutine is applied to a test acoustic-shield problem and results are demonstrated. The application of the subroutine to real-world engine-shield systems is discussed.

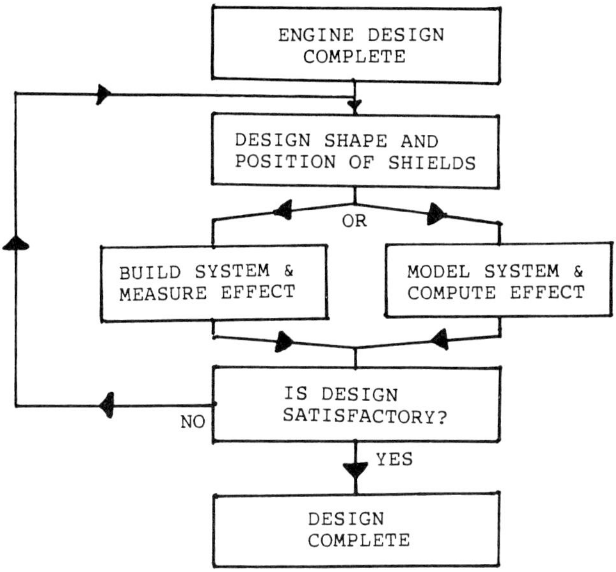

Figure 2. The Noise Shield Design Process

2 Analysis of the Engine-Shield System

The engine-shield system is a complex acoustic-mechanical system. The acoustic field arises mainly through the vibration of the engine. In order that the noise shields do not add significantly to the weight they will generally by made of light flimsy materials and hence they may vibrate significantly. The vibration of the noise shields could arise because they are mechanically coupled to the vehicle or because of acoustic coupling between the engine and shield.

The frequencies that largely determine the noise produced by the engine-shield system are the frequencies at which the engine vibration is strong, the structural resonant frequencies of the shield, and the acoustic resonant frequencies that may occur in the gap between the engine and the shield. Moreover, should resonant frequencies of any two of the engine, the shield, and the gap coincide in the frequency spectrum then the shield may well increase rather than decrease the noise from the engine. Clearly, therefore, much useful information about the engine shield system can be derived by considering the vibratory properties of the engine, the vibratory properties of the shield and the acoustic properties of the gap in isolation. However, in this paper, we will be mainly concerned with considering the problem of modelling the complete engine-shield system.

3 Computational Method

In this section the computational method employed by subroutine ABSEMGEN is described. The purpose of the subroutine is to compute the properties of the acoustic field surrounding a set of arbitrary vibrating bodies S and coupled, forced shells Γ. The method used arises from formulating the acoustic problem in the form of an integral equation and approximating the surface of the vibrating body and the shield(s) by a set of triangles. The velocity potential on the bodies, the jump in velocity potential on the shell, and the displacement of each modeshape on the shell are approximated by a constant on each triangle. Full details of the underlying formulation and method are given in [4].

The primary stage of the methods involves solving the following linear system of equations

$$\begin{bmatrix} \alpha(M_{SS} - \frac{1}{2}I_{SS}) + \beta N_{SS} & \alpha M_{S\Gamma} + \beta N_{S\Gamma} & O_{SM} \\ N_{\Gamma S} & N_{\Gamma\Gamma} & -U_{\Gamma M} \\ O_{MS} & i\rho\omega U^T_{M\Gamma} & U^T_{\Gamma M} D_{MM} \end{bmatrix} \begin{bmatrix} \underline{\hat{\varphi}}_S \\ \underline{\hat{\delta}}_\Gamma \\ \underline{\hat{V}}_M \end{bmatrix}$$

$$= \begin{bmatrix} \{\alpha L_{SS} + \beta(M_{SS} + \frac{1}{2}I_{SS})\}\underline{v}_S \\ M^t_{\Gamma S}\underline{v}_S \\ U^T_{M\Gamma}U_{\Gamma M}\underline{F}_M \end{bmatrix},$$

where the suffices S, Γ, and M denote the number of boundary elements, the number of shell elements and the number of modes used to model the vibration respectively. The matrices
$L_{SS}, M_{SS}, M^t_{SS}, N_{SS}, M_{S\Gamma}N_{S,\Gamma}, M^t_{\Gamma S}, N_{\Gamma S}$, and $N_{\Gamma\Gamma}$ arise through the discretisation of the integral operators in the boundary and shell integral equation formulation and $C_{SS} = \frac{1}{2}I_{SS}$. The matrix $U_{\Gamma M}$ describes the modeshapes of the shell, $i = \sqrt{-1}, \rho$ is the density of the acoustic medium (air), $\omega = 2\pi\nu$ where ν is the frequency in hertz. The inputs are the velocities of each of the centroids on the boundary $\underline{\nu}_S$,and the list of modal forcings $\underline{F}_M$. The outcome of computing the solution of the linear system of equations are approximations to the velocity potential at the centroids of the triangles $\underline{\hat{\varphi}}_S$, the jump in velocity potential at points on the shell $\underline{\hat{\delta}}_\Gamma$ and the modal velocity coefficients $\underline{\hat{V}}_M$. Approximations to the velocity potential can be calculated at points in the exterior by integrating over the boundary and shells.

The sound pressure at any point is straightforwardly related to the velocity potential by $P(\mathbf{p}) = i\rho\omega\varphi(\mathbf{p})$. The other acoustic properties such as the sound intensity pattern on the surface of the body and the surfaces of the shells, the sound power, and the radiation ratios can also be derived from this information.

The subroutine was written in Fortran and has the following form

SUBROUTINE ABSEMGEN(RHO,C,BVERTPTS,BELEMVERT, SELEMVERT,SVERTPTS,SDEFUPP,SMODESHAPES,MLAMBDAS, FREQUENCIES,BNORMVELS,MFORCING,EXTPTS,BINTENS, SINTENS,BPOWERS,SPOWERS,BBAFPPOW,SBAFFPOW, EXTPRESSURES)

Where the parameters RHO and C are the density and speed of sound in the acoustic medium, BVERTPS and BELEMVERT describe the triangles that make up the surfaces of the bodes, SELEMVERT, SVERTPTS and SDEFUPP describe the triangles that make up the shells, SMODESHAPES and MLAMDAS describe the dynamic properties of the shell, FREQUENCIES lists the frequencies of interest, BNORMVELS lists the normal velocities of the centroid of each triangle that makes up the boundary at each frequency and MFORCING lists the model forcing on the shells, EXTPTS lists the positions of the exterior points where the sound pressure is estimated. The output parameters are BINTENS and SINTENS which list the surface sound intensity on the surfaces of the bodies and shells at each centroid and each frequency, BPOWERS and SPOWERS which list the sound power of each body and shell at each frequency. From BBAFFPOW and SBAFFPOW the radiation ratios can be estimated. EXTPRESSURES lists the estimate of the sound pressure at the exterior points. For the engine-shield system $\underline{v}_S$ and $\underline{\hat{\varphi}}_S$ correspond to the vibration velocity and the velocity potential on the engine surface, $\underline{\hat{\delta}}_\Gamma$ corresponds to the jump in velocity across the shield, $\underline{\hat{V}}_M$ and $\underline{F}_M$ corresponds to the coefficients of modal velocity and the coefficients of direct modal forcing on the shield.

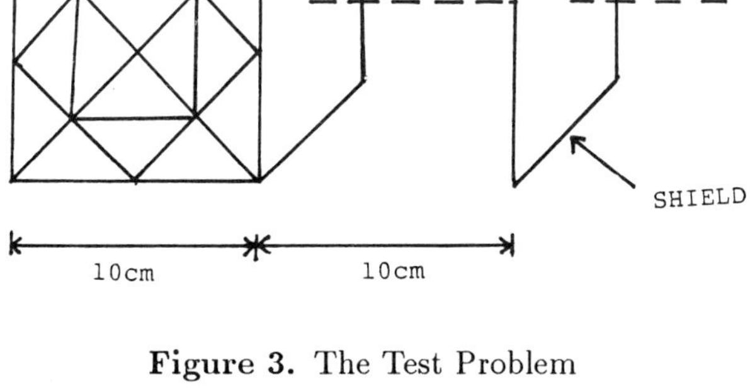

Figure 3. The Test Problem

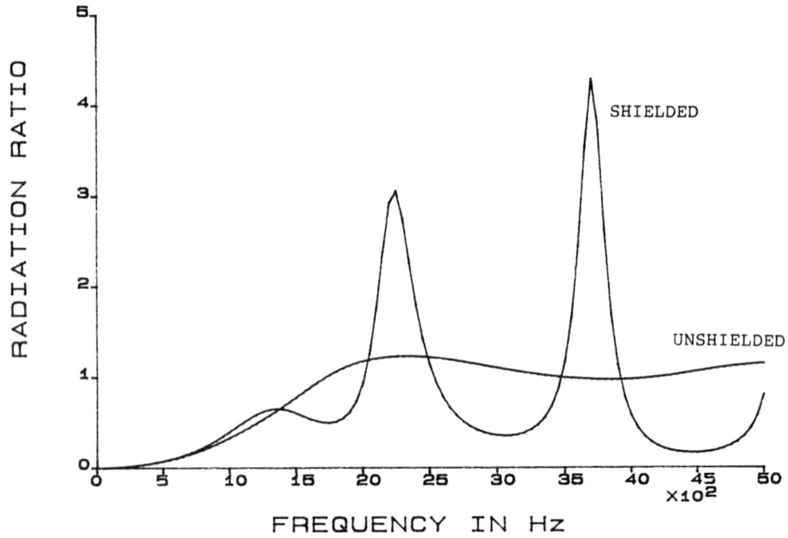

Figure 4. Radiation Ratio Curves for Both a Shielded and Unshielded Cube

4 Test problems and results

In this section some results from the application of subroutine ABSEMGEN to a simplified engine-shield problem are given. The test problems that are considered all have a 10cm cube of which one face is vibrating uniformly, the vibration of the other faces is zero. A square plate with sides of length 10cm is placed 10cm from the vibrating face and this is generally compared with the situation of the vibrating cube without the square plate. Initially the square plate is assumed rigid, although in the final test it is assumed to take the properties of a 1mm thick plate of steel. In each test the acoustic medium is air. The test problem is illustrated in figure 3.

The radiation ratios of the system with or without the plate at 10Hz, 20Hz, ... 5000Hz were computed using subroutine ABSEMGEN. The results were used to generate the radiation ratio curves shown in figure 4. The graph shows that the shield has a major effect on the radiation ratio of the vibrating cube. The radiation ratio for the shielded and unshielded cube are similar when the frequency is less than around 1300Hz. However the shield makes a significant reduction in the radiation ratio at around 1800Hz, 3000Hz and 4500Hz and significantly increased the radiation ratio at around 2200Hz and 3700Hz. The curve is similar to the one given in figure 7B(ii) in chapter 6 of reference [2].

Let us now go on to consider the effect of the shield on the sound

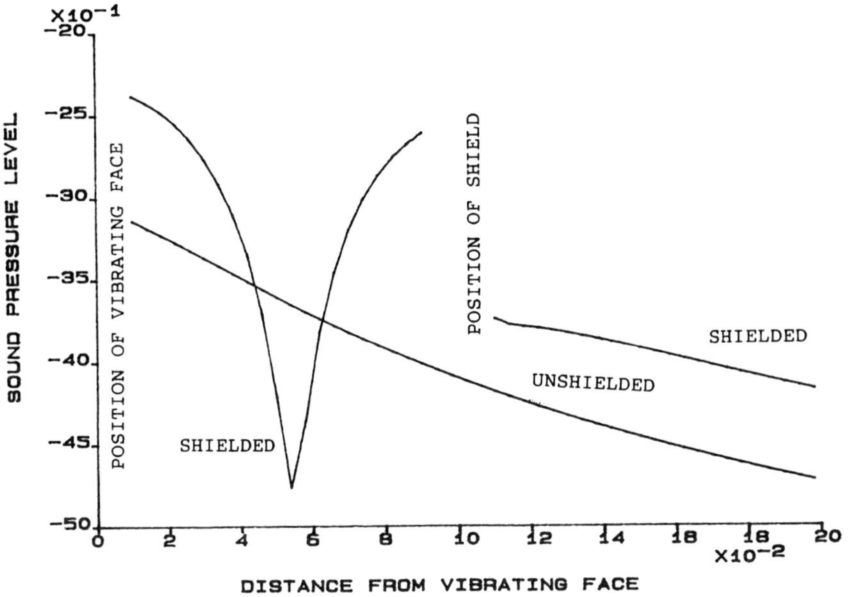

Figure 5. Effect of Shield on Sound Pressure at 2300 Hz

pressure field in front of the shield at a set of chosen frequencies. Figures 5, 6 and 7 show the sound pressure level (in this case the logarithm of the sound pressure) at points along the line from the centre of the vibrating face of the cube to the centre of the plate 2300Hz, 3050Hz, and 3700Hz.

Finally we consider the situation when the plate is non-rigid. It is assumed that the plate is a 1mm thick piece of steel that is hinged on all four sides and is assumed to have no material damping. The dynamic properties of a square plate can be found in most textbooks on vibration analysis. The first resonant frequency of the steel plate is near 490 Hz. The shield is not directly forced, though at 490Hz the steel will vibrate significantly because of coupling. Figure 8 compares the sound pressure level for the steel shield with that of the rigid shield.

5 Concluding discussion

In this paper we have shown how the boundary and shell element method in the form of subroutine ABSEMGEN can be used to compute the properties of the acoustic field surrounding a set of vibrating bodies and coupled shells. The subroutine has not yet been applied to a real-world engine shield problem.

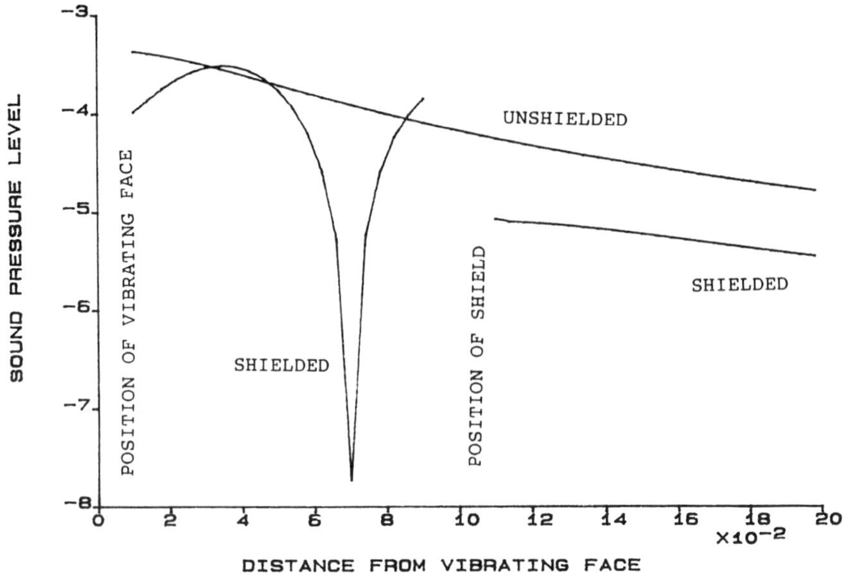

Figure 6. Effect of Shield on Sound Pressure at 3050 Hz

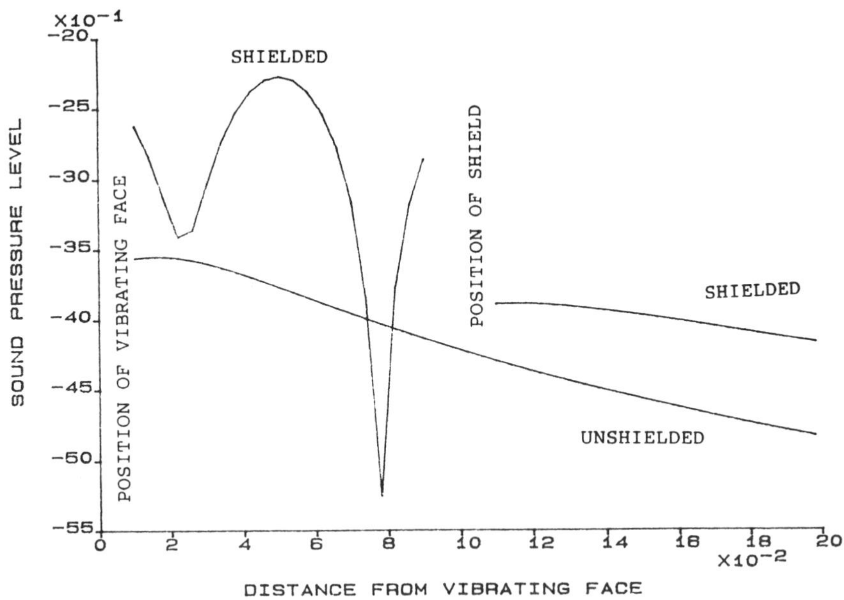

Figure 7. Effect of Shield on Sound Pressure at 3700 Hz

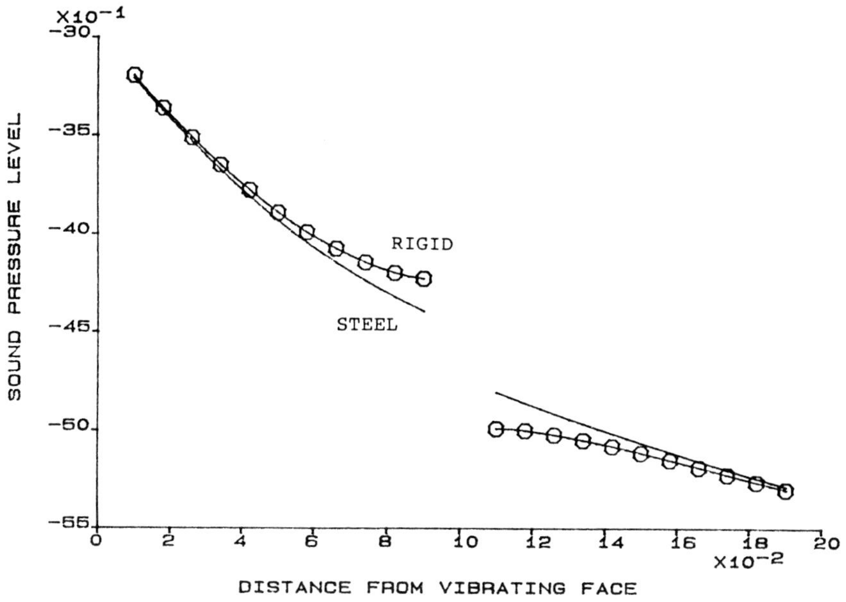

Figure 8. Comparison of Rigid and Steel Shields

At the moment experimental data for the acoustic properties of the field around an engine with an acoustic shield are not available at Ricardo. However sound pressures at microphone points in space around an engine block with a reflecting floor are available from the crankcase simulation rig project (see [1]). This situation may be modelled computationally by using boundary elements on the engine block, as described in [2] or [5], and shell elements on the floor.

References

1. Croker, M.D., (1987), Summary report of experimental work on the Ricardo crankcase simulation rigs - Part 1, Ricardo report 87/716.

2. Kirkup, S.M., (1989), Solution of exterior acoustic problems by the boundary element method, PhD thesis, Brighton Polytechnic.

3. Kirkup, S.M., (1989), BEM IV: Shell elements: their use in conjunction with boundary elements and in application to Laplace problems, Report M.18, Department of Mathematical Sciences, Brighton Polytechnic and in Proceedings of the NUMETA90 Conference, Elsevier.'

4. Kirkup, S.M., (1989), Solution of acoustic problems exterior to vibrating bodies and coupled shells.

5. Kirkup, S.M., Tyrrell, R.J., Henwood, D.J., (1989), Computational techniques for engine noise prediction, Report M.20, Department of Mathematical Sciences, Brighton Polytechnic.

6. Sung, Y.C.C. and Lalor, N., (1988), The prediction of engine noise reduction by using shields and partial enclosures and the optimization of their design, Paper C29/88. Proceedings of the Institution of Mechanical Engineers, International Conference on the Control and Refinement of Vehicle Noise.

7. Waters, P.E., (1980), The effect of noise legislation on vehicle diesel engine design, SAE 800401.

6 Acknowledgement

The authors would like to thank the directors of Ricardo Consulting Engineers Ltd and Brighton Polytechnic for funding this research.

Acoustic Analysis of Road Vehicle Silencers

K.S. Peat

University of Technology, Loughborough

Abstract This paper considers the prediction of intake and exhaust noise reduction, dealing specifically with gas noise developed at the engine which is radiated from the intake or exhaust orifice. It is shown that linear plane-wave acoustic analysis can produce accurate low frequency predictions for complex silencer designs throughout the engine speed range. A computational model, based on a transfer matrix representation of individual acoustic elements, is sufficiently fast and accurate to be used in an iterative design-analyse scheme, thereby reducing considerably the cost and time of the traditional design-manufacture-test cycle.

At frequencies below the cut-on frequency of the first non-planar mode, the plane-wave analysis is rendered inaccurate by the presence of non-planar evanescent waves which are generated at geometric discontinuities. The errors are generally small, but can be reduced by the use of discontinuity impedances developed from a three-dimensional analysis of specific discontinuities. This paper gives examples of the development and use of effective impedance corrections in a basically plane-wave model and compares the overall accuracy with respect to three-dimensional finite element analysis.

1 Introduction

The precise mathematical analysis of exhaust or intake silencer systems would require the solution of the three-dimensional, unsteady, non-linear aeroacoustic equations. Whilst such a computational analysis is technically feasible, the cost of analysis of a single system at a single engine operating condition would be prohibitive for what is, relatively speaking, a low-cost, low-technology vehicle component. The requirement of a silencer designer is for a computational analysis tool that is simple and cheap to operate such that it can be used repetitively in a design-analyse iteration towards an acceptable solution, in order to reduce both the time and cost taken by conventional cut-and-try techniques. Fortunately, the main problem area for intake and exhaust noise is the elimination of low frequency noise, because the dominant noise occurs at engine firing frequency and the first few harmonics. Furthermore the design of silencer chambers which effectively reduce high frequency noise is straightforward. At low frequencies, only plane waves can continuously propagate along uniform ducts and any non-planar waves which are formed at duct discontinuities are rapidly attenuated. Thus considerable mathematical simplification can be achieved by using one-dimensional equations for practical low frequency analysis, although the model will be in error at discontinuities of the ductwork. One-dimensional, non-linear, unsteady models are reviewed in Munjal [11]. They are generally used in the prediction of engine performance but they can also predict acoustic characteristics. Such models are still too costly to use in an iterative design-analyse cycle and require concurrent in-cylinder modelling to establish conditions at the engine ports.

Further mathematical simplifications can be achieved by using a linearised acoustic analysis and working in the frequency domain, treating the total gas flow as a superposition of a steady mean flow and noise, and analysing for discrete frequency components of the noise. Such a model is able to meet the designers requirements in terms of simplicity and cost, but it then remains to be shown whether or not the accuracy achievable is sufficient for practical purposes. The validity of the linearisation procedure is most questionable when the noise levels are highest, that is generally in the downpipe of an exhaust system for an engine operating under full load and speed. It has been shown that a linear model can produce good predictions under these most exacting conditions [12]. Errors which are attributable to the non-linear, evanescent waves which are generated at duct discontinuities are found to be small and can be reduced by the use of equivalent impedances in the one-dimensional model. The nature of these impedance 'corrections' can be ascertained from various forms of three-dimensional analysis on specific isolated discontinuities, as shown by examples in this paper.

There are several sources of intake or exhaust noise, both structure-borne, such as down-pipe ring or case noise, and airborne noise radiated from the intake/exhaust pipe orifice, which might have originated from the exhaust ports or been generated by flow through the silencer system. The analysis presented here deals only with airborne noise present at the engine end of the system and radiated from the pipe orifice. The results presented in this paper are for insertion loss, the difference between radiated noise levels of a given silencer system and some reference system attached to the same engine operating under the same conditions. It is assumed that the noise output of the engine is unaltered between the two systems, which then removes the requirement for specifying or analysing the noise levels at the valve ports. The validity of the assumption can best be judged by reference to the comparison of measured and predicted insertion loss.

2 Linear, Plane-wave Model

The most convenient method to implement a linearised plane-wave acoustic model of a general silencer system is through the transfer matrix approach [4, 2]. A transfer matrix relates unknown variables at the input of an acoustic element to similar unknown variables at the output. In the simplest case, one could relate the acoustic pressure, p, and volume velocity, U, at a single input port to similar variables at a single output port by the matrix equation.

$$\begin{bmatrix} P_{in} \\ U_{in} \end{bmatrix} = \begin{bmatrix} A\,B \\ C\,D \end{bmatrix} \begin{bmatrix} P_{out} \\ U_{out} \end{bmatrix} \tag{2.1}$$

The matrix coefficients A, B, C, D are determined with reference to the conservation equations of mass, momentum and energy. They are generally complex functions of frequency, mean flow Mach Number, temperature and geometrical details of the specific acoustic element. Details of the derivation of the transfer matrices for many basic acoustic elements are given by Munjal [11], together with a survey of earlier work in this field. Whilst simple elements, such as a uniform pipe, have only a single input and output port, it is possible and normal for elements to have multiple inputs and outputs. Thus in general one could write the transfer matrix of an ith element as

$$\left\{ \underline{\phi}^{i}_{in} \right\} = \left[\ \mathbf{T}_i \ \right] \left\{ \underline{\phi}^{i}_{out} \right\} \tag{2.2}$$

where transfer matrix $\mathbf{T}_i$ is of size $(j \times k)$ say, and relates the j input variables $\underline{\phi}_{in}$ to the k output variable $\underline{\phi}_{out}$.

Suppose a silencer system is composed of a sequence of N such elements, beginning at the engine noise source and finishing at the plane of the orifice of radiation into free space. The output variables of element i become the input variables of element $i+1$, hence

$$\left\{ \underline{\phi}^{1}_{in} \right\} = \left[\ \mathbf{T}_1 \ \right] \cdots \left[\ \mathbf{T}_i \ \right] \cdots \left[\ \mathbf{T}_N \ \right] \left\{ \underline{\phi}^{N}_{out} \right\} \tag{2.3}$$

or, following matrix multiplication,

$$\left\{ \underline{\phi}^{1}_{in} \right\} = \left[\ \mathbf{T} \ \right] \left\{ \underline{\phi}^{N}_{out} \right\} \tag{2.4}$$

where $\left[\ \mathbf{T} \ \right]$ is the overall transfer matrix. Note that, due to the frequency dependence of many of the elements of the transfer matrices, it is necessary to repeat the derivation and multiplication process for each chosen discrete frequency of analysis. In general an overall silencer has a single outlet, the orifice of noise radiation, and the engine source can be treated as a single input. The overall silencer can then be regarded schematically as shown in Figure 1.

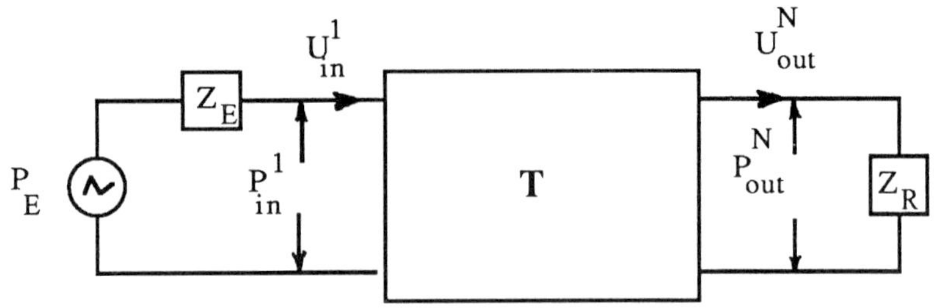

Figure 1. A Schematic Representation of a Complete System

If the source impedance, Z_E and radiation impedance, Z_R are known, then the level of the radiated noise can be related to the source pressure P_E. If a second system is then connected to the same invariant source, P_E can be eliminated to leave the ratio of radiated noise levels of the two systems, the insertion loss [2].

3 Non-Planar Effects

At low frequencies, only plane wave modes can propagate continuously through a uniform duct. A silencer is generally composed of a sequence of uniform duct sections joined by sudden discontinuities. Non-planar evanescent waves are formed at the discontinuities and they typically have a second order influence on the overall noise characteristics of a silencer. For low frequency results, an alternative to the time-consuming and costly exercise of a fully three-dimensional analysis of a complete silencer system is to analyse the effects of evanescent waves for standard types of isolated discontinuities. These effects can conveniently be expressed as effective impedances by which the basic plane-wave model needs to be adjusted for accurate analysis. Since evanescent waves decay rapidly, the effects of one discontinuity upon a successive one are generally negligible. The main disadvantage of this approach is that only specific forms of discontinuity are likely to have been analysed. However, since the effects are only second-order, it is generally possible to adopt the effective impedance of a similar discontinuity as a good approximation, as is shown in this paper.

4 Results

The analysis of effective impedances began with the form of the radiation impedance for a pipe terminating in an infinite flange [16], the impedance for a pipe radiating into free space being considered much later [8]. In both instances a formula for Z_R is obtained where $P_R/U_R = Z_R$. The effective impedance of a sudden change in cross-sectional area has also received much attention over many years [5-7, 9]. The plane wave modes each side of the discontinuity are then related by a transfer matrix of the form

$$\begin{bmatrix} P_{in} \\ U_{in} \end{bmatrix} = \begin{bmatrix} 1 & Z \\ 0 & 1 \end{bmatrix} \begin{bmatrix} P_{out} \\ U_{out} \end{bmatrix} \tag{4.1}$$

where Z is a discontinuity impedance.

The non-planar effects at the junction of a bifurcated tube have also been analysed [1, 10] and more recently [13] it has been shown that the non-planar effects of an extended inlet or outlet section can be incorporated into a basic plane-wave model by the inclusion of a discontinuity impedance Z where

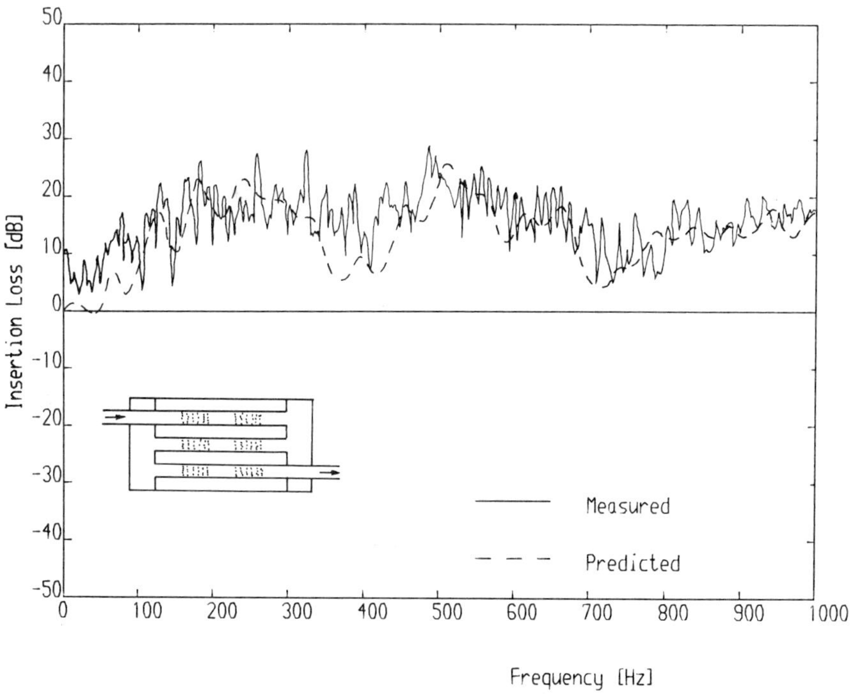

Figure 2. Insertion Loss of a Triple-Pass Silencer 1250 rev/min

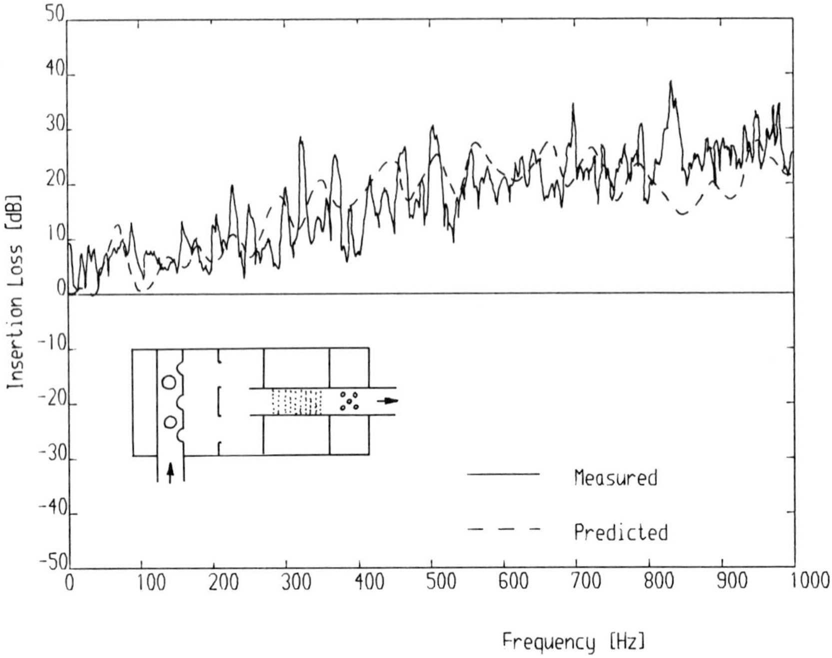

Figure 3. Insertion Loss of a Side-Entry Silencer 2750 rev/min

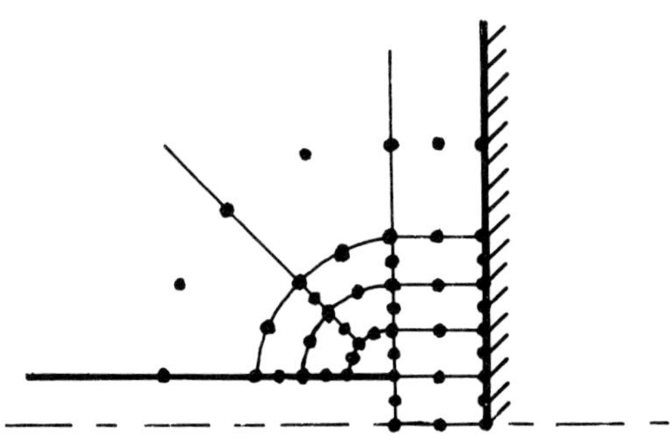

Figure 4. Finite Element Mesh for the Evaluation of Radiation Impedance from an Open Pipe Faced by an Infinite Plane Wall

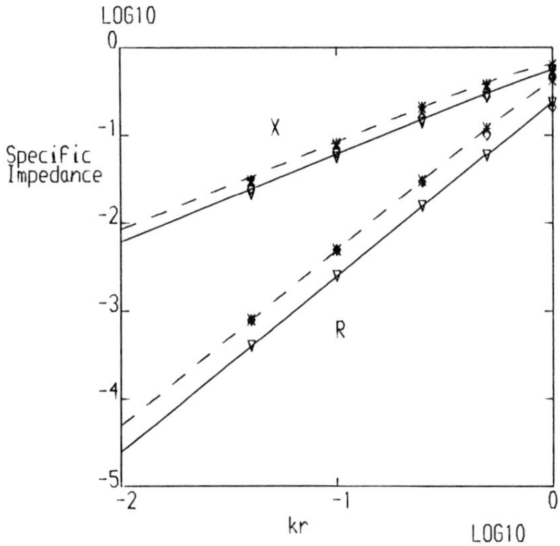

Figure 5. Radiation Impedance $Z_R = R - iX$ for the Following Radiators: (a) an open tube or radius r — analytical, ∇ finite element; (b) a tube of radius r with an infinite flange - - - analytical, $*$ finite element; (c) a tube of radius r faced by a plane wall (i) $\times$ distance r away (ii) $\diamond$ distance $2r$ away.

$$\begin{bmatrix} P_{in} \\ U_{in} \end{bmatrix} = \begin{bmatrix} 1 - iY_{AO}Z & iY_{out}Z \\ -(\frac{Y_{AO}}{Y_A})\left[1 + (\frac{\alpha^2}{1-\alpha^2})iY_{AO}Z\right] & 1 + \alpha^2 iY_{AO}Z \end{bmatrix} \begin{bmatrix} P_{out} \\ U_{out} \end{bmatrix} \tag{4.2}$$

Y_{AO} and Y_A are respectively the plane-wave admittance and characteristic admittance of the closed duct, whilst Y_{out} is the characteristic admittance of the outlet duct and α is the diameter ratio of the inlet to the outlet duct. Equations (3.1) and (3.2) refer to the zero mean flow situation. The basic transfer matrix terms are affected significantly by mean flow [11] but changes to the discontinuity impedance terms are insignificant [15].

For the simplest forms of discontinuity, the three-dimensional analysis can be performed by the Wiener-Hopf technique [6, 12] or mode-matching [5-10, 13]. In more complex situations a fully numerical procedure, such as a finite element analysis, is generally adopted [14, 15, 17].

The overall accuracy of a basically plane-wave, linear acoustic model for the analysis of the insertion loss of silencers can be judged from Figures 2 and 3. These figures show good correlation between predicted and experimental results for complex silencer systems fitted to a 5.0 litre, 6 cylinder, naturally-aspirated diesel engine operated near the two extremes of its speed range. Similar correlation has been achieved for a wide variety of exhaust and intake silencer types fitted to both naturally-aspirated and turbo-charged diesel engines [2, 3,12]. The predicted results include effective impedance terms which account for non-planar waves at geometrical discontinuities.

For intake systems, the orifice of noise radiation and air inflow is often situated close to an engine bulkhead and may be fitted with a rounded flange to reduce flow losses. A finite element analysis [14] has been used on such systems in order to evaluate the form of the effective radiation impedance. The simple finite element mesh for pipe radiation with a facing wall is shown in Figure 4, which indicates the use of infinite elements to account for radiation into the far-field. The results in Figure 5 show that as the wall is moved closer to the pipe orifice, the form of the radiation impedance gradually changes from that for radiation into free-space to that for radiation into half-space, the piston in a plane-wall form. Similar analysis for the situation with a rounded flange gives results which indicate a change of radiation impedance from free-space to half-space form as a flange is introduced. Thus it appears that the near-field geometry is the dominant effect in the form of the radiation impedance.

A further example of the use of three-dimensional analysis is given for the rain-hood of Figure 6a, a device commonly found on the intake system of large trucks. A mesh for the finite element analysis of this axisymmetric

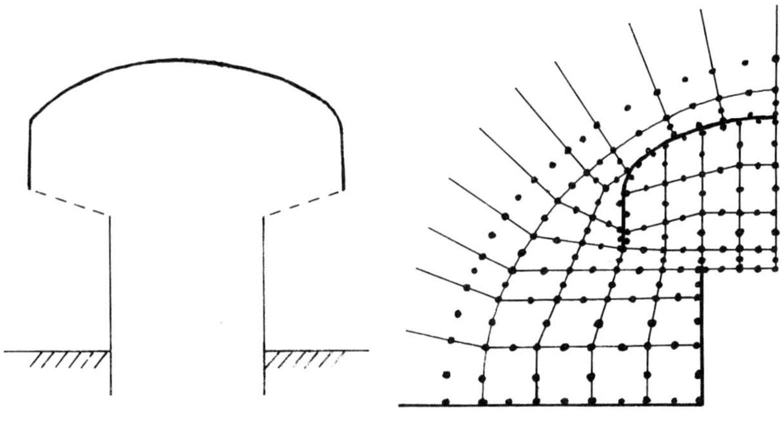

(a) (b)

Figure 6. (a) Test Arrangment for a Rain Hood (b) Finite Element Mesh for Analysis

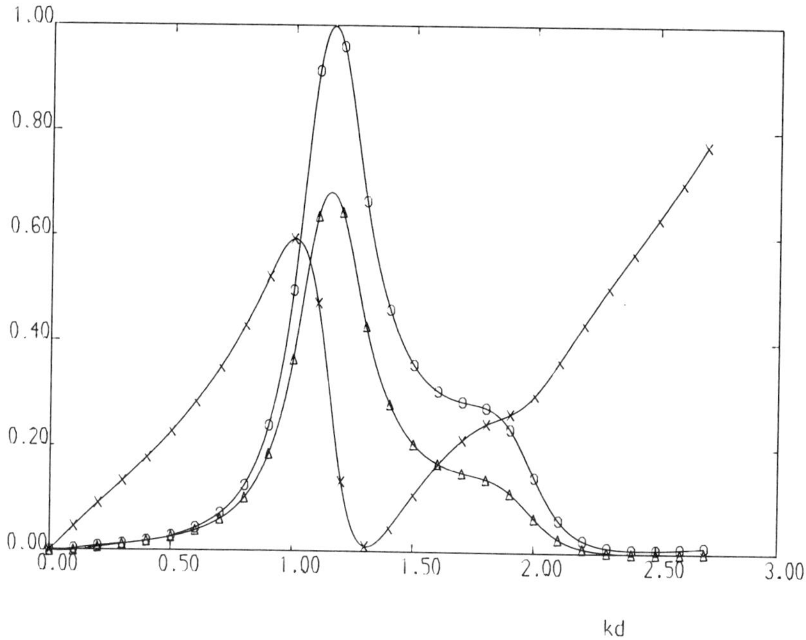

Figure 7. Finite Element Results for the Specific Load Impedance $Z = R - iX$ and the Energy Transfer of the Rain Hood: $R, (\triangle); X, (\times); [pU]_{out}/4[pU]_{in}, (0)$.

problem is shown in Figure 6b. The effective impedance at the outlet of the uniform pipe is shown in Figure 7. The impedance is more accurately termed a load impedance than a radiation impedance, since a significant amount of the acoustic energy transmitted from the main duct is reflected back into the duct by the rain hood. Thus, in order to determine insertion loss, it is necessary to evaluate in addition the energy transfer from pipe exit to radiation into free space, which is also shown in Figure 7 in the form of the ratio $(pU_{in}/(pU)_{out}$. The overall result for insertion loss with respect to a length of uniform pipe is compared with experimental data in Figure 8. For comparison, the result using plane-wave analysis with impedance corrections for non-planar waves is also shown in Figure 8. In this instance the effective impedance terms used are those for an extended inlet and radiation from a circular section pipe into half-space, the most similar 'standard' discontinuities. In reality there is a flow reversal chamber whose side-walls do not extend back even to the place of the discontinuity, and radiation is from an annulus. However the 'corrected' plane wave results are generally of acceptable accuracy and of comparative accuracy

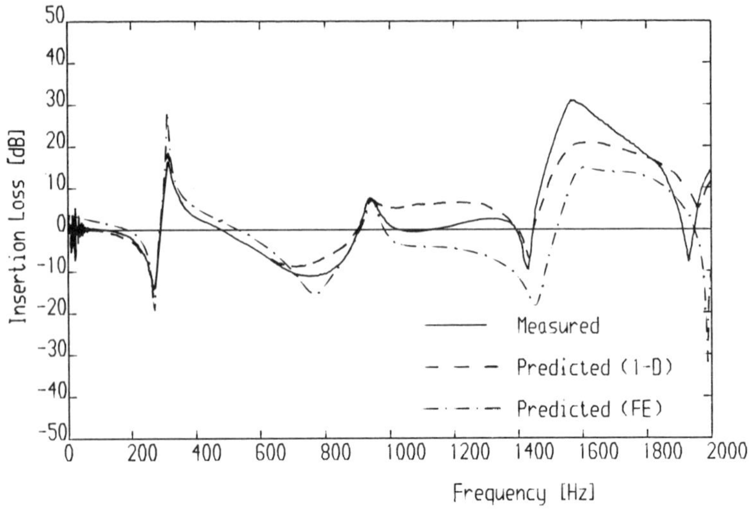

Figure 8. Insertion Loss of a Rain Hood

to the finite element results. The latter could be improved by significant mesh refinement, but already the computing cost is prohibitive for use in a design-analysis cycle, and the time required to generate the finite element mesh is equally excessive. Furthermore the above case is geometrically axisymmetric, whereas a fully three-dimensional situation would be even more disadvantageous to a finite element approach.

5 Conclusions

A simple linear, plane-wave model for the acoustic analysis of intake and exhaust silencer systems can give accurate predictions of insertion loss. The accuracy can be increased by the use of effective impedance terms to account for non-planar evanescent waves which are generated at discontinuities in the duct system. The form of the effective impedance terms can be established from the three-dimensional acoustic analysis of a 'standard' set of isolated discontinuities. The complete plane-wave model with correction terms is in a form which is very convenient for the implementation on a computer, giving a designer a quick and convenient tool for an iterative design-analysis process.

Acknowledgements

The experimental results in this paper were produced at the Motor Industry Research Association, under the supervision of Mr G.D. Callow.

The experimental work was funded jointly by the Department of Transport through the Transport and Road Research Laboratory and by a consortium of industrial members, under the QHV90 Project.

References

1. Bailin, L.L., (1951), An analysis of the effect of the discontinuity in a bifurcated circular guide upon plane longitudinal waves, Journal of Research of the National Bureau of Standards, 47, 315-335.

2. Callow, G.D. and Peat, K.S., (1988), Insertion loss of engine intake and exhaust silencers, Proc. of the I. Mech. E. Int. Conf. on Advances in the Control and Refinement of Noise, 39-46.

3. Callow, G.D. and Peat, K.S., (1988), QHV90 Project: Reduction of noise from lorry exhausts and air intakes, MIRA Report No. K125062.

4. Erikkson, L.J. and Thawani, P.T., (1985), Theory and practice in exhaust system design, SAE Paper No. 850989.

5. Hudde, H, and Letens, U., (1985), Scattering matrix of a discontinuity with a nonrigid wall in a lossless circular duct, Journal of the Acoustical Society of America, 78, 1826-1837.

6. Karal, F.C., (1953), The analogous acoustical impedance for discontinuities and constrictions of circular cross-section, Journal of the Acoustical Society of America, 25, 327-334.

7. Kergomard, J. and Garcia, A., (1987), Simple discontinuities in acoustic waveguides at low frequencies: critical analysis and formulae, Journal of Sound and Vibration, 114, 465-479.

8. Levine, M. and Schwinger, J., (1948), On the radiation of sound from an unflanged circular pipe, Phys. Rev., 73, 383-406.

9. Miles, J.W., (1944), The reflection of sound due to a change of cross-section in a cylindrical tube, Journal of the Acoustical Society of America, 17, 259-271.

10. Miles, J.W., (1947), The equivalent impedance for a bifurcated cylindrical tube, Journal of the Acoustical Society of America, 19, 579-584.

11. Munjal, M.L, (1987), "Acoustics of Ducts and Mufflets", John Wiley, New York.

12. Peat, K.S. and Callow, G.D., (1989), Acoustic analysis of lorry exhaust systems at high speeds, Internal Report of Dept. of Math. Sciences, Lough. Univ. of Tech., 1989/1.

13. Peat, K.S., The acoustical impedance at the junction of an extended inlet or outlet duct. Journal of Sound and Vibration (in press).

14. Peat, K.S., (1982), Evaluation of four-pole parameters for ducts with flow by the finite element method, Journal of Sound and Vibration, 84, 389-395.

15. Peat, K.S., (1988), The acoustical impedance at discontinuities of ducts in the presence of a mean flow, Journal of Sound and Vibration, 127, 123-132.

16. Lord Rayleigh, (1894), "The Theory of Sound", Macmillan, London.

17. Young, C.J. and Crocker, M.J., (1975), Prediction of transmission loss in mufflers by the finite element method, Journal of the Acoustical Society of America, 57, 144-148.

Car Body Design and Tooling with CAD/CAM

P.E. Bézier

Paris

Abstract The general definition of a car being given by the Management, the traditional process, from styling to production of car bodies involved about ten capital stages based on the great experience and skill of specialists such as stylists, designers, plasterers and tool setters. Nevertheless, it was very difficult to obtain a complete and accurate definition of the shape of the parts.
CAD/CAM has brought a great and total change in the functioning of many divisions: styling, R. & D., tool shop, inspection, etc.
The benefits are obvious: accuracy, lagtime and cost.

1 Introduction

At the turn of the century, when the automotive industry was just gaining momentum, mathematics played a discrete but significant role. Maybe, thermodynamics or computation of stress and strain were not carried very far, but kinematics and, most important, geometry were of primeval importance.

Since that time, the role of mathematics steadily increased. Now, the help of that science is required for the definition of objects taking part in the working of mechanical parts: engine, transmission, frame, steering device, chassis, etc. The weight of a car must be kept to a minimum according with the performance which is expected: efficiency of the engine, silent behaviour of gears, road holding, drag, all those points have been paid high attention; they can concern such phenomena as valve springs vibration, noise transmission through housings and steel plate panels, but actual computing speed was limited by that of the mechanical arithmetic machines which were, till the advent of the computer, the fastest means available.

2 Definition of Mechanical Parts

2.1 Computation

It is to be noted that, at such time, the mechanical parts of a car were the only ones to be submitted to calculation. They were mainly designed with the help of the straightedge and of the compass; for the sake of simplicity, their shape could be considered equivalent to a Boolean combination of simple volumes such as parallelepipeds, cylinders, cones, spheres and tori.

The actual shape was generally more complicated, but it was not totally defined by the drawing office and it was left to the skill and experience of craftsmen such as patternmakers or die designers to complete the definition.

2.2 Drawing and Dimensioning

Satisfactory as it was for production purposes, this type of definition was a blending of pure geometry and handicraft and, as such, could only be submitted to approximate calculation.

The definition of mechanical parts bore an important feature: vital dimensions were completed with limits and, after completion, a part could be inspected and consequently accepted or rejected as "good for scrap", leaving no place for discussion or haggling.

3 Definition of Body

3.1 Traditional Method

The design of a car body was basically different, as it only relied on drawings effected with the help of French curves and lathes.

Evidently, no mathematical definition could be assigned to the curves, even less to the surface they encompassed, and that made any attempt of calculation utterly fruitless.

In fact, car body design and production had inherited a tradition three or four hundred years old, leaving the basic principles practically unchanged; of course, some improvement existed, such as the use of lacquered steel plates or plastic sheets for tracing means, or epoxy resins for 3D models.

It was a rather difficult task since the outer shape, or "skin", should please the eye of the stylist, of the top management and, if possible at all, that of the prospective customer. In the meantime, it should have aerodynamic qualities, be easily stamped, assembled, lacquered, strongly connected with mechanical ensembles, power plant, transmission, etc, together with dashboard, upholstery, seats etc.

It is evident that, at that time, body parts could not be defined precisely and completely, and this was a real hindrance at the tooling and production stages; consequently, when the computer became available, efficient and safe, it appeared as possible to improve the definition of a car body, and soon some large companies paid attention to that problem.

The first ones belonged to the aircraft industry but automobile companies followed rapidly.

Before looking more deeply in that evolution, it seems advisable to describe briefly which was the conventional process.

Outlining

First of all, the general management sets a broad description of the car which must be produced: overall dimensions, number of seats and doors, power, speed, type of body - sedan, coupé, limousine, break, etc - and the general impression it should convey: sporting, dignified, sleek or sturdy, etc. At that stage, the styling department and the Sales division have to express their opinion.

Sketches and Small Mockups

Starting from this very sketchy description, the styling department invents a few possible solutions, some being in accordance with prevailing trend, others diverging radically from it. These solutions are expressed by drawings and by small scale models, or mockups, the proportion ranging from one fifth to one eighth, which are hand made out of clay or solid wax.

Provisory Drawing

After one solution has been selected by the general management, it is the task of the Drawing Office to trace, full scale, the shape of a so-called clay model, which will be carefully examined before taking the final decision.

After measuring the offsets of points located on the surface of the small scale model, these points are drawn onto a large scale drawing board (8 x 2 metres, i.e.; 26 x 7ft); evidently, for the sake of simplicity, these points are located in planes parallel to those of the set of coordinates of the model.

Through the points, the designers run some curves with the help of French curves and lathes, only relying on their skill and experience; during that period, they sometimes ask the stylist to express his opinion and imagine whether the 3D shape will please the eye.

This task may take weeks, if not months, to be performed; needless to say, the drawing only describes the outer shape, the "skin", of the car, making sure, in the meantime, that there is space enough in it to contain the mechanical parts, the overall dimensions of which are more or less known.

It is worth noticing that, whilst endeavouring to run a curve through given points, the designer is entitled to diverge a little from them in order to improve the smoothness of a line, provided the distance to the point

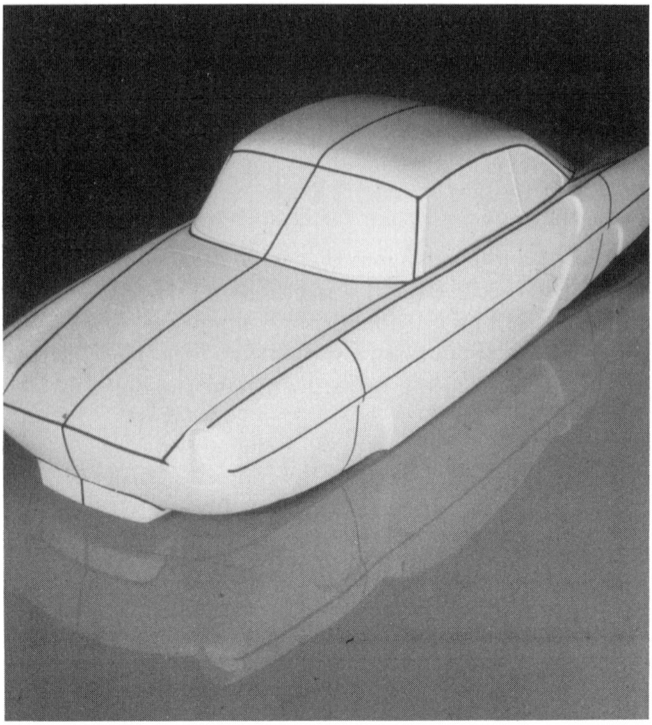

Figure 1. Mockup of the Floride car, scale 1/5

is kept within the limits, or allowance, which have been previously agreed upon.

In Europe, the distance between cross sections is commonly 100 millimeters (4 in.), and five inches in USA.

Clay Model

From the curves so traced, sections are cut in plywood sheets, which are then assembled together and form the skeleton of the so-called clay-model; some plaster is brought to fill the gaps between the plywood sections; when it is set, highly skilled plasterers remove the excess plaster to make it flush with the plywood ribs till they obtain a continuous and "pleasant" surface. During the period, the opinion of the stylist is very often requested, and the shape is modified in accordance, either locally or generally.

When the shape looks satisfactory, it is lacquered, equipped with wheels, glasses, headlamps, chrome trimmings, and bumpers and finally it looks like a real car. Again, some alterations may be required and carried out.

Figure 2. Mockup of the 4L car, full scale

Final Drawing

Since the shape of the so-called clay-model does not contain exactly the lines previously traced, measurement must take place again, in order to define a new set of points and obtain a set of drawings, which is supposed to be final, or so-called "frozen". This time, they not only define the "skin" of the car body but every inner panel, the frame and the brackets holding the mechanical ensembles, including hinges, handles, locks, screen wipers and many other implements.

This is a very difficult task, performed by designers endowed with taste, skill, experience, a very good knowledge of descriptive geometry, and above all, the ability to imagine a 3D object from the view of 2D lines and curves.

Such a task could take a large amount of time to be performed, often exceeding a semester, or a year.

During that period, the tool engineering division must express its opinion regarding the feasibility of the parts which must be stamped and assembled. This sometimes requires minor adjustments, so much that the drawings may differ slightly from the clay model shape but, from that moment on, the definition is supposedly "frozen".

Tool Design

This point being reached, the tool engineering department could embark on the tool design. A very important point is the task of "tilting" a part in order to make the stamping operation possible, and then defining the surface upon which the blank should be held and clamped for stamping; such a surface should be close to be wrapped, or developable, lest the blank be wrinkled when blocked by the upper part of the tool before being in contact with the punch [AUF.74].

Master and Patterns Building

The manufacturing and inspection of stamping tools require some 3D standard, that could also be used for inspecting stamped and assembled parts. A 2D drawing is not sufficient for this purpose, and a 3D master must be built in order to provide a means for copy-milling the press tools and for checking them as well as the parts.

This master must be built in a material that would not suffer warping or distortion as long as the car will be in production. It was carved in very expensive first class mahogany, which was replaced with araldite resin when that material became available at an affordable price.

The master was made of different blocks bolted onto a central cube, each representing a separate part of the body: bonnet, top, fenders, doors, trunk, screens, etc.

It was defined by templates representing the cross sections in the drawings, and then interpolated by patternmakers. Again, some alterations could be requested in the course of this operation.

Of course, great care had to be taken when using this master. It had to be kept, to avoid warping, in a moisture and temperature controlled underground room, and only duplicates should be used either for copy-milling, for die-spotting and for checking the stamping tools or for inspecting the stamped parts.

Tool Milling and Setting

Due to the limited accuracy of copy-milling machines, the normal practice was to leave a certain amount of material to be removed in order to erase the ridges left between the cuts performed by a toroid or ball-end cutter. That operation was performed by highly skilled tool setters, using first a pneumatic chisel, next a portable grinding wheel, then a file and a polishing stone. Of course, the shape so obtained was checked against a copy of the master but, in the meantime, attention was paid to the image of straight lamps on the surface of the polished tool; to obtain smooth lines, the operator sometimes had to diverge a little from the shape of the master copy.

Figure 3. Rough cuts on a stamping tool

The final inspection of a stamping tool did not mean much in fact. It was carried out with the help of templates made of steel plates. Had their shape been in perfect accordance with the curves in the drawings, which was far from granted, it was practically impossible to locate them accurately with respect with the coordinates system related with the tool and, regarding flexible steel plates, this was pure illusion, not to make use of harsher words.

Features

Let us sum up the description of the major steps of the process:

1. The stylist builds a small scale model or mockup.
2. A designer traces, full scale, cross sections of a volume representing the "skin" of the car body.
3. Plasterers build, full scale, a plaster-of-Paris so-called "clay model".
4. Designers trace, full scale, the car body, complete with inner panels, brackets, etc.
5. Patternmakers build a full scale master, which will be used as a standard.
6. Toolsetters finish copy-milled surfaces of the stamping tools.

The above mentioned operations share one feature: in each one, the operator is entitled to alter slightly the shape resulting from the previous ones. This means that the final product, summing up all the hard luck possibilities, may be somewhat different, be it only a little, from the initial product, not forgetting that, during the elapsed time, somebody may have unconsciously changed one's mind. Since this may have taken place in any of the six aforesaid steps, nobody can be held responsible for any detrimental drift. Hence, arguments, retouches, expenses and delay were, at that time, looked upon as inevitable.

In that scheme, mathematics played a significant part but only as geometry, without any computing. To interpolate a few curves between two opposed limits of a patch, this surface was replaced with a hyperbolic paraboloid, on which some operations could be easily performed, and then translated back onto the original surface. The result was not absolutely granted and had often to be improved in accordance with the designer's taste and experience; the help of the theory was not negligible; expressing it with equations was easy, but the mechanical arithmetic machines were too slow to be of actual help.

3.2 CAD/CAM Method

General View

For centuries, car body design and tooling methods did not suffer basic change, but everyone was conscious of the difficulties that were related with the needs, requirements and possibilities of this trade; but when computers became available, some people guessed that its incredible computing power and speed could improve much the whole system.

Available Hardware

In fact, at that time, computers were not totally available because, in the late fifties, the prevailing opinion was that a company should be equipped with a very powerful computer working for the benefit of various services or divisions, either administrative or scientifico-technical; but batch processing does not ensure that a prompt answer could be obtained in any case, and that would have been a severe limitation for any CAD system; moreover, at that time, it was considered unsafe to carry directly data through the telephone network since any parasitic or any blurred impulse could alter the significance of a character and result in a mistake. Hence, in order to overcome such an obstacle, the only solution would be to convince the management that an autonomous computer, be it of medium scale, should be allotted to the research and design division, specially for controlling some machines, and this was no easy endeavour.

Regarding tracing means, the CRT and the plotters were not very large, and stylists were reluctant to judge a curve or a drawing that would not be full scale; the large drawing machine were scarce and the milling machine-tools numerically controlled did not actually appear in Europe before 1960.

Basic Problem

The first problem to be solved was to build up a system by which the shape of a part belonging to a mechanism or to a car body could be defined with the only help of figures.

For mechanical engineers, educated and trained to work in Cartesian spaces, the first idea was to find a Cartesian expression, the ones fitting best with the computing technique being polynomials. But these proved rapidly difficult to manage, since it was often required to tilt the set of reference; hence the success of parametric definition.

As early as 1960, American aircraft companies had paid great attention to the problem, and obtained significant results; they had taken advantage of Isaac Schoenberg's works and adopted a parametric system.

The outer shape of an aircraft is somewhat simple, when compared to that of a car, and their problem was first solved with the help of third order functions, except for some regions such as the Von Karman's fillets, which were left to be defined later, when the shop hands had modelled them in aluminium sheet.

Alternatives

Progress in industry can happen according to different schemes: on the one hand, one can consider the successive steps of a process and improve those which can benefit from a discovery recently happened.

To give an example of such a case, one can examine the sequence concerning the traditional solution and think of using a measuring machine to record the offsets of some points located on the surface of a model, and improving it with a subprogram by which the distance between two consecutive points is computed and chosen according to the previous offsets instead of leaving it to the decision of the operator; another example would be leaving the definition of the slope of a curve on a recorded point to an algorithm taking into account the coordinates of two or four adjacent points.

On the other hand, one can endeavour to create a totally new system, specially devised to take full advantage of a recent invention.

The first solution is rather safe since it does not upset an existing system and leaves the possibility of restoring it without much trouble and cost; on the contrary, the second is more difficult to apply since it involves not only the purchase of new machines, but the training of people and, sometimes, an overall re-arrangement of the functioning of a service or of a shop; no doubt this is more risky, but it can bring much greater advantages.

Solutions

The solutions in existence [BAR.85] [CAGD.74] [CAS.85] [FER.64] [DAR.80] [GOR.71] [GOR.72] [MEH.69] [PAR.70] [RIE.74] [SAB.70] are adapted to the viewpoints briefly described in the preceding paragraph, some people being prone to "play it safe" and others to take chances.

Hereafter, two systems are briefly characterised, related with the two tendencies.

Coons' Problem and Solution

The first solution is, for example, that of the problem proposed to Steven Coons in the middle sixties by an automobile company: a full scale model being built, some character lines are traced on it, which delimit a network containing four-sided patches. The four boundary lines of each patch being expressed by parametric functions of rather low order, e.g. third or fifth, the problem consisted in defining the points inside each patch, ensuring in the meantime slope, or even curvature, continuity between them. In fact, this involved a previous step: the parametric definition of each boundary curve, which was the task of the designers. The system was intended to replace the work of the plasterers or patternmakers interpolating the curves previously carved in the material of the master.

Steven Coons perfectly solved the problem [COO.67] [COO.74], but his solution suffered from two limitations:

i. The cross derivatives, or twist vectors, had to be nil at the corner points, which gave the portion of the surface a pseudo-flat aspect. Steve Coons easily solved that minor problem, but the process could no more be considered as totally algorithmic since the designer was left the responsibility to choose arbitrary twist vectors so as to remove the pseudo-flats.

ii. Adjacent patches should be of similar dimensions and shape, e.g. broadly parallelograms. It could not, for instance, blend a patch resembling a square with another looking like a trapezoid.

In such a case, the accepted solution was to leave aside a part of the car, which would be dealt with by the conventional system.

It remains that in spite of these limitations the Coons' solution, algorithmic or not, is widely used throughout the world, and industry at large can pay his memory a well deserved tribute.

Direct Definition

The solution developed at Renault twenty odd years ago [BEZ,68] [BEZ.77] [BEZ.86] had a different objective: it should be used from end to end in the process running from the styling department down to the final inspection of stamped and assembled parts.

From the start it was supposed to fulfil some fundamental conditions i.e.:

i. be easily understandable by draughtsmen, designers and technicians who have learnt geometry, are extemely good at descriptive geometry but are, as a whole, only little familiar with analysis.

ii. be able to generate a great variety of curves, not only lines, circles and parabolas.

iii. be able to describe space curves as well as plane ones.

iv. give a prompt answer.

v. have an acceptable price.

Complying with those requirements, the solution can be described in the following manner:

i. The mathematical theory is based on the use of the Bernstein functions, the coefficients of which are expressed with the help of polygons and networks that somewhat mimic the shape of the corresponding curves and patches.

A text on this subject was published in 1977 as a thesis and, with some useful additions, in 1986. Up to now, B-splines and NURBS have not been implemented in the relevant software, maybe by fear that it could not be easily understood by draughtsmen, designers, stylists or technicians.

ii. Increasing the number of legs of a polygon leads automatically to expand the degree (or order) of the curves; consequently, the variety of curves is very great, even in the case of the order being practically limited to eight.

iii. When the vertices of a polygon (or network) are not coplanar, the curve (patch) is a space curve (patch); consequently, the system uses 2D as well as 3D curves (patches).

iv. The problem was not so much to have a very fast computer, but to have it available at any moment. Consequently, though it was very uncommon circa 1960, a medium size second hand computer was purchased, to be exclusively for the use of the drawing office.

Obtaining a 2D image in a short time is doubtless important, but a stylist can not pass a judgement, be it provisory, so long he has not seen and touched a full scale 3D object which has been lacquered and equipped so as to look like a real car. Consequently, it is considered compulsory to provide the drawing office with a N.C. milling machine, specially designed to carve free-form surfaces in large blocks of foam.

At the beginning, that machine was large enough to manufacture a major part of a car, say a roof, a bonnet or a flank, but now it is considered necessary to machine a complete full scale model; this means that the machine, plano-miller type, can have a capacity about 5 x 1.8 x 1.5 metres (16 x 6 x 5ft); since the material is very soft, the spindle power is only a few

Figure 4. NC carving machine

kilowatts, but the feed is about 12 metres per minute (40 ft/mn). A small scale mockup can be machined with great detail in a matter of one or two hours and a full scale car in a couple of days, which is then covered with a plastic film, lacquered and completed with all the requested equipment.

As we see, as soon as the drawing of a part looks acceptable, a solid counterpart is manufactured, giving the stylist a good opportunity rapidly to make up his mind.

When that point has been reached, the shape of the car is expressed in its entirety by figures, the drawings and 3D models having only an informative value and being some help for the people who will have to take part in the whole process.

The stylists define the skin of a car with the help of so-called "character lines", that are the boundary lines of rather large patches then subdivided to represent smaller details such as head lamps, parking lights, gougings, handles, monograms, license plates, etc. To express it briefly, it can be stated that, at the time a shape was defined by cross sections ten centimetres apart (4in), the number of patches could be over two thousands and the definition was, by all means, far from complete. Now, with the help of CAD, ninety five per cent of the surface can be expressed by sixty to eighty patches, the rest needing about three hundred very small patches.

It should be added that softwares exist which erase hidden lines of a wire

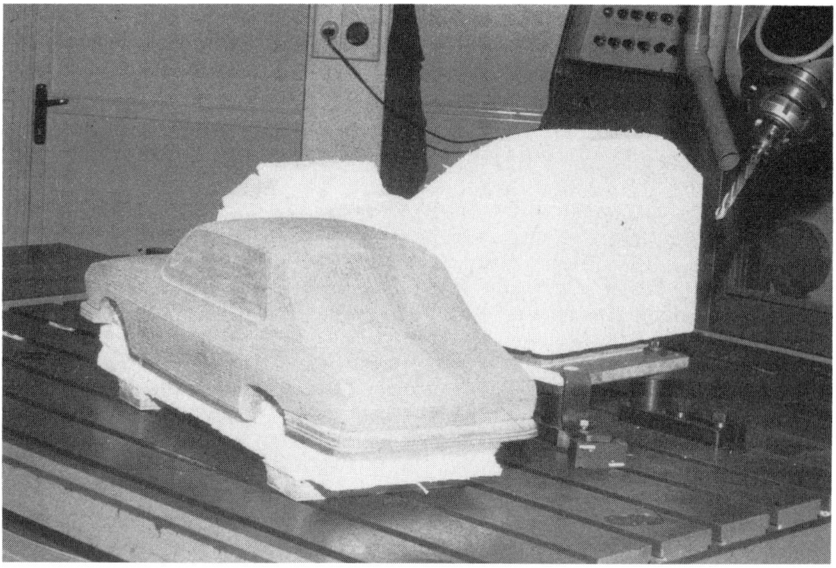

Figure 5. Clay model, scale 1/5

Figure 6. Clay model, full scale

Figure 7. Mockup model with character lines, scale 1/5

Figure 8. Same, with detailed character lines

frame drawing and colour the patches, taking into account the reflection lines and the change of shade related with the angle of light rays, which gives a very realistic aspect to a CRT image.

Starting from the numbers given by the styling department, the drawing office completes the definition of the body; one point of importance is that, owing to the possibility to define exactly the shape of each part, it is feasible to use the Finite Elements method and simulate stress, strain, vibration, buckling as well as the results of shocks, crashes and collisions; this does not preclude the building of models to be submitted to experiments, but one single experience is generally enough to confirm the result of the previous computation.

Tool Design and Manufacturing

When, at last, the definition of the body is finalised [ARA.85], the relevant information is transferred to the Methods (i.e. production engineering) division and especially to the tool-engineering and drawing office [BEZ.83]. In order to facilitate the stamping operation, it is often required to "tilt" a part, and this involved a difficult and time consuming task. Now, it is performed in a few minutes, instead of days, by the computer. Moreover, the accuracy has been much improved and the tool setting reduced in accordance.

For manufacturing the stamping tools free-form portions, copy-milling has been replaced by numerical control; the accuracy being much better, the surfaces so obtained are very close to the finished shape, and only need polishing instead of hand setting.

It should be added that if the Methods service can predict the spring-back occurring after stamping, compensation can be applied onto the shape of the tool; it is also possible, during that period, to build some tools made of compressed wood or of zinc alloy and carry out some tests, such tools being respectively able to stamp a score or a hundred parts before being worn out.

Another advantage is that the tool radius compensation be on line computed, since it must take into account the difference between rough, semi-finish and finish cuts as well as the dimension of the cutting tools which are, on the moment, avaialable at the tool crib.

For checking a finished press tool, one does not rely any more on templates, but only on offsets measured by special machines.

4 Fundamental Change

If we want to briefly sum up major changes that took place in a process endowed with a long experience, we may say that in a system making full use of the potentialities of CAD/CAM a fundamental feature is that

Figure 9. Final cuts on a stamping tool

Figure 10. Measuring machine

the basic data are carried only by figures, and are consequently accurate, invariable and undebatable. Hence

1. the stylist can work on a CRT to express his choice and then have a small scale model milled in a couple of hours, thus enabling him to try different options in a short time.

2. starting from the small scale model, the drawing office can define the "skin" of a car in a more detailed manner and have it carved in foam at the rate of one or two square metres per hour, then submitting it to the stylist for judgement. Machining a complete model, covered with a colour plastic film complete with trimming does not take more than a few days.

3. when the model has been accepted, the drawing office can embark on the complete definition of the body, issuing drawings of each part and conducting research about strength, strain, vibration, crash resilience, etc.

4. the production engineering services use the drawing office definition, the necessary transformations being accomplished by matrix and tensor computation.

5. sculpturing and inspection of press tools is performed by numerically controlled milling and measuring machines; the part of man in this work has been much reduced.

6. the overall result is a significant reduction in the time required to design, tool up and begin producing a new car body. It is admitted that this can sometimes reach fifty per cent.

5 Conclusion

Up to now, we have mainly examined problems related with hardware - CRT, computers, drawing machines, NC machine-tools, etc, and software - mathematics, data processing, etc. -.

But, as automatic as a system may be, it is invented, implemented, operated and, on the long run, controlled by man, whose role is absolutely vital.

Those who have to select a system and make a decision must have a fair knowledge of mathematics to choose correctly between the many systems available. They must evaluate the needs and requirements in accordance with their specific problem. They should consider the advantages of CAD/CAM at large, some being evident and rather easily expressed, but others lacking the property of having a precise value, i.e. delay, ease of contact with subcontractors, better connection between divisions, shops and services, etc.

To invent or create a system, mathematics are evidently of primeval importance, but it needs too a good knowledge of the mechanical as well as electric problems involved and the ability to anticipate the reactions of people who will have to take part in its functioning.

Another point should be considered with great care: it concerns some trades which are likely to decrease in importance or even to practically disappear; this is, or will be, the case for people building plaster-of-Paris "clay models", carving the masters or setting the dies and stamping tools. Other specialties will suffer a change, i.e. that of designers and draughtsmen engaged in body or tool design, and machine tool operators; in the meantime, some activities will expand: since more can be obtained from computing, mathematicians and data processing specialists will be requested to solve more problems, such as finite-elements computing, or invent new systems of curves and surfaces, or improve some aspects of already existing programmes; designers, draughtsmen, pattern-makers will have to become familiar with new aspects of their trade due to the importance of the computer.

An important change will take place in the tool shop organisation: the numerically controlled machine-tools hourly rate is greater than that of copy-milling machines; hence the need to prepare very carefully the operations so as to keep the cutting time to a minimum, this will require hiring and training new specialists, and tool setters should find there access to a better paid profession; experience proved beyond doubt that the risk to have to lay off people is practically nonexistent.

It must not be overlooked that in the most automatic system man remains one of the most important components; this is a lesson that engineers engaged in automatisms have learnt ages ago.

References

1. [ARA.85] Araki, H. and Amano, Y., (1985), Replacement of physical models by highly accurate numerical models for manufacturing stamping dies in the integrated CAD/CAM system. PROLAMAT.

2. [AUF.74] Auffray and Pichouron, (June 1984), Définition des surfaces développables s'appuyant sur une courbe. Application *á* la détermination des surfaces des outils d'emboutisage. Mémoire du C.N.A.M.

3. [BAR.85] Barnhill, R.E., (1983), Surfaces in computer-aided geometric design; a survey with new results. CAGD sept. 1985.

4. [BEZ.83] Bézier , P.E., (1983), Unisurf, from design to tool shop. Proceedings of C.A.P.E. 83 meeting. North Holland.

5. [BEZ.68] Bézier, P.E., (April 1968), Emploi de la commande numérique pour l'étude des carrosseries et des outillages. Journal de la Société des Ingéieurs de l'Automobile.

6. [BEZ.77] Bézier, P.E., (1977), Essai de définition numérique des courbes et surfaces expérimentales. Thesis, Paris Univ.

7. [BEZ.86] (Bézier, P.E., (1986), Mathématiques et CAO (Tome 4). HERMES publication.

8. [CAD.84] (1984), Proceedings of the Bristol Congress.

9. [CAGD.74] (1974), Proceedings of the Salt Lake City meeting, Academic Press.

10. [CAS.85] de Casteljau, P., (1985), Mathématiques et CAO (Tome 2) HERMES publications.

11. [COO.67] Coons, S.A., (June 1967), Surfaces for computer-aided design of space forms. Technical report MAC-TR-41. M.I.T. Cambridge Mass.

12. [COO.74] Coons, S.A., (1974), Surface patches and B-splines curves. CAGD.

13. [FER.64] Ferguson, J., (April 1964), Multivariable Curve Interpolation. Journal of the Association for Computing Machinery.

14. [FOR.80] Forrest, A.R., (1980), Recent work on geometric algorithms, MMCGD.

15. [GOR.71] Gordon, W. and Riesenfeld, R., (April 1974), Bernstein-Bézier methods for the computer-aided design of free-form curves and surfaces. JACM.

16. [GOR.72] Gordon, W.J. and Riesenfeld, R., (1974), B-splines and surfaces, CAGD.

17. [MEH.69] Mehlum, E., (December 1969), Curve and surface fitting based on variational criteria for smoothness. Central Institute for Industrial Research. Oslo.

18. [PAR.70] Parizot, S. and et Bouffard-Vercelli, R., Application de la commande numérique *á* l'étude et *á* la réalisation d'une corrosserie. Encyclopédie de la Carrosserie. Edition des Compagnons du Tour de France.

19. [RIE.74] Riesenfeld, R., (March 1974), Application of b-spline approximation to geometric problems of computer-aided design. Doctoral dissertation. Computer science division, University of Utah, Salt Lake City.

20. [SAB.70] Sabin, M., (July 1970), Parametric splines in tension. British Aircraft Corporation.

Surface Data Exchange Problems and Solutions

R.J. Goult

Cranfield Institute of Technology

Abstract This paper surveys the current standards for geometric data exchange with particular reference to the types of free form surface used in automotive design. Some of the problems arising in IGES and VDA-FS data exchanges are highlighted, and the development of the new I.S.O. draft proposal (STEP) for product data exchange is described. This is intended to supersede IGES, VDA-FS and SET, and is published for the first time in early 1989. The provisions made in this specification for the exchange of curve and surface data are described, and the author shows the extent to which these solve some of the earlier problems.
The limitations of the neutral file concept for surface data exchange between basically incompatible system representations are described, and the author shows some effective methods of curve and surface approximations which enable the data to be transmitted with a tolerance defined by the user. The two methods which have been developed are designed to maintain continuity conditions at curve and surface boundaries, and this is a significant feature in the CAD context.
The parametric evaluator concept, by which geometric information can be communicated between a CAD system and an application system, is described. The application of this concept to problems in machining complex sculptured surfaces is illustrated.

1 Introduction

During the past decade the use of computers in design and manufacture in the automotive industry has expanded dramatically. This growth of computing has helped to reduce the lead time from initial design to production, but has brought with it some new problems. Not the least of these is the problem of communicating geometric data between computer systems. The problem is particularly acute for data concerned with the description of external body panels where the geometry can be quite complex, and loss of continuity resulting from the communication will be very obvious in the finished product.

The geometric data communication requirement arises at many stages in the design and manufacturing process, including; design studio to manufacturer, model making for wind tunnel testing, communication with subcontractors, and the communication of design data to NC machining processes. The geometric data communication requirement arises at many stages in the design and manufacturing process including design studio to manufacturer, model making for wind tunnel testing, communication with subcontractors and the communication of design data to NC machining processes. The traditional method of communication is by drawings, these are potentially ambiguous, and involve potential errors in the re-creation of the geometry in the new system. Clearly what is required is some method of directly communicating the geometric data in magnetic or electronic form without human re-interpretation. The problem is compounded by the different forms of representation of 3-D curve and surface data used by the available CAD systems, and the proliferation of different systems in use. A motor manufacturer within our Esprit CAD Interfaces project found no less than 7 different systems in use within the company, and more than 40 others being used by suppliers and sub-contractors. For surface data, the forms of internal representation being used include piecewise polynomial, Bezier, B-Spline, rational and procedural. Some of these representations can be exactly converted to other forms, but in many cases a complete exchange of geometric data is impossible.

The types of possible communication channel between systems are via direct translators, via some standard neutral file format, or via parametric evaluators. Direct system to system translators were the first to be developed, they enable a maximal communication between systems, but have the disadvantages of requiring a separate translator for each pair of communicating systems, and of obsolescence as the systems change. The standard neutral file format has the advantage of only requiring one pre-processor and one post processor for each system supported (2N rather than $N(N-1)$ for direct translation), but the neutral file format itself may inhibit the data communication. As will be shown in section 4, parametric evaluators can provide an effective communication channel between a design system and application software, but cannot truly offer geometric data exchange.

2 Neutral File Specifications

The use of a neutral file as a transfer mechanism requires, a pre-processor to convert the data from the native format of the sending system to the neutral file format, and a post-processor to convert the neutral file data to the internal representation used by the receiving system. The data types which can be satisfactorily communicated are therefore limited by the system capabilities, the neutral file format itself, and by the pre- and post-processors. The unattainable ideal neutral file would be sufficiently flexible as to permit the communication of all types of geometric data currently used in CAD systems, and would use the most stable forms of representation. Even the ideal neutral file specification cannot enable the complete exchange of surface data between systems using incompatible representations, some form of approximation must take place in these cases. IGES (Initial Graphics Exchange Specification) was the first attempt to define a universally acceptable neutral file for geometric data exchange. When version 1.0 was published as an ANSI standard in 1981 its scope was limited to the exchange of 2 dimensional drawing information. IGES uses an 80 column card image format for the ASCII data used to represent geometry, and some of the files can be quite cumbersome. Further entities have been added to extend the capabilities of IGES as it has evolved, and the latest version IGES 4.0, [1], supports finite element and electrical applications as well as 3 dimensional surfaces and CSG solids. IGES is now the most widely used neutral file exchange medium, but problems have been encountered with unstable forms of representation, ambiguous entity specifications, and incompatibilities between processors.

The early compatibility problems were mainly caused by different CAD system vendors supporting different subsets of the IGES entities. The formal validation of processors has only recently started at the CAD CAM Data Exchange Technical Centre in Leeds, and a similar process has yet to start in the USA. The first general purpose entity for surface data in IGES was the spline surface, which was restricted to piecewise parametric cubic surfaces, this was later supplemented by the Rational B-Spline surface entity. Two other neutral file specifications, which have become national standards and were introduced with the objective of overcoming some of the shortcomings of IGES 1.0, are VDA-FS and SET.

VDA-FS [2] is a German national standard initiated within the automative industry with the restricted objective of providing an IGES compatible format for the communication of 3-dimensional curve and surface data. Parametric curves and surfaces are communicated by their explicit polynomial coefficients, but without the limitation of degree which occurs in the IGES spline surface. A processor validation service was established in Germany at the time when VDA-FS was standardised.

SET [3] is a French national standard which continues to evolve. It was primarily developed within the aerospace industry and is used in a number of joint European projects. One of the objectives of SET was to provide a more compact format than the card image of IGES, and it was unusual in intending to provide archiving capabilities as well as data exchange. The general purpose surface entity provides for explicit rational representation of unlimited degree.

At the end of 1984, the International Standards Organisation recognised the need for a standard for product data exchange, and after a review of the existing national standards decided that none of these could be extended to meet the requirements. A working group, TC 184 SC4/WG1, was established to develop a proposal for the new standard. The scope of this work is much broader than the earlier standards since it is intended to cover all aspects of product data for the entire life cycle, and a wide range of applications. The acronym STEP (Standard for the Exchange of Product Data) was coined for this standard which will be identical to the PDES specification being developed in the US by the IGES organisation. Many countries have been involved in the STEP development, and the major technical contributions have been provided by the IGES/PDES organisation and partners in the European Esprit CAD*I project. The first draft proposal for STEP was published, [4], for international review early this year, and is currently being revised to take account of the review comments. STEP has been heavily influenced by experience gained with the earlier standards, a formal language EXPRESS is used for the specification to remove the potential for ambiguity in interpretation, and attention has been paid to the needs for efficiency and stability of representation.

The entities in STEP for the communication of curve and surface data pay particular attention to stability, and most geometry is defined in terms of the basic entities of points and directions. An investigation undertaken within the CAD*I project, [5], showed that for parametric curves and surfaces the Bezier representation is more stable in the presence of rounding errors than the explicit coefficient representation. Table 1 shows some of the results obtained. As a consequence of this work, the B-Spline curve and surface entities, which also permit the communication of Bezier control points, were selected for the communication of parametric curves and surfaces in STEP.

3 Approximation Problems

The entities in STEP provide, after appropriate mathematical conversion, a stable method of communicating the geometric description of parametric curves and surfaces of polynomial, Bezier and B-Spline form including rational versions of these. This ensures that where two systems are mathematically compatible, data can be exchanged reliably, but no neutral file can overcome system incompatibility problems. Table 2 shows some of the potential problems and suggests some appropriate approximation methods. Perhaps the most significant problem is that of degree reduction, since many current CAD systems use parametric polynomial surfaces, but differ widely in the maximum permissible degree which ranges from 3 to 20 or more. Other problems occur in transferring data from rational to polynomial systems, in communicating data describing procedural surfaces such as those found in the Ford PDGS system, and in offset curves and

Table 1. Comparison of transmission errors after truncation to 6 significant figures of polynomial approximations to simple curve shapes when communicated by polynomial coefficients, Bezier control points or uniform B spline control points

Curve Shape	Degree	Polynomial Error / Bezier Error	B Spline Error / Bezier Error
	2	1.00	1.00
	3	1.81	1.49
Semi-circle	4	2.14	1.77
	5	1.81	.76
	6	18.50	135.16
	7	1.05	5.60
	8	1.87	8.05
	2	.74	1.00
	3	6.60	1.53
Rectangle	4	34.54	3.36
	5	2.71	17.91
	6	61.97	93.04
	7	188.64	910.57
	8	762.47	9202.46
	2	.79	1.31
	3	2.13	2.81
Exponential	4	1.16	3.34
	5	1.62	6.19
	6	3.22	12.29
	7	3.21	40.66
	8	2.29	64.66

surfaces. In all the above cases, a complete data transfer is impossible, but satisfactory communication can still be achieved if a suitable approximation of a simpler type can be found to the original curve or surface. Ideally, the approximation should be able to achieve a closeness of fit which meets a user-defined tolerance, and should maintain continuity with adjacent curves or surfaces. Both of these criteria are met by the Chebyshev and Orthogonal polynomial methods, [6,7,8], described in outline in the following sections.

3.1 Generalised Chebyshev Method

In the classical method of Chebyshev economisation the degree of a polynomial function $P_n(x)$ of degree n is reduced to $n-1$ after subtracting

Table 2. Data conversion approximation methods for curve and surface data transfer. Systems shown in ( )

Source	Destination			
	Low Degree (Catia)	High Degree (Strim 100)	Rational (Integraph)	Piecewise Low Degree (Medusa)
Low Degree (Catia)	Exact	Exact	Exact	
High Degree (Strim 100)	Chebyshev or Orthogonal	Exact		Chebyshev or Orthogonal
Rational (Integraph)	Orthogonal	Orthogonal	Exact	Orthogonal
Piece-wise Low Degree (Medusa)		Orthogonal		

the multiple of the Chebyshev polynomial of degree n which eliminates the leading coefficient. The properties of the Chebyshev polynomials ensure that the resulting function is the closest function of degree $n-1$ in the sense that

$$e(x) = \max \mid P_n(x) - P_{n-1}(x) \mid \tag{3.1}$$

is minimised in the range $-1 \leq x \leq 1$. For a parametric polynomial curve, the x, y and z components can be treated as independent polynomial functions of t, usually over the range [0,1]. With appropriate change of range the classical method could be applied to produce a close approximation of lower degree. The disadvantage of this approach is that since the Chebyshev polynomials are non zero at the ends of the range the approximating curve will deviate from the original curve at the end points. Watkins and Worsey, [9], overcome this problem by re-adjusting the end points after the degree reduction process has been completed, an alternative is to replace the Chebyshev polynomials by constrained Chebyshev polynomials. Using the interval [0,1] as standard range, families of constrained Chebyshev polynomials can be defined as equi-oscillatory polynomials with multiple roots of order M at $t = 0$ and $t = 1$. Treating the 3 components of the polynomial curve $\underline{r}_n(t)$ as independent functions, the subtraction of appropriate multiples of the constrained Chebyshev polynomials from $\underline{r}_n(t)$ will produce an approximation $\underline{r}_{n-1}(t)$ which matches $\underline{r}_n(t)$ and its first $m-1$ derivatives, and is the closest such function in the minimax sense described above. As with the classical method the magnitude of the error can be pre-determined from the coefficient required for the constrained Chebyshev polynomial. For multiple degree reduction the process described here can be repeated one degree at a time until the target degree is attained. The predicted errors at each stage can be accumulated to give an upper bound on the error, which is not usually attained.

Lachance, [7], and Sherar, [8], have computer constrained Chebyshev polynomials of degrees up to 21 and for $m = 1, 2$ or 3, these are considered to be sufficient for most practical applications. These authors have successfully generalised the method in order to reduce the degree of biparametric polynomial surfaces.

If $C_m^n(t)$ denotes the generalised Chebyshev polynomial of degree n with m-fold constaints at $t = 0$ and $t = 1$ the degree reduction of a biparametric surface is achieved as follows

$$\text{Let} \quad (u, v) = \sum_{i=0}^{m} \sum_{j=0}^{m} A_{ij} u^j v^i \tag{3.2}$$

where $A_{ij}, 0 \leq i \leq m, 0 \leq j \leq n$, are the tensor coefficients. Two alternative representations which separate the two variables are

$$r(u, v) = \sum_{i-0}^{m} P_i(u) v^i \tag{3.3}$$

and

$$r(u, v) = \sum_{j=0}^{n} P_j(v) u^j \tag{3.4}$$

For any pair of positive integers (l, k) with $2km$ and $2ln$, a constrained Chebyshev economisation is defined by the equation

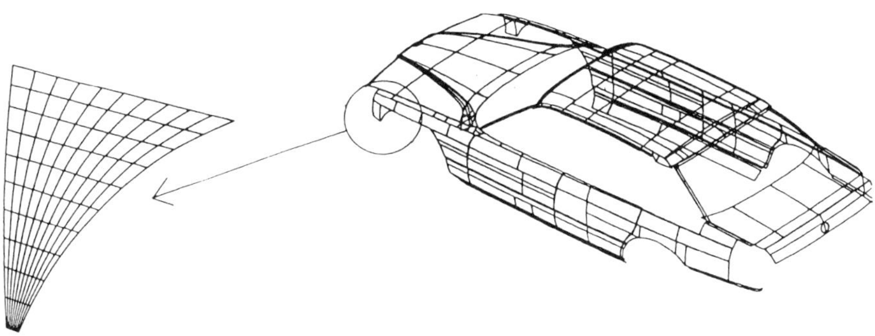

Figure 1.

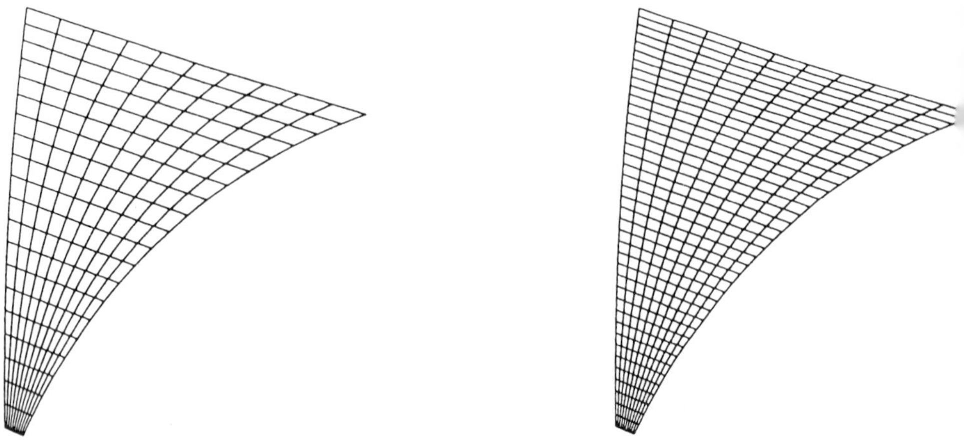

Tolerance = 1.0 mm

No Of Patches = 2

Tolerance = 0.1 mm

No Of Patches = 4

Figure 2.

$$s(u,v) = r(u,v) - P_m(u)C_m^k(v) - P_n(v)C_n^l(u) + A_{mn}C_m^k(v)C_n^1(u) \quad (3.5)$$

It can be verified that $s(u,v)$ is a parametric surface of degrees at most $n-1$ and $m-1$ in u and v respectively. The properties of the constrained Chebyshev polynomials ensure that at the patch corners, $s(u,v)$ matches $r(u,v)$ in value and in derivatives according to the selected values of k and l.

Along the boundaries, $u = 0$ or 1, and $v = 0$ or 1, and the result obtained is identical to that given by the degree reduction of the corresponding boundary curve. This property ensures that for adjacent surfaces any pre-existing continuity up to that corresponding to the constraints on the Chebyshev functions will be maintained in the approximating surfaces. As with curves, the process can be repeated any number of times until the target degrees are attained. A simple modification of formula (3.4) enables degree reduction to be made in u or v only. An important property of this method is that an upper bound for the maximum error in the approximation can be computed from the coefficients used. As a practical surface approximation algorithm, the error bound can be utilised, together with an automatic range bisection strategy, to sub-divide the surface into patches in order to obtain an approximating surface of target degree, with specified continuity conditions which meets any user defined tolerance. Figures 1 and 2 show some of the results obtained using this algorithm.

3.2 Orthogonal Approximation

The method described in section 3.1 can only be used when the original curve or surface is of parametric polynomial form. A method was therefore sought which was suitable for more general type of curve and surface definition. By modifying the error minimisation criteria from absolute deviation to mean least squares error, orthogonal polynomials provide an appropriate basis for the approximation. The classical basis of this type is the set of Legendre polynomials, but the use of these makes it impossible to maintain boundary continuity conditions. By analogy with the constrained Chebyshev polynomials, a set of orthogonal polynomials is defined to have multiple roots at $t = 0$ and $t = 1$ and to be mutually orthogonal in the sense that,

$$\int_0^1 P_m(t)P_n(t)dt = 0, \tag{3.6}$$

where $P_m(t)$ and $P_n(t)$ are distinct polynomials in this set. If the polynomials are normalised such that,

$$\int_0^1 (P_n(t))^2 dt = 1, \tag{3.7}$$

then the polynomials of degree n which minimises,

$$e(t) = \int_0^1 (f(t) - p(t))^2 dt, \tag{3.8}$$

and satisfies the constraints at $t = 0$, $t = 1$ is given by

$$p(t) = \sum_{i=1}^{m} a_i P_i(t), \quad \text{where} \quad a_i = \int_0^1 (f(t)p_i(t)dt \tag{3.9}$$

In order to apply these results to curve approximation, as a first step it is necessary to obtain a curve of the required type which coincides with the original curve (and if necessary its derivatives) at the end points. If only C_o continuity is required, then this first approximation curve can be the straight line joining the end points and the orthogonal polynomials used will have the simple constraints $P_i(0) = P_i(1) = 0$. For tangential constraints, the first approximation can be a Ferguson or Bezier curve, and the orthogonal polynomials will satisfy higher order constraints.

If $\underline{r}(t)$ denotes the original parametric curve, and $\underline{r_o}(t)$ is the first approximation obtained from the end conditions, then the orthogonal polynomials are used as defined by (3.9) to find the best least squares approximation to,

$$\underline{S}(t) = \underline{r}(t) - \underline{r}_o(t). \tag{3.10}$$

If this approximation is $\sum_{i=1}^{m} \underline{a}_i P_i(t)$,
then the required approximation to $\underline{r}(t)$ is,

$$\underline{r}_o(t) + \sum_{i=1}^{m} \underline{a}_i P_i(t). \tag{3.11}$$

As with the Chebyshev method, it is possible to obtain reliable estimates for the approximation error, in this case the mean least squares error, from the coefficients. It should be noted that the main computation in obtaining this approximation is given by (3.9), and thus depends upon the evaluation of an integral, this makes the method suitable for use with any integrable function. In practice, this integral is evaluated numerically, but the oscillatory nature of the orthogonal basis functions means that some care is

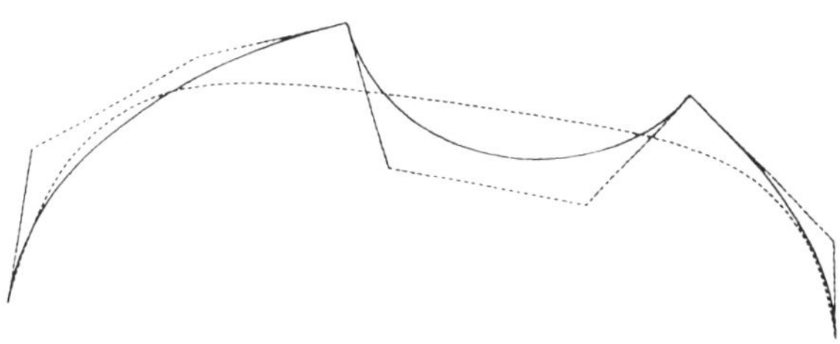

Simple Constrained Degree 4 Approximation

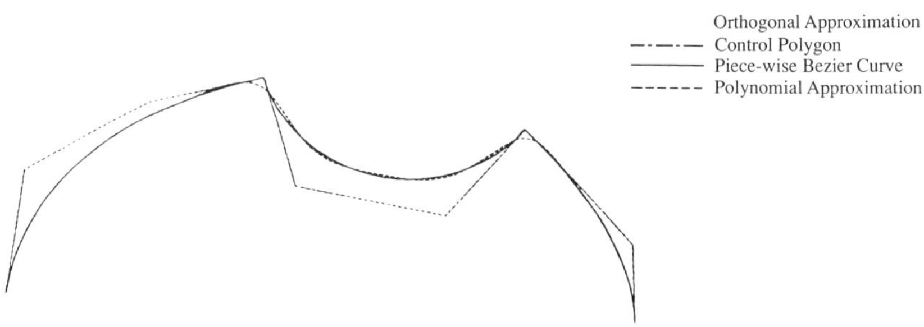

Tangent Constraint Degree 20 Approximation

Figure 3.

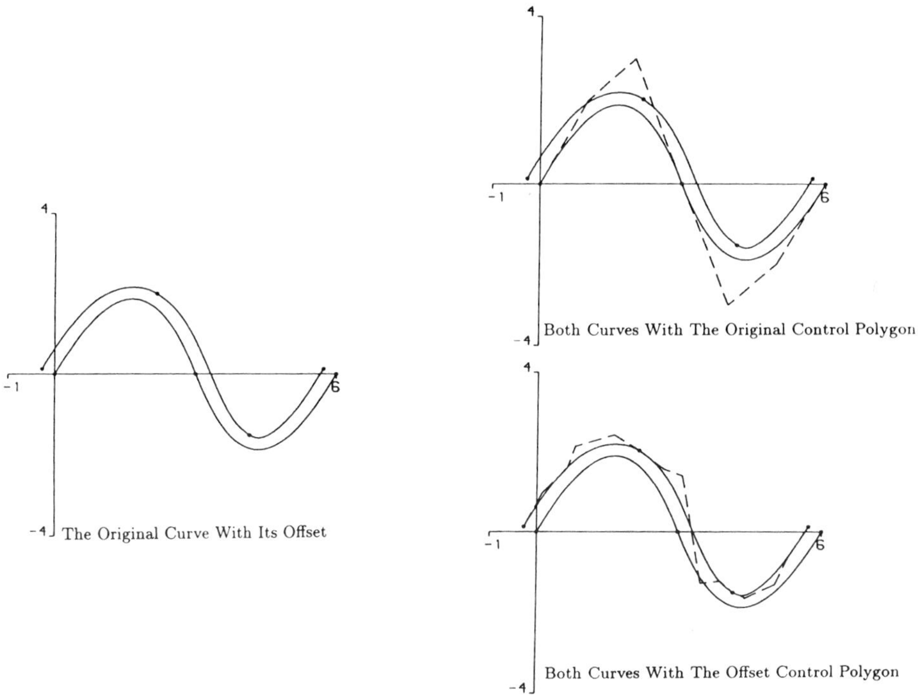

Figure 4.

needed at this stage. The method has been sucessfully applied to approximating rational curves, offset curves, other types of non-polynomial curve, and for approximation by a high order polynomial to piecewise polynomials curves of lower degree. Figures 3 and 4 show some sample results.

For surfaces the method can be generalised and applied as a two stage process. In the first stage, the boundary curves are individually approximated as described above. A first surface approximation, which can be a Coons' blend, is then constructed to interpolate the 4 boundary approximation curves. A tensor product surface of constrained orthogonal polynomials is then computed to approximate the difference between the original surface and the first approximation. Once more error estimates are obtainable, and the method ensures that adjacent surfaces will share a common boundary approximation curve.

4 Parametric Evaluators

Parametric evaluators provide a form of applications interface to CAD surface design systems. The concept differs significantly from the neutral file approach in that a complete and explicit surface definition is never communicated. For each type of surface communicated a 'black box' of software is provided to perform the surface evaluation function, this is accompanied by data to define the parameters of the particular surface being evaluated. All surfaces must be parametric of the form $\underline{r}(u, v)$, and the evaluator is a function or procedure, which for the input of parameters u and v, returns the values of the point $\underline{r}(u, v)$ and, usually, the first and second partial derivatives at this point. The data returned is sufficient for many applications, including surface display (for which point and surface normal are required), NC machining computations, surface offset calculations, and the computation of geodesics on the surface. One significant advantage of parametric evaluators is that the surface data is always, in theory, communicated exactly with none of the errors associated with the type of approximation process described in section 3.

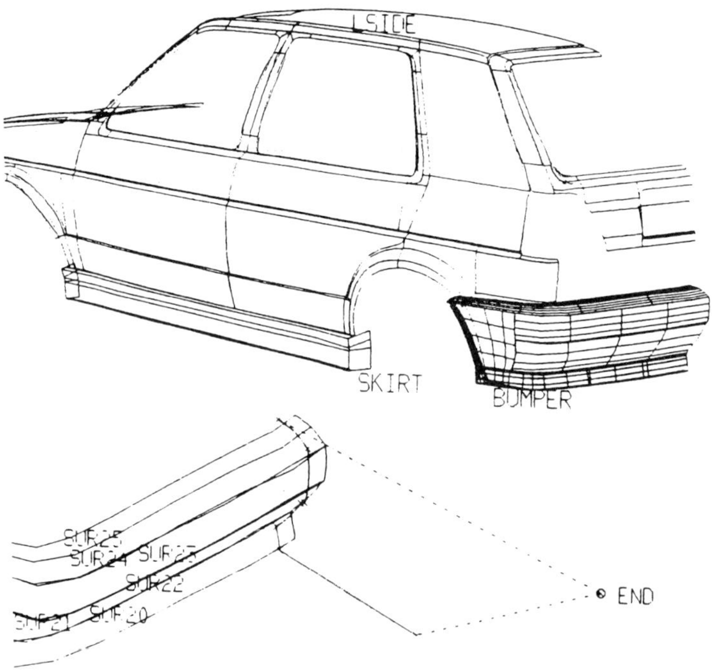

Figure 5.

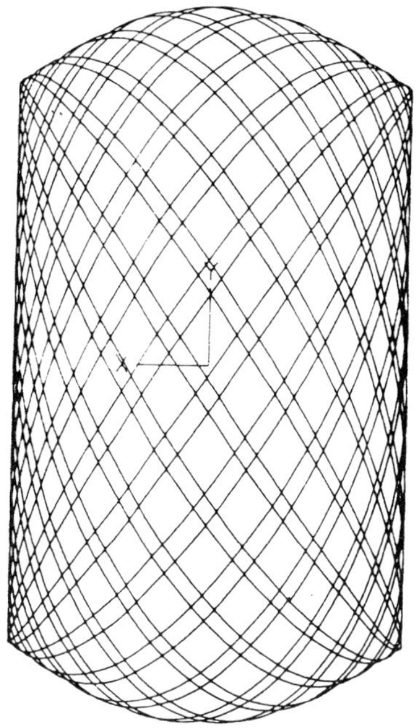

Figure 6.

The major disadvantage of parametric evaluators is that the application software must be specially written to make use of the data in this form. Successful parametric evaluator interfaces have been developed at Cranfield to the APT IV Sculptured Surfaces system for NC machining, and to the companion CASPA graphics display system (Figure 5). A parametric evaluator based package to compute geodesic paths on a surface for filament winding applications has also been developed (Figure 6). In the United States, Northrop and General Dynamics have also used the parametric evaluator concept for joint projects involving geometric definitions from incompatible surface design systems. For their applications, the confidentiality of the geometric definition when communicated in this way is considered to be a major advantage.

The inclusion of parametric evaluators in a later version of STEP is being proposed, but there are some significant problems to be overcome. In order to standardise the concept an agreed standard format for curve and surface data to be used by the evaluators has to be devised, and the not insignificant problems of communicating evaluators between different computers and operating system have to be solved.

5 Acknowledgements

I would like to thank Esprit project 322 (CAD*I) for funding much of the work described in this paper. Particular thanks are also due to Peter Sherar for providing information on Chebyshev approximation, and to Elizabeth McLellan and John Hamilton for providing illustrations.

References

1. "IGES, (1988), Initial Graphics Exchange Specification, Version 4.0", National Institute of Standards and Technology. Gaithersberg, Maryland.

2. "Verbund der Automilindustrie - Flachen Schmittstelle, DIN", (1984).

3. "Standard D'Exchange et de Transport (S.E.T.), (1985), AFNOR Z-68-300", Afnor, Paris.

4. "First Draft Proposal for STEP, (1989), DP10303", NSIT, Gaithersberg, Maryland.

5. "Lachance, M.A., (1987), "Some Experiments on Transferring Polynomial Data" in ESPRIT project 322: CAD*I, Status Report 3, Kernforschungszentrum Karlsruhe.

6. Goult, R.J., (1988), "Parametric Curve and Surface Approximation", in D.C. Handscomb (ed.) "The Mathematics of Surfaces III", Oxford University Press.

7. Lachance, M.A., (1988), "Chebyshev Economisation for Parametric Surfaces", Computer Aided Geometric Design 4,3.

8. Sherar, P.A., (1987), "Constrained Chebyshev Economisation for Parametric Curves and surfaces". Cranfield MSc Thesis.

9. Watkins, M.A., Worsey, A.J., "Degree Reduction of Bezier Curves", Computer aided Design 20, 7, 398-405.

Feature-Based Design of Automobile Surfaces

J.C. Cavendish and S.P. Marin

General Motors Research Laboratories, Michigan

Abstract We describe a new feature-based mathematical formulation for computer aided *design* and *representation* of surfaces. This formulation requires for input only limited information describing the specific geometry of a feature (pocket, channel, rib, etc), and produces on output a composite surface with continuous normals which provides a complete mathematical description of the shape. We describe techniques which are flexible enough to permit interactive design and easy modification of complex multi-featured surfaces, and which significantly simplify down-stream surface applications such as NC-machining.

1 Introduction

In this paper we propose a new feature-based CAD approach to surface design which is motivated by viewing the design process as one of *surface assembly* in which a number of known component surfaces are combined along user-specified boundaries with a given base surface in order to define a final composite surface containing desired geometric features; for example, pockets. With the aid of a novel strategy for recursively blending the component surfaces to the base surface, we are able to design and represent complicated multi-featured surfaces through a sequence of simple operations. We begin our exposition with an emphasis on *single* feature design.

1.1 Single Feature Design

Most often, a single feature involves a specified *secondary* surface, S_1, (for example, the floor of a pocket) which must be joined to a *primary* surface, S_0, with a smooth transition occurring between two curves. One of these two curves, $\tilde{C}_0$, is specified on S_0 to start the transition and the second curve, $\tilde{C}_1$, is given on S_1 to complete the transition (see Figure 1). We will assume that both S_o and S_1 are representable in explicit form; that is, S_0 and S_1 can be described by functions $z = f_0(x, y)$ and $z = f_1(x, y)$ respectively. (The important case where S_O or S_1 are parametrically defined is reasonably straight forward to deal with and presents no essential

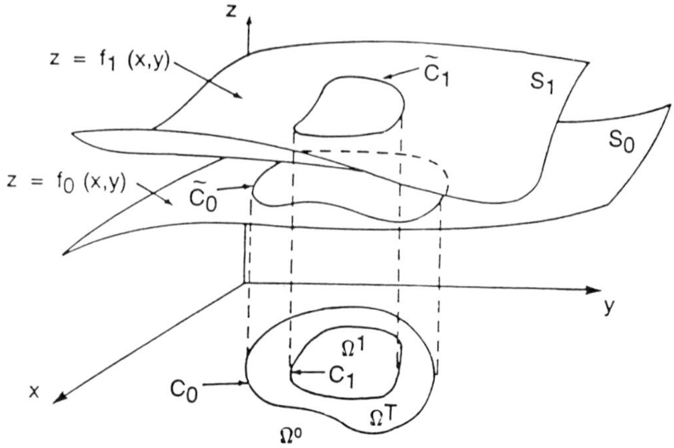

Figure 1. Primary and secondary surfaces prior to pocket construction

roadblocks to the method described here).

Note that the plan view projections of $\tilde{C}_0$ and $\tilde{C}_1$ partition the xy-plane into three disjoint regions: the region Ω^0 outside C_0, the region Ω^1 inside the curve C_1 and a region Ω^T between these two curves. We wish to design a new surface, P_1 which is equal to S_0 on Ω^0, equal to S_1 on Ω^1, and which makes a smooth transition between S_0 and S_1 in the region Ω^T. Our strategy is to construct first a smooth function $z = \Phi(x, y)$ which we call a *transition function* and which satisfies the related problem:

$$\Phi(x,y) = \begin{cases} 0 & if \;\; (x,y) \;\epsilon\, \Omega^0 \\ \epsilon(0,1) & if \;\; (x,y) \;\epsilon\, \Omega^T \\ 1 & if \;\; (x,y) \;\epsilon\, \Omega^1 \end{cases} \tag{1.1}$$

In (1.1), the notation $\Phi(x, y)\epsilon\,(0, 1)$ is meant to suggest that Φ increases monotonically from 0 to 1 in Ω^T. The precise definition of Φ on Ω^T will be described shortly. Once $\Phi(x, y)$ has been defined, we can construct the desired single feature surface, P_1, by means of the explicit representation $z = g_1(x, y)$ where

$$g_1(x,y) = (1 - \Phi(x,y))f_0(x,y) + \Phi(x,y)f_1(x,y) \tag{1.2}$$

1.2 Multiple Feature Design

Generalization of the single feature design (1.2) to a multi-feature design is straightforward. We proceed as follows if the intent is to design a geometry involving a base surface $S_0 : z = f_0(x, y)$ with N features whose tops or bottoms are defined by component surfaces $S_i : z = f_i(x, y), i = 1, 2, \ldots, N$. In order to integrate the *ith* feature into the design, input is entered to define the transition function $\Phi_i(x, y)$ corresponding to the shape of the *ith* feature. In turn, Φ_i is constructed from input data used to define Ω_i^T, the transition region appropriate for the *ith* feature. We remark that Ω_i^T depends on the plan view boundaries of the *ith* feature and not on any of the component surfaces. Next, define N intermediate surfaces, $P_i : z = g_i(x, y), i = 1, 2, \ldots, N$ via the recurrence:

$$g_i(x, y) = (1 - \Phi_i(x, y))g_{i-1}(x, y) + \Phi_i(x, y)f_i(x, y), 1 \leq i \leq N \qquad (1.3)$$

where $g_0(x, y) = f_0(x, y)$. The desired N-featured surface is P_N represented by $z = g_N(x, y)$.

1.3 The Definition of Pocket Boundaries

We describe next how pocket boundaries are specified in the plan view xy-plane. We begin by inputting a sequence of vertices to define the approximate shape of the inner boundary curve, C_1 for the *ith* feature. For example, if a pocket is to be a rectangular shape - with rounded corners - we input the (x, y)-coordinates of four vertices, (x_i, y_i) $1 \leq i \leq 4$, to define a piecewise linear approximation to the intended inner curve. Then, to control the degree of rounding at each corner, we input four radii, $(r_i, 1 \leq i \leq 4)$. The sharp corners are automatically replaced by smoothly fitted circular arcs with designated radii of curvature. We obtain the outer curve, C_0, as a *radial offset* to the inner curve just constructed. This is done by displacing each point on the smoothed inner curve, C_1 by a given amount R, an input offset distance, in the direction of the unit outward normal to C_1. This construct for defining feature boundary curves in plan view is useful for both convex shapes and shapes with reentrant corners, provided the offset, R, is small enough. Figure 2 illustrates these geometric constructions.

1.4 The Definition of $\Phi_i(x, y)$

Before concluding this paper with an illustration of a multi-feature surface constructed by implementing equation (1.3), we first define the *ith* transition function $\Phi_i(x, y)$ in the blending region, Ω_i^T. Recall that Ω_i^T is constructed by first defining an inner feature curve, C_1, composed of a smooth

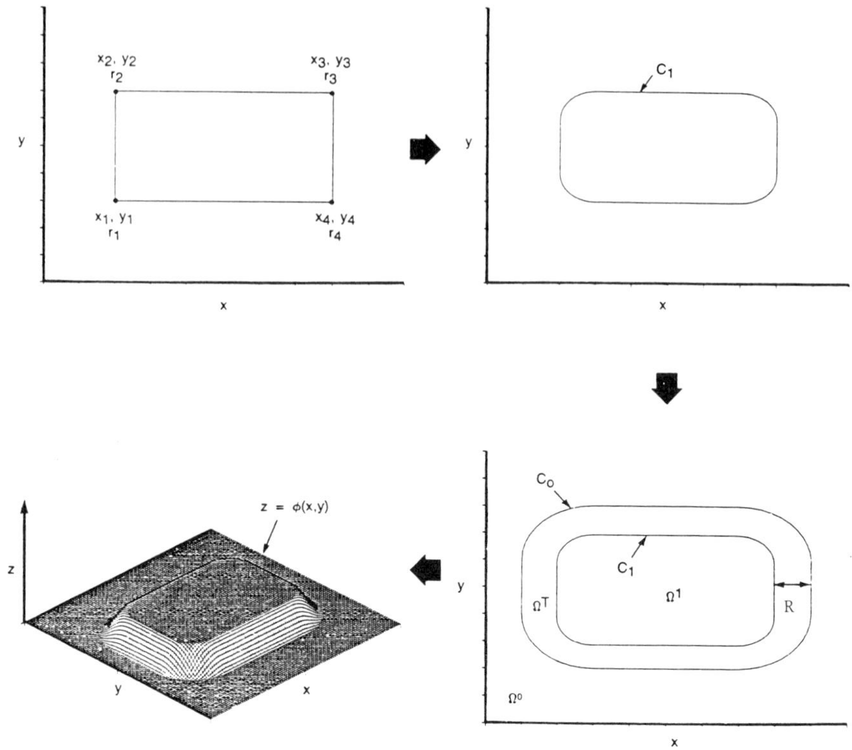

Figure 2. Construction of feature boundary curves in plan view

union of straight lines and circular arcs and described parametrically by $C_1 : (x(s), y(s)), 0 \leq s \leq L$. From this curve we can generate an entire family of well defined, smooth radial offset curves, say $C_{1,r} : x(r,s), y(r,s)$, by moving outward along the normals to C_1 a positive distance, r. In this notation, the original inner boundary curve, C_1 corresponds to the member $C_{1,0}$. In this paper we have assumed that the outer curve, C_0 is defined by $C_0 = C_{1,R}$ for some offset $R > 0$. Note that the family $C_{1,r} : 0 \leq r \leq R$ covers Ω_i^T and establish a smooth one-to-one correspondence between the cartesian coordinates $(x,y)\epsilon\Omega_i^T$ and the radial/axial coordinates $(r,s)\epsilon[0,R] \times [0,L]$. A family of simple transition functions can now be defined by introducing $z = h(r)$ where $h(r)$ is a simple univariate function; for example, a cubic Hermite polynomial satisfying end point conditions $h(0) = 1, h'(0) = h(R) = h'(R) = 0$. For each fixed value of s, the slice of the surface $S : x = x(r,s), y = y(r,s), z = h(r)$ will then be a plane cubic curve. With this, $\Phi(x,y)$ assumes the definition $\Phi(x,y) = h(r(x,y))$ for $(x,y)\epsilon\Omega^T$ in equation (1.3). Figure 2 displays a rectangular shaped

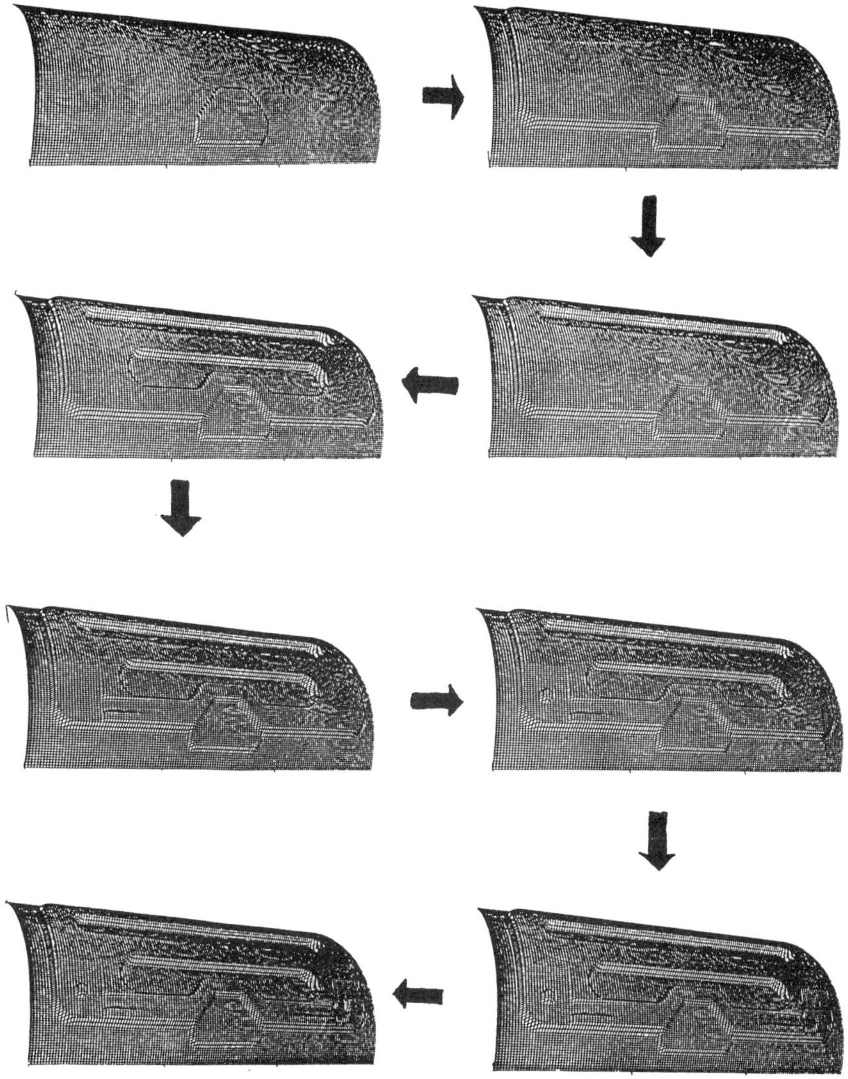

Figure 3. An eight-feature surface design

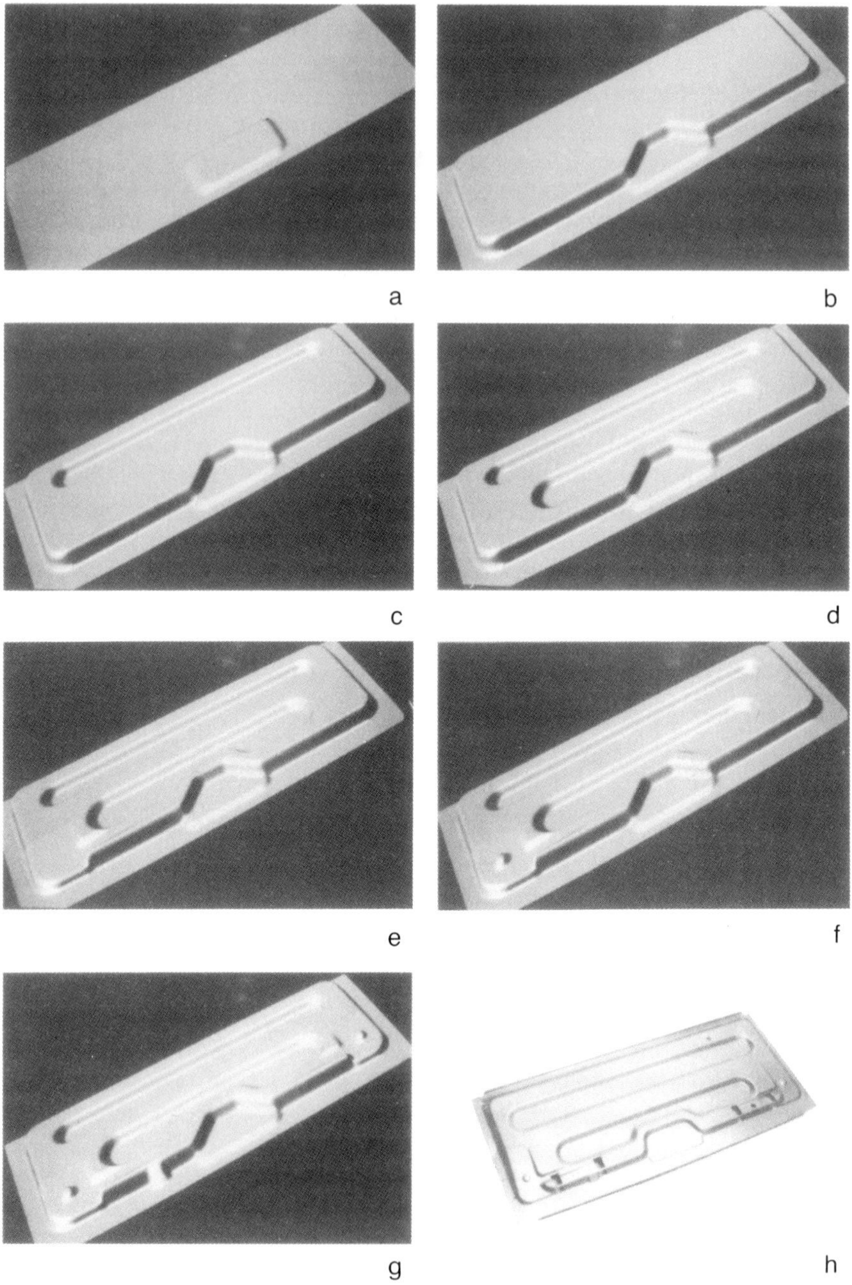

Figure 4. Comparison of eight-feature surface defined by methods presented here (a-g) and actual automobile inner panel (h)

transition function which blends together the surfaces $f_1(x, y) = 1$ and $f_0(x, y) = 0$ (see equation (1.2)).

2 An Example

We conclude this paper with an 8-feature surface design effort shown in Figure 3. For this exercise, all component surfaces are concentric cylinders or inclined planes. All input data needed to define pocket boundaries were entered interactively via a PC-AT interfaced to an IBM-3090 mainframe in less than 15 minutes of console time. Evaluation of the surface on a 150x150 rectangular grid for display purpose requires approximately 30 seconds of terminal time. Modifications to the final surface (for example, addition of a new pocket, deletion of an existing feature) require only seconds to implement. Finally, in Figure 4(a-g) we show the same eight-feature design displayed in Figure 3, but now all component surfaces are simple planes. Figure 4h is a photograph of the actual automobile surface (a so-called inner panel) from which we extracted the input data (for example, radii, wall angles, offset depths) needed to execute the mathematical algorithms presented in this paper and which produced the surface in 4g.

Simulation of the In-cylinder Flow in a Small Loop-Scavenged Two-Stroke Engine

R.J.R. Johns and M.H. Sandford[1]

Ricardo Consulting Engineers

Abstract The current interest in two-stroke engines for automotive applications has highlighted a lack of detailed knowledge of the in-cylinder flow; such information can be obtained using computational fluid dynamics (CFD). The paper provides some background on engine CFD and describes how CFD may be applied to loop-scavenged two-stroke engines. The system of partial differential equations for the conservation of mass, momentum and energy in a form suitable for engine flow calculations is presented, together with additional equations which allow the incorporation of turbulence effects. To illustrate the information that can be obtained, results are presented of the scavenge flow in a small motorcycle engine.

1 Preamble

Recent developments, particularly in fuel-injection systems, have led to renewed interest in two-stroke gasoline engines for automotive applications. However, the attractions of these engines (high power-to-weight and power-to-volume ratios, relative simplicity and low cost) must be tempered with the necessity of conforming to emissions legislation. In the past, two stroke engines have not been renowned for having the cleanest of exhausts. In addition to the blue smoke often seen in two-stroke exhausts resulting from lubricating oil, substantial contributions to unburnt hydrocarbons also result from short-circuiting of the charge (fresh fuel-air mixture which passes from the transfer ports directly out of the exhaust during scavenging) and from light load misfire. Although the short-circuiting problem can be alleviated by using in-cylinder fuel injection, other difficulties arise, such as achieving adequate fuel-air mixing in the period prior to ignition and providing adequate mixture strength in the vicinity of the spark plug. This

[1]M.H. Sandford now with Lotus Cars

has highlighted the lack of a detailed understanding of the in-cylinder processes, especially during the scavenging and compression periods.

Traditionally, performance of a scavenging arrangement has been assessed experimentally. For example, Jante [1] devised a technique for loop scavenged engines using pressure measurements obtained from pitot tubes in a motored engine with the cylinder head removed. This has been evaluated and refined by Blair and Kenny [2] who concluded that, when data from the Jante method is analysed correctly, it is capable of differentiating between good and poor scavenging systems. Dedeoglu [3] used two different fluids in a single shot rig equipped with a glass cylinder to visualize the mixing between 'air' and 'exhaust' in a loop scavenged gas engine. To aid visualisation the liquids were dyed and to monitor the temporal variation of 'exhaust' composition the capacitance changes due to the different dielectric constants of the liquids were recorded.

Whilst these and other techniques have proved extremely valuable in developing two-stroke engines to their current state, they do not provide the detailed information that is required for further refinement of the engine. In particular, knowledge about the time- and space-resolved velocity and species concentrations within the cylinder are vital in understanding the in-cylinder flow and mixing processes, in order to address problems such as those described above. In principle, this information can be obtained from optical diagnostic techniques, such as Laser Doppler Anemometry (LDA) and Laser Induced Fluorescence, however these are either too time-consuming, expensive or difficult to use on a routine basis, as required for engine development.

An alternative means of obtaining detailed information is to model the in-cylinder flow by solving numerically the fundamental differential equations describing fluid motion. This approach has increased in popularity during the last few years as numerical methods and the associated software have advanced and computer power has become cheaper, such that solutions to realistic problems can now be achieved at moderate cost.

Probably the first detailed simulation of a loop scavenged engine was that of Sher [4] who used a two-dimensional representation of the cylinder and porting arrangement. Around the same time, Johns [5] described how the three-dimensional flow in a loop scavenged two-stroke engine could be calculated by adapting the mesh arrangement developed for diesel engine flow calculations; the present work closely follows this approach. The first calculations of the flow in two-stroke engines for realistic conditions are those of Adachi et al [6], Cartellieri and Johns [7] and Johns [8] where the flow in uniflow scavenged engines was computed. Of particular interest here, the role of swirl was identified in controlling the size of a region of recirculating flow resulting from separation of the flow entering the cylinder and detrimental to scavenging performance.

Subsequent work by Sweeney et al [9] and Smyth [10] has focused on loop scavenged engines. In the former paper, [9], the authors computed the scavenging efficiency for a range of different cylinder configurations and compared these to measurements. Although the computations showed marginally higher scavenging efficiencies than those measured, the ranking of the cylinders was in agreement. The disagreement was attributed, in part, to uncertainties in the boundary conditions, in particular the velocity profile of the jet issuing from the transfer ports. This aspect was improved in the latter paper, [10], where measured velocity profiles were used to characterise the transfer port outflow. Here, it was found that the exit angle deviated significantly from the design direction of the port. In the elevation plane, the deviation from the design direction varied between 34° for a 3mm port opening to 14° for a fully open port, whilst in the horizontal plane the corresponding deviations were 24.5° and 10.4° respectively. Comparison between the results from computations using the measured velocity profiles with those using an assumption of a uniform profile aligned with the design direction of the port shows that the latter underestimates the amount of short-circuited charge, by incorrectly directing the flow too far towards the rear of the cylinder. Agreement with measured axial velocities in the cylinder of a motored engine was moderate.

2 Contents of Paper

Firstly, in section 3, the salient features of the mathematical model for in-cylinder flow are presented. To avoid unnecessary repetition, discussion is limited to the more important aspects; further details can be found in the references. The aspects particular to two-stroke engines, such as control of the grid for opening and closing of ports are discussed in section 4, together with details relating to the engine used to illustrate the computations. Results of the calculation are discussed in section 5 and the work is summarised in section 6.

3 Mathematical Description

Analysis of the flow is based upon solving numerically a system of partial differential equations which describe the conservation of mass, momentum and energy (see, for example, Bird et al [11] for a detailed derivation of the equations). However, engine flows are invariably turbulent and, although these equations are valid for both laminar and turbulent flows, the difficulty in economically resolving the small scale fluid motion at high Reynolds numbers dictates that the effects of turbulence are modelled rather than directly simulated. This is achieved by: replacing instantaneous quantities appearing in the differential equations by an ensemble-averaged mean and a fluctuation about that mean; ensemble-averaging the resulting equations to produce various correlations of means and turbulent fluctuations; replacing the unknown correlations by the dependent variables of additional equations of quantities which are related in simple algebraic fashion to such additional variables; and generating differential equations for the new variables in which unknown terms can be either ignored or, with reasonable assumptions, expressed in terms of known quantities. This process, referred to as turbulence modelling, is discussed in, for example, Launder and Spalding [12]. The resulting equations are discussed in more detail below.

3.1 Mean Flow Equations

The continuity and momentum equations, in Cartesian tensor form, can be expressed as:

$$\frac{\partial \rho}{\partial t} + \frac{\partial \rho U_i}{\partial x_i} = O \tag{3.1}$$

$$\frac{\partial \rho U_j}{\partial t} + \frac{\partial \rho U_i U_j}{\partial x_i} = -\frac{\partial p}{\partial x_j} - \frac{\partial \tau_{ij}}{\partial x_i} \tag{3.2}$$

where $\rho(x,t)$ is the averaged density at position x and time t; U is the averaged velocity, p the pressure and τ_{ij} the stress tensor. The unknown stresses are modelled in terms of an isotropic visocity μ_{eff} and the strain rate:

$$\tau_{ij} = -\mu_{eff}\left(\frac{\partial U_i}{\partial x_j} + \frac{\partial U_j}{\partial x_i}\right) + \frac{2}{3}\left(\rho k + \mu_{eff}\frac{\partial U_\ell}{\partial x_\ell}\right)\delta_{ij} \tag{3.3}$$

The viscosity is related, in turn, to two scalar quantities characterising the turbulence, namely the turbulent kinetic energy k and its dissipation rate ϵ, for which conservation equations may be derived:

$$\frac{\partial \rho k}{\partial t} + \frac{\partial \rho U_i k}{\partial x_i} = \frac{\partial}{\partial x_i}\left(\frac{\mu_{eff}}{\sigma_k}\frac{\partial k}{\partial x_i}\right) + G - \rho\epsilon \tag{3.4}$$

$$\frac{\partial \rho\epsilon}{\partial t} + \frac{\partial \rho U_i \epsilon}{\partial x_i} = \frac{\partial}{\partial x_i}\left(\frac{\mu_{eff}}{\sigma_\epsilon}\frac{\partial \epsilon}{\partial x_i}\right) + \frac{\epsilon}{k}(C_{\epsilon 1} - C_{\epsilon 2}\rho\epsilon) \tag{3.5}$$

The turbulence generation rate, G, is given by:

$$G = -\tau_{ij}(\partial U_i/\partial x_j) \tag{3.6}$$

In addition to the above, other scalar quantities may also be of importance. For example, in engines, thermal effects such as compression, heat transfer and mixing processes cannot be ignored, necessitating solution of an energy equation. Species concentrations are also of interest, especially when mixing between streams of different compositions occurs, as during two-stroke scavenging. These can be accounted for by solving additional differential equations which account for conservation of the quantity in question. These equations can be expressed in a general form as:

$$\frac{\partial \rho \phi}{\partial t} + \frac{\partial \rho U_i \phi}{\partial x_i} = \frac{\partial}{\partial x_i}(\frac{\mu_{eff}}{\sigma_\phi}\frac{\partial \phi}{\partial x_i}) + s_\phi \tag{3.7}$$

where ϕ is the dependent variable in question and s_ϕ is a source or sink of ϕ.

3.2 Coordinate Tranformation

Two coordinate transformations are applied to the above equations to account for the separate effects of curved surfaces and boundary motion. Here, general orthogonal coordinates are used to represent the curved boundary shape and the Cartesian form of the equations can be transformed rigorously using the method described by Pope [13]. Uni-directional boundary motion, such as that resulting from movement of the piston, can be accounted for by transforming the curvilinear coordinates into a system which has time-varying metric coefficients (Johns [5]). After transformation, the general scalar equation (3.7) can be written as:

$$\frac{1}{\ell_q}\frac{\partial}{\partial t}(\ell_q \rho \phi) + \nabla.\{\rho \underline{\tilde{U}} \phi - \frac{\mu_{eff}}{\sigma_\phi}\frac{1}{\ell_j}\frac{\partial \phi}{\partial \xi}\} = s_\phi \tag{3.8}$$

where ℓ is a metric coefficient, $\ell_q = \ell_1 \ell_2 \ell_3, \xi$ is a coordinate direction and $\tilde{U}$ is the velocity of the fluid relative to that of the coordinate system.

3.3 Numerical Solution

Each term in the differential equation is replaced by an approximate algebraic expression to produce an equation which relates the dependent variable at each grid cell to its six neighbours in the orthogonal directions and the old time level value. Implicit differencing is used for the source terms and the equations, which are diagonally dominant, are solved using standard techniques.

The use of general orthogonal coordinates dictates that the coordinate system, which is problem-dependent and defined by the mesh, must be generated a priori. A three-dimensional curvilinear-orthogonal mesh is not only difficult to generate for a general specification of boundary shape, it is also unnecessary for many in-cylinder geometries as the hexahedra produced by extrusion or rotation of a two-dimensional mesh is often sufficient. A two-dimensional mesh can be easily obtained by solving numerically a pair of Laplace equations for the physical coordinates of the mesh line intersections, these being coupled by the boundary conditions [14].

3.4 Boundary Conditions

Different types of boundaries can exist in engine flow calculations, including walls, inlets, outlets and various forms of symmetry. The practices adopted at each are described briefly below.

Walls

The appropriate terms in the difference equations for cells immediately adjacent to the boundary are replaced by expressions obtained by assuming that the flow is locally a one-dimensional turbulent Couette flow:

$$\tau_w = (U_p/y_p)\mu y_p^+/\ell n(E y_p^+) \tag{3.9}$$

$$y_p^+ = \rho(k_p C_\mu^{\frac{1}{2}})^{\frac{1}{2}}(y_p/\mu) \tag{3.10}$$

$$\frac{\partial k}{\partial y} = O \tag{3.11}$$

$$\epsilon = (C_\mu^{\frac{1}{2}} k_p)^{\frac{3}{2}}/\kappa y_p \tag{3.12}$$

$$\dot{q}_w = -(\mu y_p^+/\sigma_h)\left\{(1/\kappa)\ell n(E y_p^+) + P\right\}^{-1}\frac{dTw}{dy} \tag{3.13}$$

where τ_w is the wall shear stress, y_p and U_p the normal distance and tangential velocity, y_p^+ the dimensionless distance, $\dot{q}_w$ the heat flux, E and κ are constants and P a laminar sublayer resistance factor.

Inlets and Outlets

Inflow boundaries have a major influence on the internal flow whilst the effect of outflow boundaries, though not unimportant, is considerably weaker and localised to the region in the immediate vicinity of the boundary. The upwind differencing of the convective terms in the equations reflects this and it is, therefore, the temporal and spatial variations of the velocity and all scalar quantities at inlets which mainly determine the in-cylinder flow structure and fuel/air - products distribution.

Here, it is assumed that the flow through the transfer ports is quasi-steady, in which case the velocity profiles obtained under steady flow conditions and which are dependent only upon port opening, may be scaled by the instantaneous flow rate to produce time-dependent, "quasi-steady" profiles which may be inserted directly as boundary conditions. The validity of this approach has not been assessed in two-stroke engines, however comparison between steady and unsteady velocity profiles in the valve gap of four-stroke engines has been encouraging [15]. The instantaneous flow rate may be obtained from a gas dynamics simulation of the intake and exhaust systems in which the following is assumed: the flow in the pipes is unsteady and one-dimensional; inflow to and outflow from the cylinder obey an orifice equation which relates the flow rate to the pressure drop; the cylinder conditions may be determined from simultaneous solution of the zero-dimensional form of the continuity and energy equations.

Symmetry

Many instances occur in engine flow calculations where it is possible to take advantage of symmetry to reduce computer resources. Usually this is axial-symmetry, which can reduce a problem from three to two dimensions. The loop-scavenged two-stroke is symmetric about a plane containing the cylinder axis and a line joining the centre of the exhaust port with a point diametrically opposite on the cylinder. By applying a zero flow condition for the velocity component normal to this surface and a zero gradient condition for the other velocity components and scalar quantities, the required computational resources are halved.

4 Engine Geometry and Computational Details

The engine chosen to demonstrate the computations is a conventional loop-scavenged, two-stroke motorcycle engine of 175 cc capacity. Although the cylinder capacity is smaller than would be found for an automotive powerplant, this engine was chosen because of the availability of experimental data, including port effective areas, LDA measurements of the transfer port exit flow and engine performance data, including air flow. The main details of the engine and operating condition are given below:
Bore - 61.5 mm
Stroke - 58 mm
Capacity - 175 cc
Speed - 3500 rev/min
Load - 5.3 bar BMEP
Equivalence Ratio - 1.15 inflow
Exhaust Port Opening/Closing - 105°/255°
Transfer Port Opening/Closing - 119°/241°

The computational mesh is shown in figure 1 with the piston at BDC. This contains 30 x 14 x 25 (axial x radial x circumferential) cells in the cylinder and 8 x 9 x 25 cells in the combustion chamber located in the cylinder head, giving a total of 12300 active computational cells in the solution domain. However, the number of cells actually used varies during the computation according to whether the exhaust and transfer ports are open or closed. The sequence of events is as follows: calculation is started at 90° with both ports closed and axial planes A to B operational; at the time of exhaust opening, planes B to C are introduced and these expand with the piston motion to fill the exhaust opening, plane B remaining fixed at the top of the exhaust and plane C 'attached' to the piston; similarly, when the transfer port opens, plane C remains fixed at the top of the transfer port whilst planes C to D expand. Closing of the ports is effected with a reverse process, whereby the respective planes disappear at the appropriate time in the cycle.

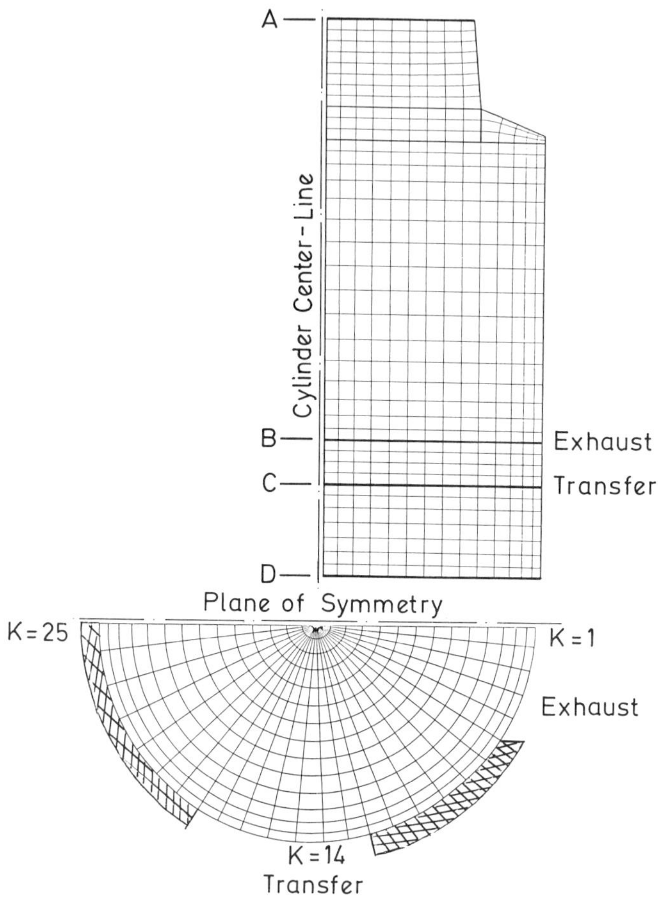

Figure 1. Computational Mesh

The transfer port geometry was very similar to that of Smyth et al [10] and LDA velocity measurements made at Ricardo produced similar deviations from the design flow direction. The data of Smyth et al is more comprehensive than the latter and has been used here. This is summarised in figure 2, which shows the variation with port opening of the deviation angles from the design flow direction in both horizontal and elevation planes. The inflow boundary condition was determined by scaling a uniform velocity profile in a direction given by the components of figure 2 with the instantaneous flow rate obtained from the cycle simulation, discussed below.

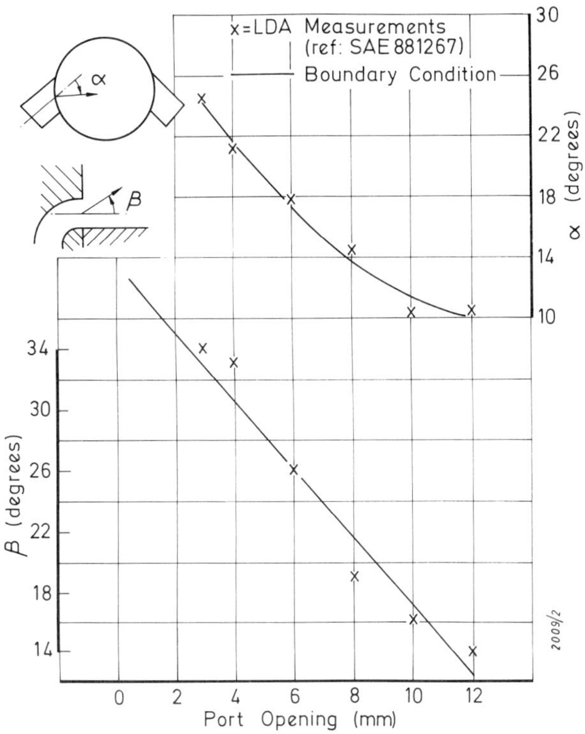

Figure 2. Variation of Transfer Port Efflux Angles with Port Opening

5 Results

5.1 Cycle Simulation

The cycle simulation required computation of 24 cycles to converge to a cyclically-independent solution, that is, a solution that is free from the effects of estimated initial conditions. This high number is attributed to the total volume of the inlet, crankcase and exhaust systems being approximately 18 times the swept volume and the coupling between the inlet and exhaust systems.

The cycle simulation program provided both the initial thermodynamic conditions and boundary conditions at the exhaust and transfer ports. The predicted exhaust and transfer port mass flow rates are shown in figure 3, where the sign convention is positive into the cylinder. The total air flow rate is within 1% of that measured and the characteristics are similar to those reported by Benson et al [16] for a loop-scavenged two-stroke. In figure 3 it can be seen that the initial exhaust blowdown pulse is attenuated at 120° by the opening of the transfer port, where the flow for the first 15 degrees is in the wrong direction, that is, into the crankcase. As the cylinder pressure drops, the transfer flow becomes established and scavenging of the cylinder occurs. Immediately before BDC(180°), the exhaust flow is aided by the rarefaction wave which has been reflected from expansion of the initial pressure wave in the divergent part of the exhaust pipe. This is closely followed by a weak positive, or charging, pulse at 190° resulting from reflection of the blowdown pulse at the step change in area at the silencer - tailpipe junction. The effect of the early arrival of the charging pulse is to increase the cylinder pressure and cause reverse flow through the transfer passage into the crankcase. The subsequent oscillations in mass flow result from interactions between the inlet - crankcase - cylinder - exhaust system.

The initial conditions of gas pressure and temperature obtained from the cycle simulation were 6 bar and 2100 K respectively at 90° crank-angle. The mass flow rates shown in figure 3 were used directly to scale the velocity profile. The reverse flow through the transfer ports into the crankcase during the early part of scavenging was accounted for in the gas composition boundary condition by assuming that the combustion products 'pushed' into the crankcase were subsequently expelled before fresh fuel/air mixture flowed into the cylinder.

5.2 In-Cylinder Flow Analysis

Figure 4 shows the velocity, equivalence ratio (fuel-air ratio normalised by the stoichiometric fuel-air ratio) and temperature fields at 145° in the symmetry plane (planes K = 25/1), the orthogonal plane through the transfer port (plane K = 14 and its mirrored counterpart) and various diametral planes. The view, which is from the side and slightly above, also shows the outline of the exhaust and transfer port openings. The 7 diametral planes in which velocities are depicted correspond to those of the scalar quantities. It can be seen that the exhaust outflow is well established and that a 'tongue' of inflowing fresh charge has nearly penetrated to the symmetry plane. Separation of the inflow jet as it enters the cylinder produces two regions of recirculating flow at 'A' and 'B'. The effect of the cooler inflow on the temperature field can also clearly be seen in producing temperatures near the transfer port of 400 K whilst the combustion products are at nearly 1900 K. Around the area indicated 'C' there is a direct path between the inflow and the exhaust and it is here that short-circuiting of the charge first occurs. The turbulent exchange coefficients for heat and mass transfer are identical (unity Lewis number) and, except where heat transfer effects dominate near boundaries, there is close correspondence between the equivalence ratio and temperature fields.

By 155° the opposing inflow jets have reached the symmetry plane, producing a stagnation point at 'D', seen in figure 5. The formation of a distinct 'loop' can be seen in plane K = 25/1, although growth of the recirculation centred at 'A' has resulted in some combustion products being trapped near the cylinder wall above the transfer ports, thus inhibiting scavenging of this region. It is observed that the fresh charge has reached the exhaust at 'C' and, from this time on, the exhaust outflow contains a proportion of unburnt fuel.

The area of recirculating flow denoted 'A' continues to grow until it fills most of the cylinder, seen at 170° in figure 6. The residual combustion

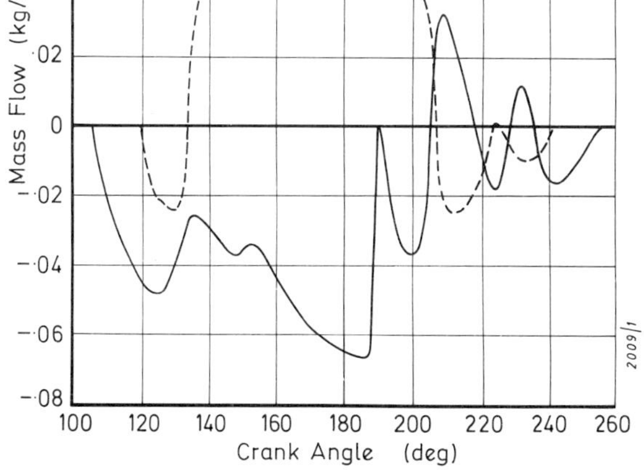

Figure 3. Exhaust and Transfer Port Mass Flow Rates

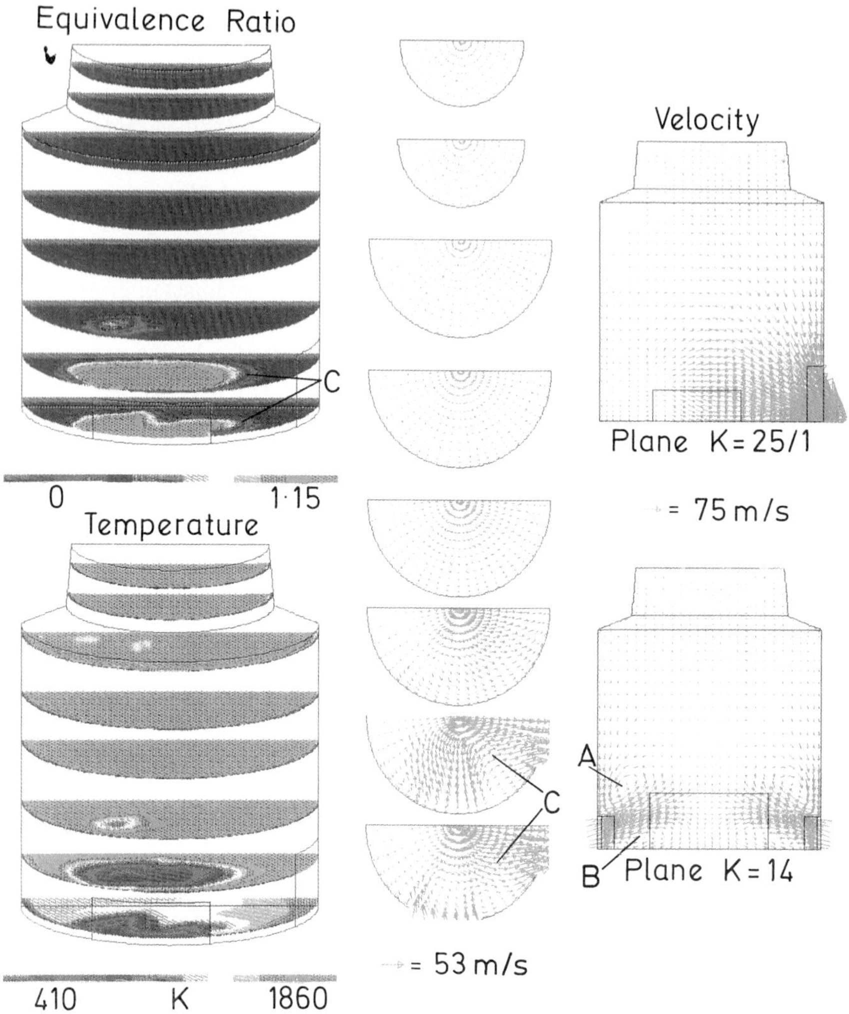

Figure 4. Flow Structure at 145°

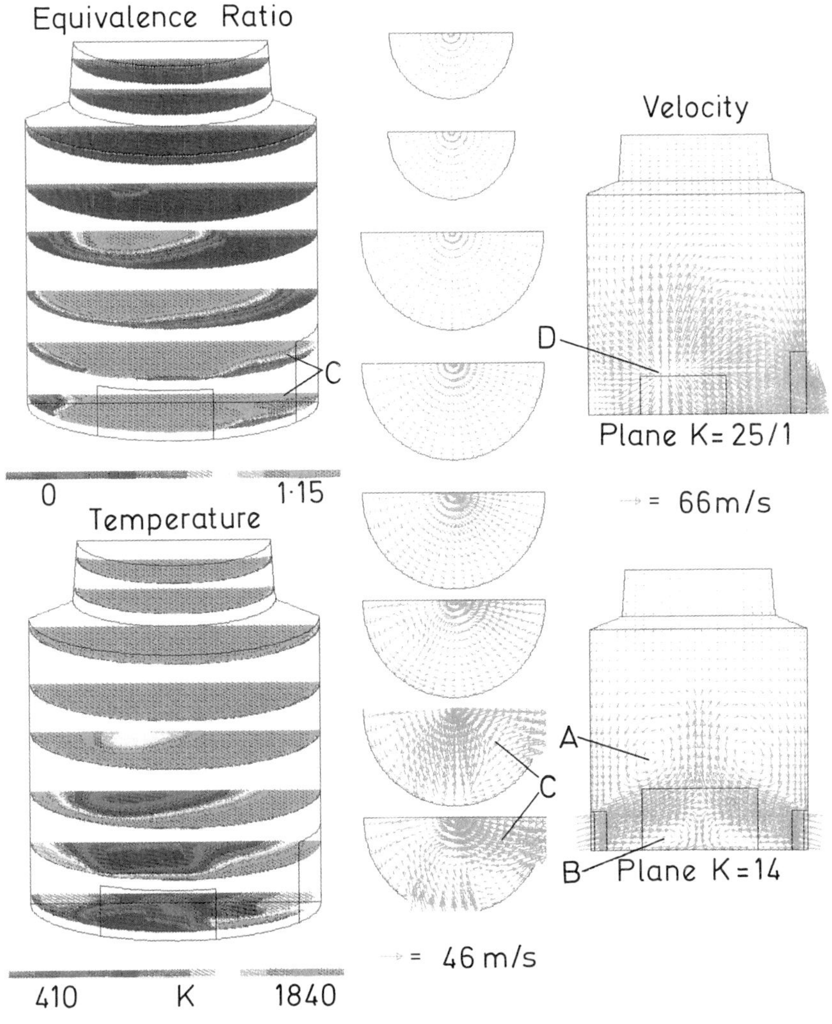

Figure 5. Flow Structure at 155°

products occupy a volume between the transfer and exhaust ports and extending to the combustion chamber in the cylinder head.

By 190° the exhaust flow is zero and , although most of the scavenging has been accomplished, there is a substantial zone containing hot combustion products between the transfer and exhaust ports. This can be clearly seen in the temperature field in figure 7 where the trapped residual gas has

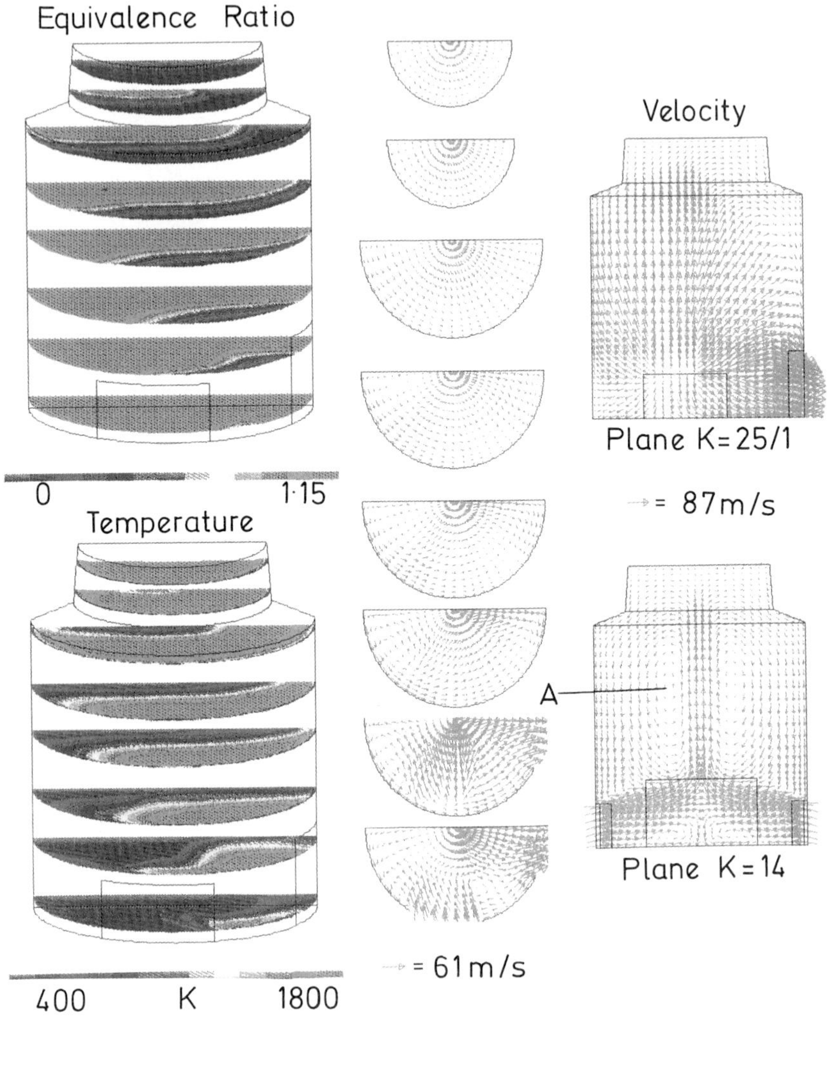

Figure 6. Flow Structure at 170°

a temperature of 1620 K, indicating that very little dilution has occurred up to this time. The region of recirculating flow at 'E', generated by impingement of the flow on the cylinder wall above the exhaust, appears to be largely responsible for the subsequent dissipative mixing of this zone.

The maximum temperature has reduced considerably by 210° (figure 8) and is confined to a small area just above the piston. Further up in

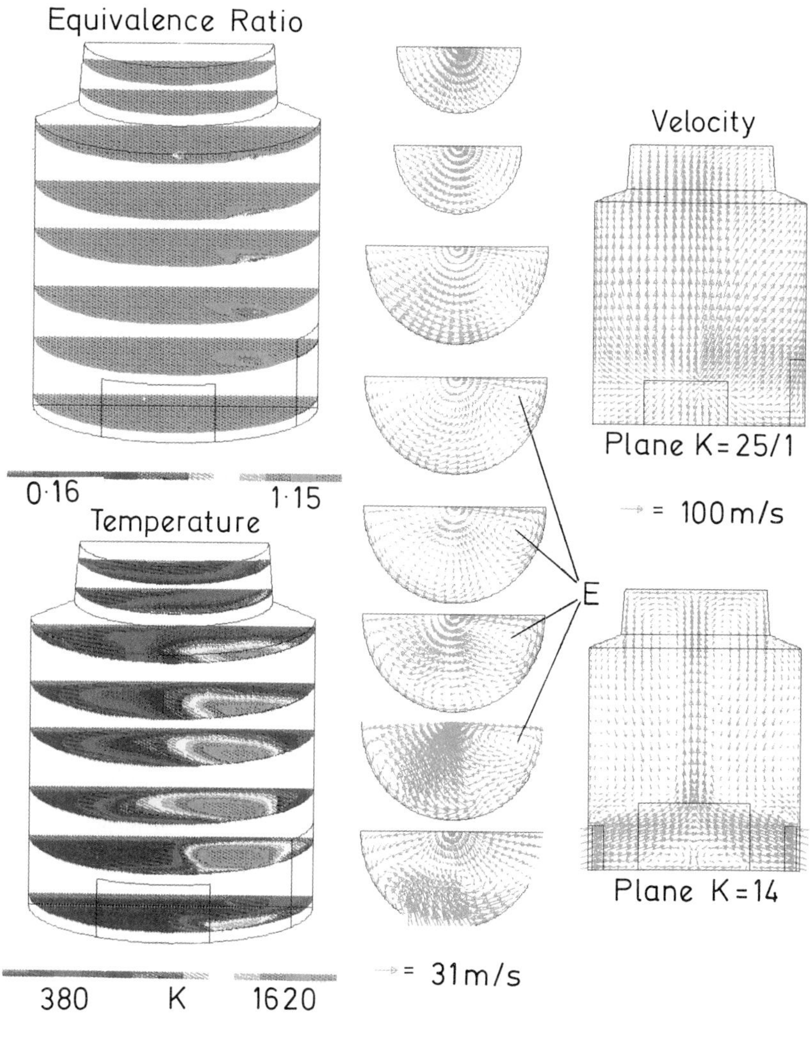

Figure 7. Flow Structure at 190°

the cylinder, the combustion products and surrounding fresh charge have mixed such that a well defined 'core' of hot products no longer exists. The reverse flow from the exhaust pulse, which reaches a maximum at this time, generates a substantial flow directed towards the back of the cylinder. This 'pushes' the remaining combustion products away from the exhaust, seen in figure 9 at 230°, although mixing continues to dilute the exhaust concentration, evidenced by a further drop in temperature to 820 K. A mechanism such as this may be responsible for causing autoignition later in the cycle in racing engines and evidenced by damage at the rear of the piston. Substantial reverse flow into the transfer ports occurs during the last part of scavenging.

6 Discussion and Summary

A number of general points are made below concerning calculation of the in-cylinder flow in two-stroke engines.

The advances in numerical methods and computer software and hardware allow realistic flow calculations in loop-scavenged two-stroke engines. From the above analysis, it can be seen that although the in-cylinder flow in two-stroke engines is extremely complicated, not only can phenomena such as short-circuiting be seen but also quantified and the controlling mechanisms understood.

Although time consuming, more extensive comparison with experimental data is required to determine the accuracy of the flow calculations. The assumption of quasi-steadiness has been applied with reasonable success to specifying boundary conditions at the inlet valve in four-stroke engines, however the validity of this has not yet been assessed for the outflow velocity profiles from transfer ports.

To a large extent, the flow structure is determined by the inflow and outflow boundary conditions as it is these which provide the driving force. However, the strong interactions between the inlet, crankcase, cylinder and exhaust dictate that simultaneous modelling of the entire system is really required. Methods for inlet and exhaust modelling are already well established and current efforts at Ricardo are directed towards integrating multi-dimensional in-cylinder flow calculations with one-dimensional gas dynamics software [17] to allow simultaneous solution of the equation systems pertaining to each.

7 Acknowledgements

The authors would like to thank the Directors of Ricardo for permission to publish this paper.

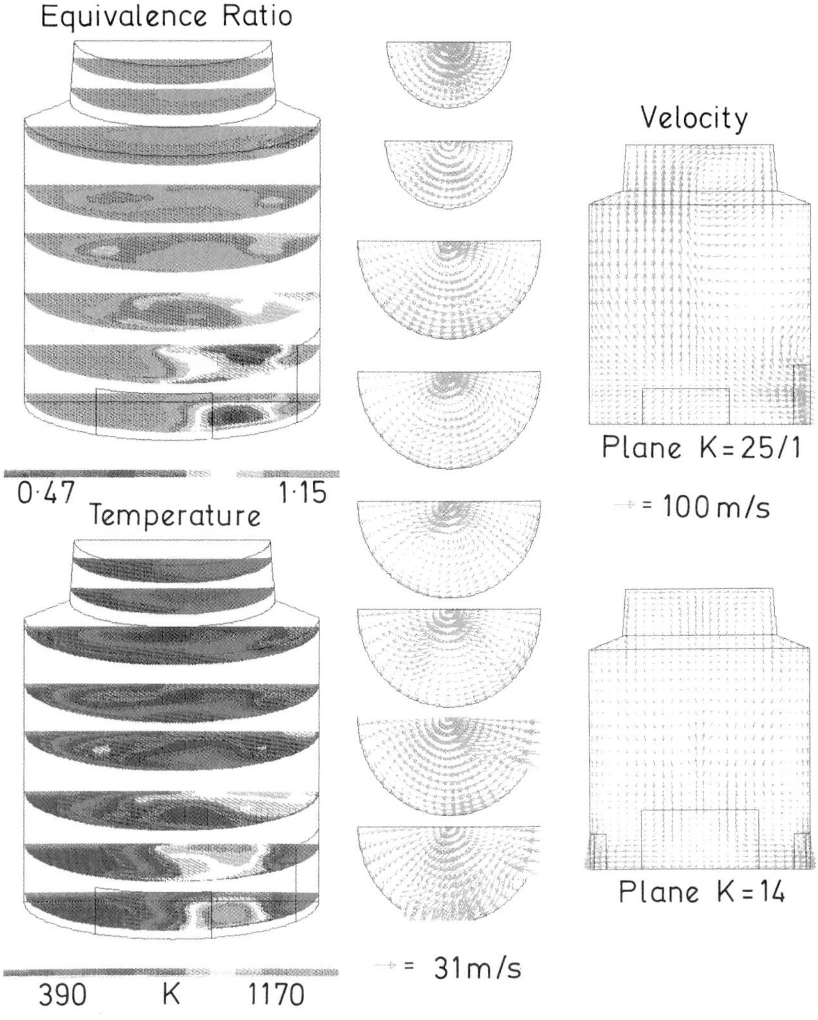

Figure 8. Flow Structure at 210°

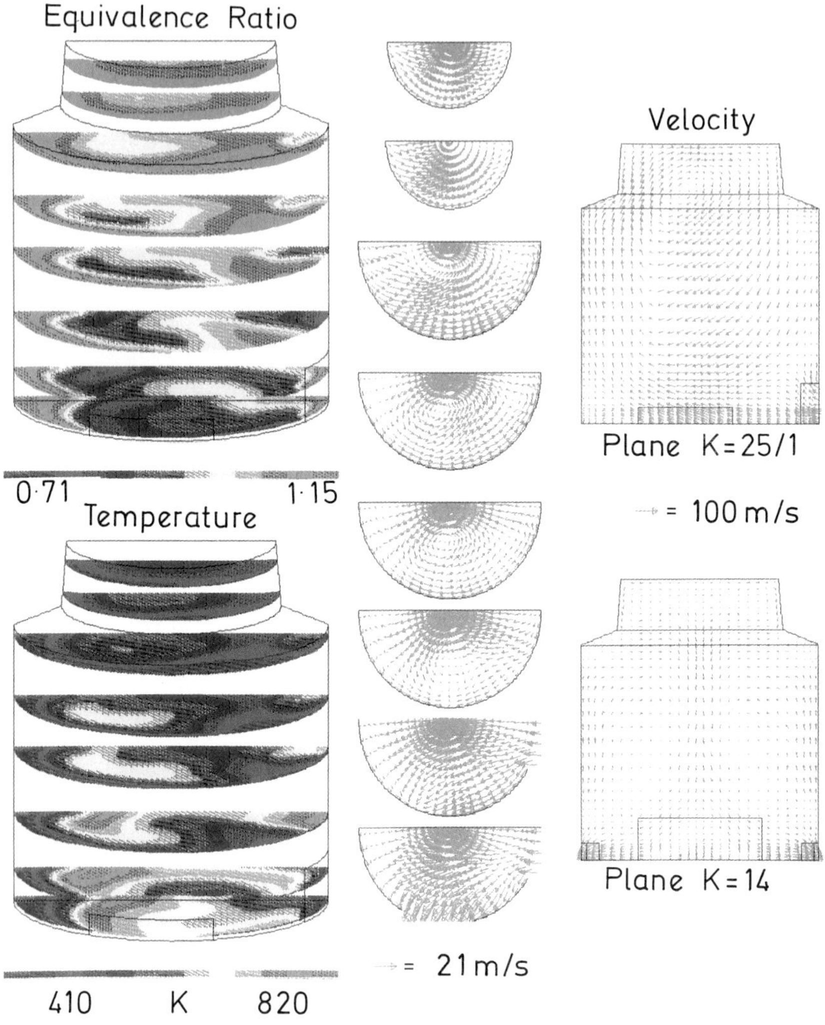

Figure 9. Flow Structure at 230°

References

1. Jante, A., (1968), Scavenging Flow Ducts of 3-Cylinder 2-Stroke Cycle Engines, SAE paper 680468 and SAE Trans., vol. 77.

2. Blair, G.P. and Kenny, R.G., (1980), Further Developments in Scavenging Analysis for Two-Cycle Engines, SAE paper 800038.

3. Dedeoglu, N., (1971), Scavenging Model Solves Problems in Gas Burning Engines, SAE paper 710579.

4. Sher, E., (1979), "Theoretical and Experimental Investigation of the Scavenge Process in a Two-Stroke Cycle Engine", Ph.D Thesis, University of Manchester Institute of Science and Technology (UMIST), England.

5. Johns, R.J.R., (1980), "Prediction of Flow in Diesel Engine Cylinders", Ph.D. Thesis, University of London, England.

6. Adachi, T., Kato, S. and Tsujimura, H., (1982), Numerical Analysis of Scavenging Flow in Two-Cycle Internal Combustion Engine, ASME conference on Flow in Internal Combustion Engines, Phoenex.

7. Cartellieri, W. and Johns, R.J.R., (1983), Multidimensional Modelling of Engine Processes: progress and Prospects, paper D8.5, Proceedings of the 15th International Congress on Combustion Engines (CIMAC), Paris.

8. Johns, R.J.R., (1984), A Unified Method for Calculating Engine Flows, ASME paper 84-DGP-18.

9. Sweeney, M.E.G., Swann, G.B.G., Kenny, R.G. and Blair, G.P., (1985), Computational Fluid Dynamics Applied to Two-Stroke Engine Scavenging, SAE paper 851519.

10. Smyth, J.G., Kenny, R.G. and Blair, G.P., (1988), Steady Flow Analysis of the Scavenging Process in a Loop Scavenged Two-Stroke Cycle Engine - A Theoretical and Experimental Study, SAE paper 881267.

11. Bird, R.B., Stewart, W.E. and Lightfoot, E.N., (1966), "Transport Phenomena", Wiley, New York.

12. Launder, B.E. and Spalding, D.B., (1972), "Mathematical Models of Turbulence", Academic Press, London.

13. Pope, S.B., (1978), The Calculation of Turbulent Recirculating Flows in General Orthogonal Coordinates, J. Comp. Phys. vol. 26, 197-217.

14. Gosman, A.D. and Johns, R.J.R., (1979), The Generation of Curvilinear-Orthogonal Grids for the Solution of Problems in Fluid Mechanics, Imp, Coll. rept. FS/79/23.

15. Bicen, A.E., Vafidis, C. and Whitelaw, J.H., (Sept 1985), Steady and Unsteady Airflow Through the Intake Valve of a Reciprocating Engine, Journal of Fluids Engineering, Trans. of ASME, vol. 107, 413-420.

16. Benson, R.S., Baruah, P.C. and Whelan, C., (1975), Simulation Model for a Crankcase Compression Two-Stroke Spark Ignition Engine Including Intake and Exhaust Systems, Proc. I. Mech. E., vol. 189, 7/75.

17. Morel, T., Flemming, M. and LaPointe, L., 'Measurement and Simulation of Manifold Pressure Dynamics in a Chrysler 2.2 liter Sl Engine', to be presented at SAE 1990, Detroit.

Simulating Flame Propagation and 'Knock' in Spark Ignition Engines

R.G. Myhill and W.P. Sweetenham

AEA Technology, Harwell Laboratory

Abstract This paper describes simulation techniques for the processes within a cylinder of a spark ignition engine. The model includes a 'two zone' model of the gas thermodynamics, a simple flame propagation and combustion model and a detailed model of the chemical reactions within the unburnt gas. The aim is to study the chemical reactions that lead to 'knock'.

The model leads to a set of simultaneous ordinary differential equations with a large range of timescales. These are solved using the stiff O.D.E. solver FACSIMILE. Careful limiting of integration step size is needed to ensure that the simulation does not run past the end of one phase of the combustion process before changing smoothly to the next phase. Results from the model give good agreement with data from engine experiments.

1 Introduction

Many problems in chemical kinetics and combustion involve processes with widely differing time scales. The simulation of these processes leads naturally to systems of differential equations which are stiff. The FACSIMILE program [1] has been designed specifically to handle systems of stiff ordinary differential equations and incorporates robust numerical integration techniques based upon the implicit backward-difference formulation. Many practical problems are large and the resulting linearised equations are sparse. FACSIMILE incorporates a sparse linear algebra package and hence is able to exploit the structure of the equations automatically. FACSIMILE and its predecessors have been refined over many years of application, at Harwell and elsewhere, to challenging practical problems. As a result of the experience a number of features have been incorporated into the software to aid the solution of these problems.

Several of these problems have arisen from chemical kinetics systems associated with the automotive industry, particularly for the chemistry of combustion and the atmospheric chemistry of the resultant pollution. For example Cox [2] has developed a model for autoignition processes, knock, based upon the Shell Thornton model [3]. The parameter fitting option of FACSIMILE was used to derive some of the basic unknown parameters in the model, such as the rate coefficients, from experiments in a rapid compression machine. Derwent and Hough [4] have used the software to investigate the effect of different vehicle emission scenarios upon atmospheric pollution. The software is being used currently to model the processes in a catalytic converter, to derive a detailed model of butane combustion and to investigate NO_x formation.

This paper illustrates how the features of FACSIMILE have been used in the simulation of combustion processes during the compression and expansion strokes of an internal combustion operating in the Otto cycle. It also describes some of the mathematical analysis used in the calculation of geometrical quantities that are required by the model. This work is being carried out at Harwell on behalf of the Harwell Petrol-Engine Working Party, whose members are Ford, Jaguar Cars, Elf and the Lucas Group, together with support from the vehicle Markets Division of the Department of Trade and Industry. Results of the model can be found in [5].

The work is primarily concerned with the autoignition chemistry and the effects of the thermodynamics within the cylinder on this. Consequently a detailed description of the fluid flow taking place in the cylinder is not required. A simplified two zone model with as few free parameters as possible is used. The two zones are assumed to consist of homogeneous regions of unburnt and burnt gas respectively. The model only covers the compression and expansion strokes of the cycle.

The model contains four interdependent mathematical sub-models:

- The thermodynamic model which describes the thermodynamic properties such as pressure, volume and temperature of the enclosed gases. The gases present in the cylinder are allowed to have temperature dependent heat capacities. The model includes heat gained from the autoignition chemistry and combustion as well as heat lost to the cylinder walls.

- The heat loss model for the rate of heat loss to the cylinder walls.

- The autoignition model describing the chemical reactions occurring in the unburnt gas which can lead to engine knock. It acts as a source of heat for the thermodynamic model but the corresponding changes in the composition of the unburnt gas are *not* included in the thermodynamic model. That is the composition of the unburnt gas is frozen with respect to the thermodynamic model. The thermodynamic model in turn feeds back into the autoignition model via the calculated temperature of the unburnt gas.

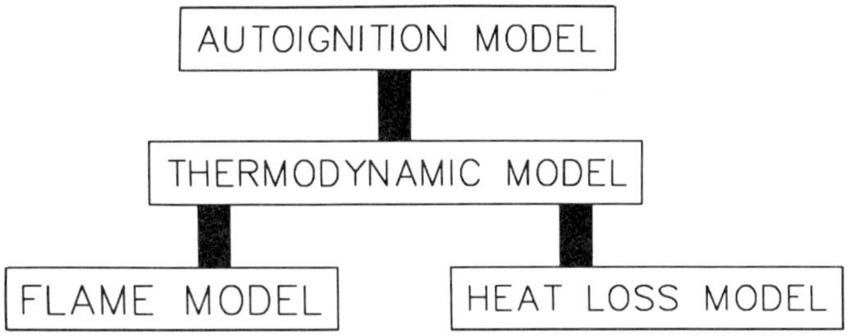

Figure 1. The Four Interdependent Mathematical Models

- The flame model for the evolution of the flame front after combustion has commenced. It is based on the turbulent flame model of Keck [6]. It includes a number of simplifying assumptions about the flame geometry. This model is used to provide the rates of burning and entrainment. Again there is a two way interaction between this model and the thermodynamic model.

Figure 1 shows the inter-relationship between the four models.

The coding of the problem in the FACSIMILE language means that it is a relatively simple task to change the emperical models used in the autoignition, flame and heat loss sub-models without altering the underlying thermodynamic model. Thus the model can be updated as new theoretical models and experimental evidence becomes available.

2 The Mathematical Model

2.1 The Thermodynamic Model

In reality the flame front has a very complex geometry which may even have a disconnected structure (see Figure 2). This means that pockets of unburnt gas may be entrained in the region of burnt gas. However for simplicity the mathematical model assumes that all the burnt gas is contained in an expanding truncated sphere centred on the spark plug. This region of burnt gas is known as zone 2. The remaining region, zone 1, contains unburnt gas. This is assumed to consist of both entrained and unentrained gas. Both zones are assumed to contain homogeneous mixtures of gases. The effects of the actual complex geometry of the flame front on the rates at which gas is burnt and entrained are included in the flame model. It is assumed that the pressure, P, is uniform throughout the cylinder.

The thermodynamic system consists of the two volumes of burnt and unburnt gas contained in the cylinder and does not include the cylinder

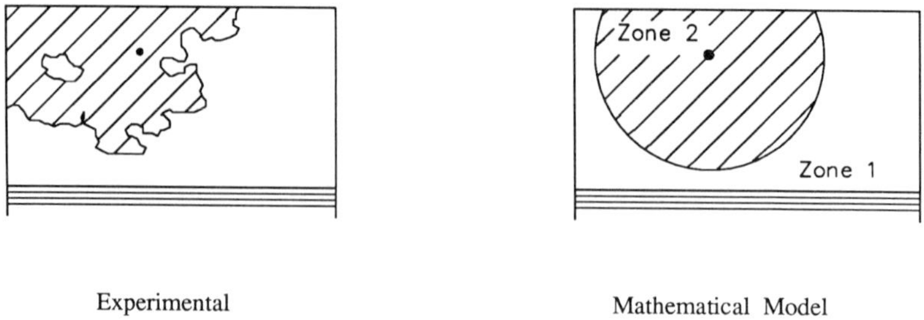

Figure 2. Comparison of Actual and Model Flame Geometries

walls or piston head. The state of the system is therefore specified by six time dependent variables.

P, the pressure in the cylinder,

T_1, the temperature of the unburnt zone, zone 1,

T_2, the temperature of the burnt zone, zone 2,

r_2, the radius of the zone of burnt gas, zone 2,

F, mass fraction of the gas that has been burned,

q, mass fraction of unburnt gas entrained in the burnt gas zone.

Notice that the mass fraction of unburnt unentrained gas is $1 - F - q$ by the conservation of mass. To describe the time evolution of the system we need a set of six simultaneous differential equations. It is assumed that all the processes are **quasi-static** so that the equation of state for an ideal gas,

$$PV = nRT \tag{2.1}$$

can be assumed to hold throughout the cycle.

2.2 Derivation of the rate equations for pressure and the gas zone temperatures

We derive the temperature equations by applying the First Law of Thermodynamics in each zone separately. In zone 1 there is no combustion so that,

$$\frac{dT_1}{dt} = \left(\frac{1}{c_p^1}(\frac{dQ_1}{dt} + V_1\frac{dP}{dt}\right), \tag{2.2}$$

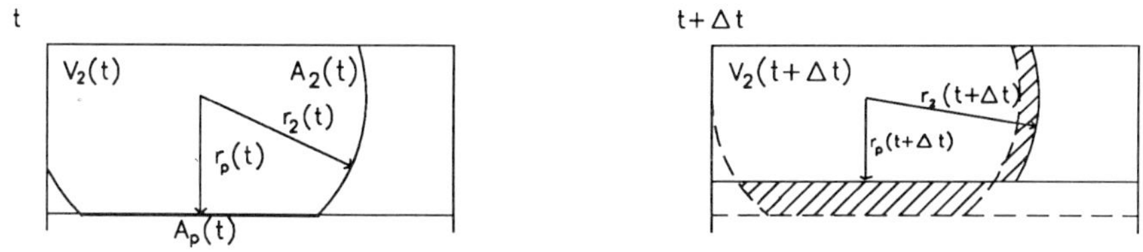

Figure 3. Diagram Showing the Volume of Burnt Gas at Times t and $t + \Delta t$

where
$\frac{dQ_1}{dt}$ is the rate of heat input to zone 1,
c_p^1 is the temperature dependent, constant pressure heat capacity of the zone 1 gas.

Similarly for zone 2 where we have combustion,

$$\frac{dT_2}{dt} = \frac{1}{c_p^2}\left(\frac{dQ_2}{dt} - \frac{dF}{dt}\Delta H + V_2\frac{dP}{dt}\right), \tag{2.3}$$

where ΔH is the enthalpy of combustion.

The model assumes that the gases in both zones are ideal so that differentiating the ideal gas law, $PV = nRT$, and using (2.2) and (2.3) we have,

$$\frac{dP}{dt} = \frac{-c_p^1c_p^2R}{c_v^2c_p^1V_2 + c_v^1c_p^2V_1}\left\{\frac{P}{R}\frac{dV}{dt} - T_1\frac{dn_1}{dt} - T_2\frac{dn_2}{dt} + \frac{\Delta H}{c_p^2}\frac{dF}{dt} - \frac{1}{c_p^1}\frac{dQ_1}{dt} - \frac{1}{c_p^2}\frac{dQ_2}{dt}\right\} \tag{2.4}$$

where
$\frac{dV}{dt}$ is the total rate of change of cylinder volume due to the motion of the piston.
$\frac{dn_i}{dt}$ is the molar rate of production of gas in zone i,
$c_v^i = c_p^i - R$ is the temperature dependent, constant volume heat capacity of the zone i gas,
R is the universal gas constant.

2.3 Derivation of the burnt gas radius equation

The radius of zone 2, z_2, is determined by the volume of burnt gas. Explicitly we use the relation between the rate of change of r_2 and the rate of change of the volume V_2 to determine the time dependency of r_2. Consider the truncated sphere containing the volume of burnt gas V_2 at a time t and short time $t + \Delta t$ afterwards. This is shown in Figure 3.

A_p is the area of the burnt zone in contact with the piston head whilst A_c denotes the total area of the piston head. r_{pst} is the radius of the circle of intersection of the spherical burnt zone with a plane passing through and parallel to the piston head. r_p is the perpendicular distance from the spark-plug to the piston head. A_2 denotes the area of the burnt zone (zone 2) not in contact with the piston and cylinder heads or the cylinder walls and r_2 is the radius of this truncates sphere. To first order in Δt,

$$V_2(t+\Delta t) - V_2(t) = [r_p(t+\Delta t) - r_p(t)]A_p + A_2[r_2(t+\Delta t) - r_2(t)],$$

$$= \frac{1}{A_c}\frac{dV}{dt}\Delta t A_p + A_2[r_2(t+\Delta t) - r_2(t)],$$

$$\therefore \frac{dV_2}{dt} = \frac{A_p}{A_c}\frac{dV}{dt} + A_2\frac{dr_2}{dt}.$$

Our equation for the rate of change of r_2 is therefore

$$\frac{dr_2}{dt} = \frac{1}{A_2}\left(\frac{dV_2}{dt} - \frac{A_p}{A_c}\frac{dV}{dt}\right) \tag{2.5}$$

2.4 The Heat Loss Model

The heat transfer to the walls is due to radiation and forced convection. It is assumed that the heat transfer due to radiation is negligible, a reasonable assumption for a spark fired internal combustion engine. Any radiative heat transfer that does actually occur is absorbed into the convective heat transfer equation by suitably adjusting the heat transfer coefficient to correctly describe the experimental data. It is assumed that there is no heat transfer across the flame front.

The rate of heat transfer is assumed to be proportional to the product of the surface area of a gas zone in contact with the cylinder walls and the temperature difference between that zone and the wall. This gives the following heat transfer equations.

$$\frac{dQ_n}{dt} = -K_n(T_n - T_E)A_n \quad n = 1, 2, \tag{2.6}$$

where T_n is the temperature of the gas in zone n, T_E is the temperature of the engine wall, A_n is the surface area of zone n in contact with the cylinder walls and K_n is a heat transfer coefficient measured in $WK^{-1}cm^{-2}$ which is determined from the experimental data using FACSIMILE's parameter fitting facility.

This is an empirical model which brings in three unknown parameters that need to be determined experimentally, the two heat transfer coefficients and the temperature of the engine wall. Many alternative and more complex empirical and theoretical models have been put forward for this term (see for example [7]).

2.5 The Autoignition Model

The autoignition model is intended to simulate the chemcial reactions which can lead to engine knock. It includes 31 different chemical reactions involving 24 chemical species. The physical conditions of pressure and temperature in the unburnt gas (zone 1) in which the chemical reactions are occurring is supplied by the thermodynamic model. In addition to modelling the progress of the autoignition reactions the autoignition model calculates the rate of heat release from these chemical reactions. This in turn is fed back into the thermodynamic model in the form of the heat rate, H_1, which depends on the rates of two key reactions in the autoignition model involving iso-octane and n-heptane respectively. The flow of heat into zone 1 due to the pre-ignition chemistry is modelled by adding the term H_1 to the first zone's heat transfer equation,

$$\frac{dQ_1}{dt} = K_1(T_1 - T_E)A_1 + H_1 \tag{2.7}$$

The chemical model is based on the Shell Thornton model [3] which describes autoignition in terms of a hydrocarbon oxidation mechanism appropriate for alkane-type fuels types with more than five carbon atoms, in the temperature region 700-850 K. This has been refined and extended in the light of the expanding body of kinetics information on the type of elementary reactions involved [2]. The overall combustion reactions are:

$$C_7H_{16} + 11O_2 = 7CO_2 + 8H_2O,$$
$$C_8H_{18} + 12.5O_2 = 8CO_2 + 9H_2O.$$

2.6 The Flame Model

The flame model used is based on the **turbulent flame model** given by Keck [6]. This is also known as the **eddy entrainment model**. Recall that the need for a flame model was to describe the rate at which gas is burned and entrained in the cylinder.

The mathematical model is formulated in the following way. Firstly another truncated sphere centred on the spark plug is introduced. This is the imaginary flame envelope which has a radius r_f that is generally greater that the radius r_2 of the zone of burnt gas, zone 2. The rates of burning and entrainment depend on the flame surface area. To describe the system we need equations describing the rate of change of F, q and r_f with time.

Equation for r_f

The flame radius is related to the radius of the zone of burnt gas, r_2. When r_2 is small $r_2 \approx r_f$. As r_2 grows r_f will grow much faster due to the increase in turbulence and the corresponding subsequent increase in the area of the flame. When r_2 becomes large the burning achieves a quasi-steady state. The empirical relation between r_f and r_2 given in [6] is,

$$r_f = r_2 + \frac{u_T l_T}{s_l}\left(1 - exp\left(-\left[\frac{V_1 F r_2}{V_2(1-F)3l_T}\right]^2\right)\right), \tag{2.8}$$

where V_i are the volumes of gases in the two zones.

The turbulent flame model has three parameters that must be determined by experimental measurement;

s_l : a **laminar flame speed** which determines the rate of burning at the flame front.

u_T : a **characteristic speed** which determines the rate at which unburnt gas is entrained within the zone of burnt gas.

l_T : a **characteristic length of an eddy** (or a **characteristic time** $\tau_B = l_T/s_l$) which determines the rate at which entrained gas is burned.

Given a set of experimental data points consisting of temperature and pressure points measured throughout the engine cycle FACSIMILE can be used to give the values of these parameters that best fit the model.

Equation for F

Firstly gas is burned at the flame front at a rate proportional to the area of the flame front A_f, the density of the unburnt gas and the laminar flame speed s_l. Secondly unburnt entrained gas is burned at a rate proportional to the mass of entrained unburnt gas and inversely proportional to the characteristic time τ_B,

$$\frac{dF}{dt} = \frac{(1-F)}{V_1} A_f s_l + \frac{q}{\tau_B} \tag{2.9}$$

Equation for q

The imaginary flame front is assumed to propagate at a characteristic speed u_T which is greater than the laminar flame speed s_l. Thus the flame entrains unburnt gas at a rate proportional to u_T,the density of the unburnt gas and the exposed area of the flame front A_f. At the same time entrained gas is being burned at a rate proportional to the mass of the entrained gas and inversely proportional to the characteristic time τ_B. Thus

$$\frac{dq}{dt} = \frac{(1-F)}{V_1} A_f u_T - \frac{q}{\tau_B}. \tag{2.10}$$

2.7 The Four Stages of Combustion

The combustion process can be divided into four distinct phases:

Stage 1: Adiabatic Compression

The first stage of combustion consists of the compression of the initial charge of air-fuel mixing up until the spark plug is fired. During this first stage there is only one zone of unburnt gas, zone 1, and no burning. As well as undergoing compression the gas is losing heat to the surrounding cylinder walls and gaining heat from the autoignition chemistry taking place within the gas.

Stage 2: Ignition and Homogeneous burn

This is an intermediate stage that occurs between the ignition of the air-fuel mixture by the spark and the development of a spherical flame front. At the commencement of this phase a small spherical region of gas, centred on the spark plug, is singled out to be burnt during the phase. The contents of this region are assumed to burn homogeneously at a constant rate. It is a necessary step as the next stage does not allow for an initial flame development.

Stage 3: Development and Propagation of a Flame Front

This stage describes the part of the cycle when there is a propagating flame front in the cylinder. It starts with a spherical flame whose radius is equal to that of the burnt gas zone after the homogeneous burning has finished. The flame then expands rapidly and turbulently. This propagation is described by the turbulent flame model described below. The phase ceases when the flame area becomes negligible.

Stage 4: Final Burning of Entrapped Gas

This stage covers the final burning of the entrapped gas and the subsequent expansion and cooling of the burnt gas. At the end of the previous stage there is still some unburnt unentrapped gas left as the imaginary flame front reaches the cylinder walls before all the unburnt gas has been either burnt or entrapped. At the commencement of the phase any such residual gas is instantly transferred to the region of entrapped gas so that it is finally burnt.

3 Geometry Calculations

A number of calculations need to be performed relating to the interaction of the flame, burnt gas zone and cylinder geometries. These are performed in a set of Fortran subroutines which are called by FACSIMILE. The required calculations are the volumes of gas in burnt and unburnt zones, the areas of burnt and unburnt zones in contact with cylinder walls, the area of flame surface, the area of burnt zone surface, and the area of piston head in contact with the burnt zone. Mathematically these reduce to

- Area of sphere enclosed in a cylinder: $S_{sphere} \cap V_{cylinder}$.
- Area of cylinder enclosed in a sphere: $S_{cylinder} \cap V_{sphere}$.
- Area of cylinder outside a sphere: $S_{cylinder} \cap V'_{sphere}$.
- Volume of sphere enclosed in a cylinder: $V_{sphere} \cap V_{cylinder}$.
- Volume of cylinder outside a sphere: $V'_{sphere} \cap V_{cylinder}$.
- Circular area of plane inside a sphere: $S_{piston} \cap V_{sphere}$.

3.1 Geometrical Configuration

Currently the model describes the intersection of a sphere with a right cylinder. Five parameters are needed to describe the system; the radius, r_c, and length, L, of the cylinder, the radius of the sphere, r_s, and the position of the centre of the sphere (assumed to be the spark plug) that is its displacement from the axis of the cylinder, r_d, and its distance below the top of the cylinder, r_a. The general configuration is displayed in Figure 4.

3.2 Area and Volume Calculations

The area and volume calculations can be performed by considering the corresponding arc-lengths and areas in a horizontal slice through the cylinder and integrating these numerically over the relevant height range, see Figure 5. The exceptions are the area calculations on the top of the bottom of the cylinder which are performed directly.

The intersection of a horizontal plane with the sphere and the cylinder is in general two circles. Special cases occur when either of these circles is completely enclosed by the other. In the plane the intersection of a surface with a volume reduces to the intersection of a line with an area. Likewise the intersection of two volumes becomes the intersection of two areas.

We shall first look at the general case shown in Figure 6.

$S_{cylinder} \cap V_{sphere}$: $2r_c\phi$

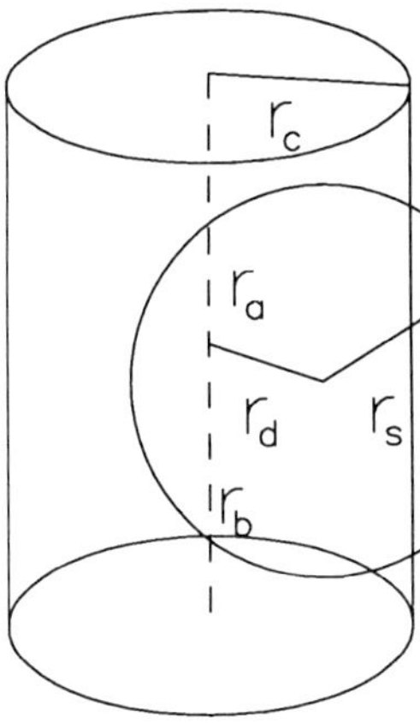

$$L = r_a + r_b \,.$$

Figure 4. Geometrical Configuration of Sphere and Cylinder

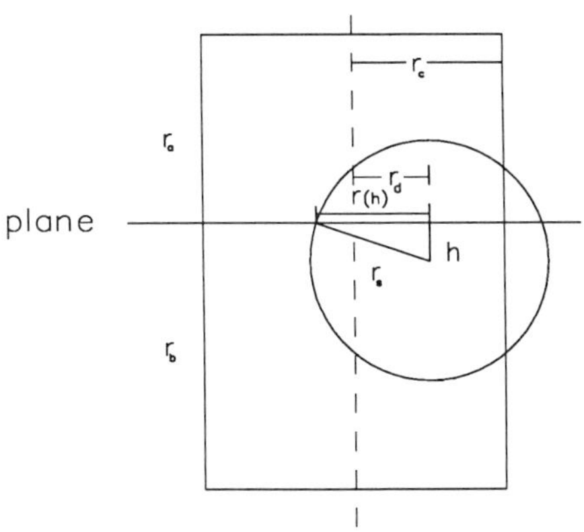

$$r(h) = \sqrt{r_s^2 - h^2} \quad -r_s \leq h \leq r_s \,,$$

Figure 5. Vertical Section Through the Sphere and Cylinder

$S_{sphere} \cap V_{cylinder}$: $\quad 2r(h)\theta$

$V_{sphere} \cap V_{cylinder}$: $\quad r^2(h)(\theta - cos\theta sin\theta) + r_c^2(\phi - cos\phi sin\phi)$

$V'_{sphere} \cap V_{cylinder}$: $\quad \pi r_c - r^2(h)(\theta - cos\theta sin\theta) - r_c^2(\phi - cos\phi sin\phi)$

where $\theta = cos^{-1}\left(\frac{r^2(h)+r_d^2-r_c^2}{2r(h)r_d}\right), \phi = cos^{-1}\left(\frac{r_c^2+r_d^2-r(h)^2}{2r_c r_d}\right)$.

The first special case occurs when the sphere surface is totally enclosed within the cylinder volume. This occurs when $r(h) \leq r_c - r_d$. The correct formulae may be obtained by considering the limiting case when $\theta = \pi$ and

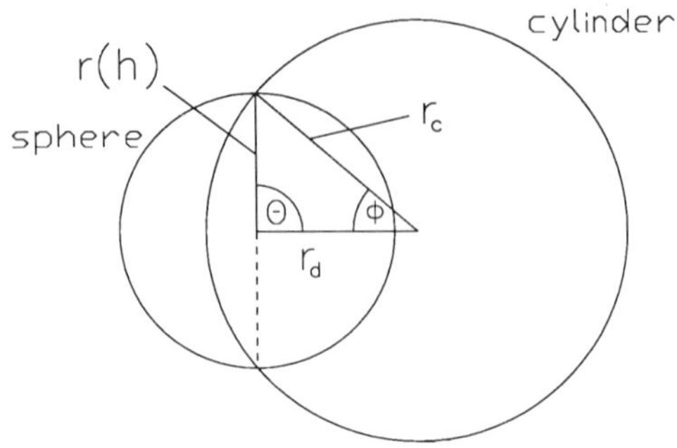

Figure 6. Horizontal Section Through the Sphere and Cylinder

$\phi = 0$. The second special case occurs when the cylinder surface is totally enclosed within the sphere volume. This occurs when $r(h) \geq r_c + r_d$. The correct formulae may be obtained by considering the limiting case when $\theta = 0$ and $\phi = \pi$.

The expressions for the area of the intersection of a plane with the intersection of the sphere and the cylinder may be used to calculate the area of the burnt zone in contact with the piston head, A_p, by taking the slice through piston head surface.

The numerical integration is performed by the Harwell Subroutine Library routine QA05. This Fortran routine integrates a function calculated by a Fortran subroutine to a user specified absolute accuracy. There are no problems with the integrations as there are no singularities in the functions to be integrated. In the quadrature care must be taken to ensure that the accuracy is such that it looks to the O.D.E. solver as a smooth function. The adaptive quadrature used in QA05 can introduce discontinuities which can give rise to difficulties for a linear multi-step method with automatic error control, particularly when higher order formulae are being used.

In general the cylinder and the sphere can be divided vertically into five distinct regions where the different cases occur. These are shown in Figure 7.

If $r_s \leq r_c + r_d$ then the middle, second special case region will be absent. If $r_s \leq r_c - r_d$ then the middle three regions will not be present and there will be just one region in which the first special case holds. In fact in this case explicit analytic expressions may be used to calculate the surface area and the volume of the, possibly truncated, sphere.

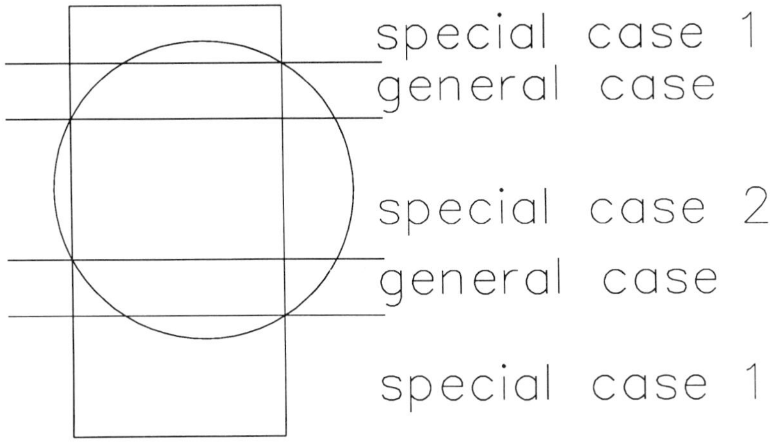

Figure 7. Vertical Section Showing Regions of Different Cases

4 Action Points

4.1 Integrating up to a Specified Point

In normal use FACSIMILE integrates a set of initial valued ordinary differential equations, choosing its own time steps. To request output or stop the calculation, the user provides a smooth function of the variables to be integrated and time. The user then specifies that actions, such as terminating integration are to be taken when the provided function passes through given values. After FACSIMILE has taken a time step, it evaluates the function, using the end of step values for the variables and time. If the function has passed through one or more of the specified values during the last step, FACSIMILE uses its polynomial interpolation functions for the variables to interpolate back to the value of time, within the step, at which the function given by the user took the specified values.

Having calculated the intermediate values of time of the completed step, FACSIMILE then takes the actions specified by the user in chronological order. This method of specifying action points is flexible and works efficiently. However, it requires the functions defining the differential equations to be evaluated beyond the end of the range of integration.

4.2 End of Burning at the Flame Front

At the end of stage three of the simulation, the radius of the burnt zone and the position of the piston are such that the sphere just encloses the cylinder, and the surface area of the burnt zone is zero. The differential equation for the rate of change of the radius of the burnt zone, (2.4), involves dividing by the surface area of the burnt zone, so the rate of change is undefined after that point in time. This needs special action to be taken by the integrator to prevent division by zero. Action points are provided in FACSIMILE to prevent difficulties of this kind.

4.3 Integrating up to an Action Point

To solve the problem, after each completed integration step, the interpolation formulae for the variables as functions of time are used to evaluate the user's function at the end of the next step. Before the next step is taken, the code knows the first step size that will be attempted and so the interpolation formulae are extrapolated to that point. If the user's function is predicted to reach an action point during the next step, then the code interpolates through the step about to be taken to find at what time the action point is expected to be reached. The code then sets a maximum step size so that that point cannot be passed.

The code also needs an action point for completed steps to be specified that will stop integration just before the equations become undefined.

4.4 Using Action Points Calculated on Predicted Steps

In the code for the engine, the function used to control the steps is the ratio of the volume of unburnt gas over the total cylinder volume. The maximum step size is set so that, based on predicted steps, that function should never pass through zero. The code stops that phase of the calculation once a step has been completed in which the function dropped to 0.001.

5 Parameter Fitting

FACSIMILE has a parameter fitting facility that takes experimental data corresponding to quantities calculated within a model written in the FACSIMILE language. The parameter fitting process also needs to be given a list of parameters from the model whose values are not known *a priori* and rough estimates of those parameters' values. The parameter fitting facility automatically finds values for the specified parameters which give best agreement with the experimental data. This process is performed under the control of routine VA05 from the Harwell Subroutine Library. VA05 is a robust non-linear least squares routine.

The experimental data required by FACSIMILE comprise values for the observed quantities together with estimates for the experimental error. After finding the closest fit, FACSIMILE determines whether the fit is within the error estimates and whether the fit is suspiciously too close. FACSIMILE then tests which of the fitted parameters have their values accurately determined by the experimental data and what range the parameter values can lie within while still giving a reasonably good fit.

The basic kinetics data for the autoignition chemistry were determined from experimental results from a rapid compression machine using FACSIMILE's parameter fitting facility. This facility was then used on experimental measurements of pressure and temperature from a real engine to fit the additional model parameters such as the heat transfer coefficients, laminar flame speed, and the characteristic speed and length.

6 Comparison with Experiment

Here we summarise the results of comparing the model with engine experiments; further details were presented at the Third International Conference on Numerical Combustion, Antibes, May 1989 [5]. Experiments were carried out at Harwell on a 2 litre production engine running at 1500 rpm, with ignition at 26° before top dead centre, in which gas temperatures and pressures were measured. Knowledge of the engine gave the geometric information needed. FACSIMILE's parameter fitting facility was used to find values of the burning rate parameters and the heat transfer coefficient which gave best agreement with the experimental measurements. There was good agreement found between experiment and simulation for the pressures and for the temperature of the unburnt gas. The temperature of the burnt gas was over predicted.

The values fitted were all concerned with the thermodynamic model. Running the simulation with the values that gave the best fit, the chemical model predicted rapid auto-ignition and under these conditions the engine does knock. The engine does not knock when the ignition point is moved to 22° before top dead centre. When the simulation was run again with all other parameters unchanged and ignition retarded to 22° before top dead centre, the chemical model did not predict rapid autoignition.

Further simulations were carried out to compare the qualitative behaviour of the model when parameters such as engine speed and air/fuel ratio were altered. In each case, the model's knock predictions varied qualitatively in agreement with experimental experience. We lacked the experimental evidence for a quantitative comparison.

7 Conclusions

This is a simple model which enables the study of auto-ignition chemistry and the effects of various engine parameters upon it. It is also valuable in interpretation of the experimental data. The model is capable of extension easily by the inclusion of different and possibly more complex sub-models; flame, auto-ignition and heat loss models. The geometry calculation could be extended so that the model would be able to cope with more realistic engine geometries. The main limitations of the model are the heat loss model and the wrinkled nature of the flame front.

FACSIMILE produces a measure of how stiff the system of differential equations supplied to it are. This consists of an estimate of the largest eigenvalue of the Jacobian multiplied by the current timestep. The eigenvalue gives an estimate of the fastest time rate in the problem whilst the timestep gives an estimate of how fast the actual solution is varying. The dimensionless stiffness ratio for the engine model varies between about 10^6 and 10^8.

8 Acknowledgements

The autoignition chemistry model was provided by R.A. Cox. The simulation program was written at Harwell by J. Cole and M. Norgett. Further work was done by the present authors.

References

1. Curtis, A.R. and Sweetenham, W.P., (February 1988), "FACSIMILE/CHEKMAT User's Manual", Harwell Report AERE R. 12805.

2. Cox, R.A. and Cole, J.A., (1985), Comb. Flame, 60, 109.

3. Halstead, M.P., Kirsch, L.J., Prothero, A. and Quinn, C.P., (1975), Proc. Roy. Soc. Lond. A, 364-515.

4. Derwent, R.G. and Hough, A.M., (October 1987), *The impact of emission reduction scenarios on photochemical ozone and other pollutants formed downwind of London*, Harwell Report AERE R. 12859.

5. Cox, R.A., Myhill, R.G. and Sweetenham, W.P., (May 1989), *Knock Prediction in Spark Ignition Engines* Proceedings of the 'Third International Conference on Numerical Combustion'. Antibes, France.

6. Keck, J.C., (1982), *Turbulent Flame Structure and Speed in Spark-Ignition Engines*, Proceedings of the Nineteenth International Symposium on Combustion, 1451-1466.

7. Annand, W.J.D., (1986), *Heat transfer in the cylinder and porting*, in "The Thermodynamics and Gas Dynamics of Internal Combustion Engines." pp773-805. Ed. J.H. Horlock and D.E. Winterborne. Oxford University Press.

Mathematical Modelling of Vehicle Heat Flows

A. Veshagh, E. Bright and R. Moffatt

University of Warwick

Abstract Design and development of the new generation of fuel efficient power units and their ancillary equipment require an integrated tool for maximum energy utilisation and overall cost optimisation. As more stringent requirements are imposed on engine cooling and passenger compartment heating and air-conditioning systems, vehicle heat flows have to be studied in greater detail to enable the designer to meet the demands set by engine performance, passenger comfort and cost consideration. An integrated design tool is needed to predict the overall system performance and optimum solution.
This paper describes the development of a mathematical model of the vehicle thermal systems consisting of components influencing or being influenced by transfer of heat. The model consists of a suite of FORTRAN programs which predicts heat gains and losses in the heat paths from engine to passenger compartment, and allows the designer to evaluate and optimise his design.

1 Introduction

The design of heating and cooling systems for today's vehicle requires not only knowledge of local temperatures, velocities and heat transfer coefficients but also knowledge of the interaction between the engine and various components affecting flow of heat. To ensure that vehicle overall energy generation and utilisation is efficient and cost effective, the designer needs to consider the total car as a thermodynamic system to determine the effect of design parameters affecting the efficient use and balance of energy generated by the power unit and dissipated or absorbed by other system components.

The design of vehicle heating and cooling systems, however, involves a lot of development work to match the heating/cooling requirements of passenger compartment and engine to those of radiator and heater matrix. Designers have generally employed traditional methods based on empirical correlations and measured data with inadequate consideration of the whole vehicle as a single thermodynamic system. While such techniques have been adequate to date, the introduction of stringent fuel economy and emission standards and market requirements have changed the scene by requiring more efficient and rigorous techniques to achieve acceptable heating and cooling systems. The designer, therefore, needs a tool for identifying and assessing the efficiency of various design options and to enable him to specify an optimum and cost effective system for each application.

The development of modern computers has had a great impact on heat transfer analysis and related design calculations. A computer model of the vehicle heating and cooling systems can be an invaluable tool as it may be used to predict performance trends of individual components and also to size and specify them at early stages of design. Computer modelling of vehicle thermal systems can also be used to assess and determine the degree of insulation and energy utilisation in the whole vehicle to optimise the design, predict problems, and to a certain extent, direct development programmes.

This paper reports on the development of a computer model at the Advanced Technology Centre in collaboration with the Rover Group, for thermal analysis of vehicle heating and cooling systems and their interaction with other vehicle components. The model was developed for parametric study of the vehicle system by systematically varying parameters such as vehicle speed, throttle position, etc. to study their effect on thermal performance and interaction of various heating and cooling components such as radiator, heater matrix and passenger compartment. Furthermore, the engine cooling and passenger compartment heating/cooling requirements can be predicted at the concept stage of design, reducing the cost of developing a refined design.

The approach followed in this work is to develop thermodynamic and heat transfer models for the various system components and couple them to develop a simulation model of the whole vehicle system. Computer model validation against available experimental data is described and the use of the model is illustrated by predicting the thermal performance of a Rover 800 vehicle heating and cooling systems under two operating conditions.

2 Benefits

There are a number of opportunities in the design of vehicle heating and cooling systems which may be analysed by employing a simulation model. These vary from the study of the effect of varying ambient conditions through to the design of protective systems for damage mode (eg frost ... etc) which may occur in the system. Benefits offered by the vehicle model include the study of:

- packaging and weight constraints (eg different designs of radiator and heater matrices) and system performance.

- reduced parasitic losses (i.e. small fan and water pump)
- coolant and oil temperatures under various operating conditions
- effect of insulation and glazing on interior temperature etc.

3 Modelling Approach

The approach taken for this work is to model each component as a separate entity and assemble them together into a vehicle model. The overall model is based on a modular arrangement of separate sub-models and data files to simulate the vehicle and the behaviour of its components.

Mathematical models of the engine, cooling system, passenger compartment and other ancillary equipment were formulated and calibrated to experimental data. Two approaches i.e. analytical and empirical were taken in most cases for modelling the behaviour of each component. Component models were assembled into the vehicle model to represent the overall system and also used for validation purposes on a component-by-component basis.

Component models were developed such that they can also be employed for performance prediction of the component they represent on a stand-alone basis as well as representing part of the vehicle model. Each component model was first validated against experimental data and then integrated into the vehicle model.

3.1 Engine model

The engine model is based on performance and heat balance data obtained experimentally as shown in Figures 1 and 2. The main input data consists of engine speed and demanded power or throttle position. Engine characteristics are represented by a series of numerical arrays for each engine operating condition starting from full load condition through to idle condition. For any set of input data, engine performance and heat balance parameters (e.g. output power, heat loss to coolant, ... etc.) are calculated by linear interpolation between stored data points.

An analytical model was also developed for the prediction of engine performance characteristics from basic design data using a step-by-step cycle simulation method. The engine cycle was based on the energy conservation and perfect gas; two differential equations are therefore obtained as follows:

$$\frac{dP}{d\Theta} = k\frac{P}{V}\frac{dV}{d\Theta} + (k-1)\frac{1}{V}\frac{dQ}{d\Theta}$$

and

$$\frac{dW}{d\Theta} = P\frac{dV}{d\Theta}$$

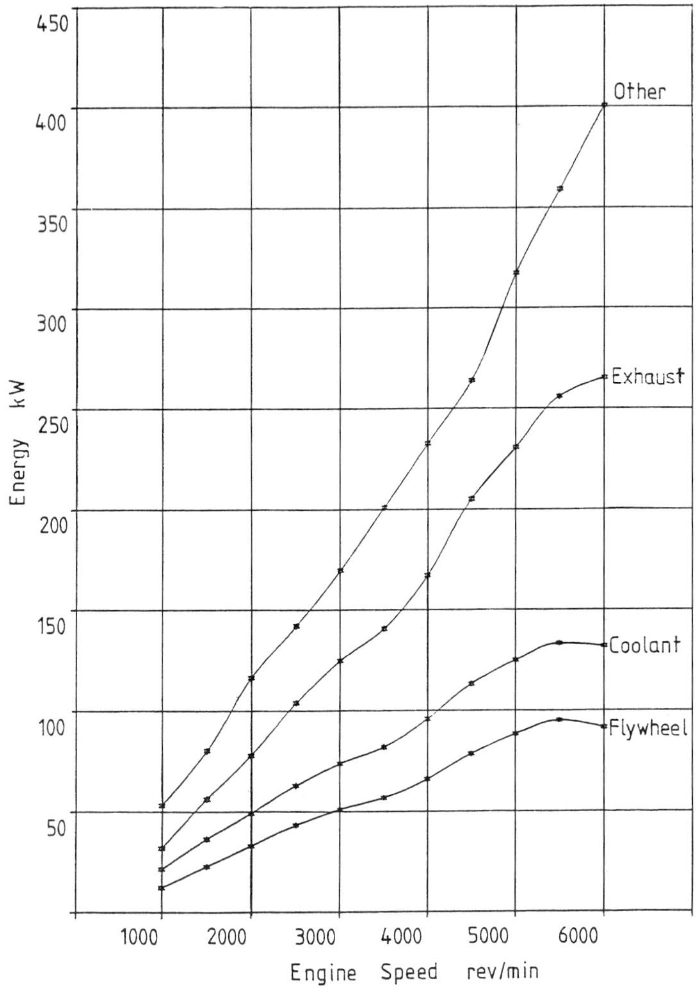

Figure 1. M16i Engine Heat Balance - As Tuned

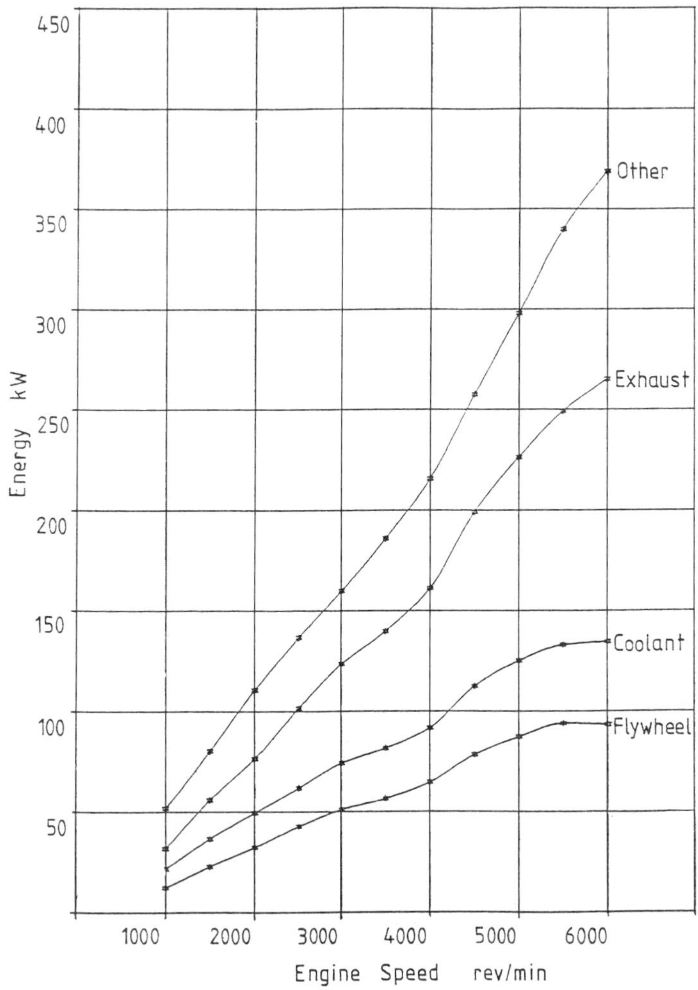

Figure 2. M16i Engine Heat Balance - Fully Optimised

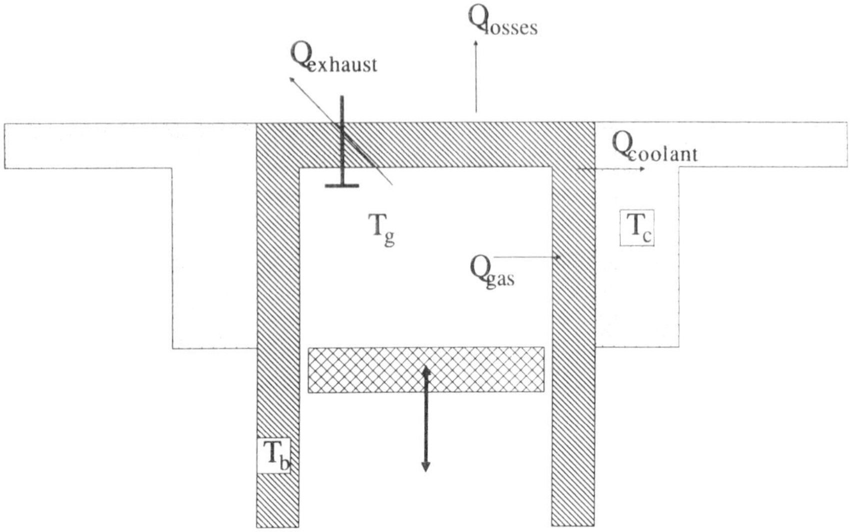

Figure 3. Engine Model

These have to be solved simultaneously to obtain temperature and pressure variations with respect to time.

Fourth order Runge-Kutta method was used to solve for gas temperature and pressure variations during the cycle. The governing differential equations were solved step-by-step; no major instability or any stiffness in the solutions were observed.

The combustion process in this model was represented by a simple heat release process. In order to match the results with experimental data and select the combustion model parameters, a simple curve fitting technique was employed.

The total engine heat balance, as illustrated in Figure 3, is represented by:

$$\dot{Q}_{\text{fuel}} = \dot{Q}_{\text{power}} + \dot{Q}_{\text{exh}} + \dot{Q}_{\text{oil}} + \dot{Q}_{\text{block}} + \dot{Q}_{\text{coolant}} + \dot{Q}_{\text{rad}} + \dot{Q}_{\text{conv}}$$

In order to simulate the thermal response of the engine block and coolant circuit, the first law of thermodynamics was applied in its unsteady form and a differential equation for the temperature variation of each system was obtained:

$$M_B C_B \frac{dT_B}{dt} = (\dot{Q}_{\text{gas}} - \dot{Q}_{\text{coolant}} - \dot{Q}_{\text{losses}}) \text{ for engine block}$$

$$M_C C_C \frac{dT_C}{dt} = (\dot{Q}_{\text{coolant}} - \dot{Q}_{\text{in}}) \text{ for coolant circuit}$$

where

$$\dot{Q}_{\text{gas}} = U g A_B (T_{\text{gas}} - T_B)$$

$$\dot{Q}_{\text{coolant}} = U_C A_B (T_B - T_C)$$

$$\dot{Q}_{\text{losses}} = \dot{Q}_{\text{conv}} + \dot{Q}_{\text{rad}}$$

$$\dot{Q}_{\text{in}} = M_C C_C (T_C - T_{\text{in}}.$$

The above equations can be solved simultaneously to predict the thermal response of the engine block and coolant circuit.

3.2 Cooling system model

The model of cooling system represents the transfer of heat to and from the cooling circuit as shown in Figure 4. This model consists of the models of engine, radiator and heater matrix which are linked together to simulate the thermodynamic behaviour of the system. The flow of coolant through the circuit is simulated by using waterpump and thermostat characteristics as shown in Figure 5.

Coolant flow rate and temperature at various points in the circuit are calculated based on heat balance principles.

The core of the model is the heat balance loop which iterates on the value of mass flow rate to satisfy the overall heat balance for the circuit (see Figure 6). Starting from the assumed value for the engine outlet temperature, radiator and heater flow rates are estimated using thermostat characteristics. Heat loss by coolant when flowing through the radiator and heater is calculated based on measured characteristics. An iteration loop is then employed to solve successively for the values of engine outlet temperature until the following energy balance is satisfied.

$$\dot{Q}_{\text{engine}} = \dot{Q}_{\text{rad}} + \dot{Q}_{\text{heater}} + \dot{Q}_{\text{losses}}$$

Linear interpolation is used to predict the performance of the radiator and heater between available data points. It is assumed that the heat gain by the coolant is directly from the engine and that there is no other heat

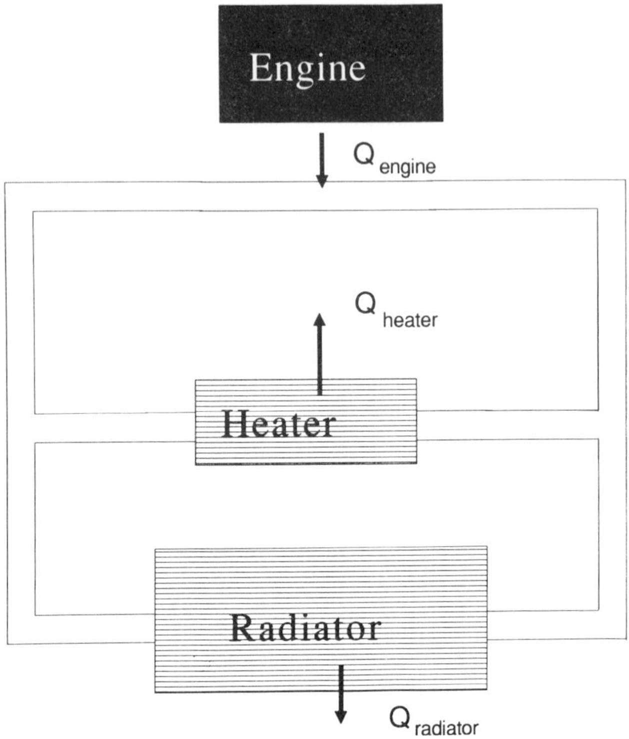

Figure 4. Transfer of Heat to and from Cooling Circuit

source in the system. The effect of pressure losses due to the flow of coolant is assumed to be negligible.

Radiator calculations are based on the assumption that air flow velocity through the radiator is uniform and directly related to vehicle speed. The effect and interaction of fan with the system is not at this stage modelled. A simple model is used to increase air flow when the fan is switched on. The thermostat model assumes a linear relationship between flow rate and temperature when the thermostat is partially open.

3.3 Passenger compartment model

This model represents the passenger compartment as a single zone surrounded by a uniform shell (Figure 7). Heat transfer calculations are based on heat supplied by the vehicle heater as well as those from the occupants,

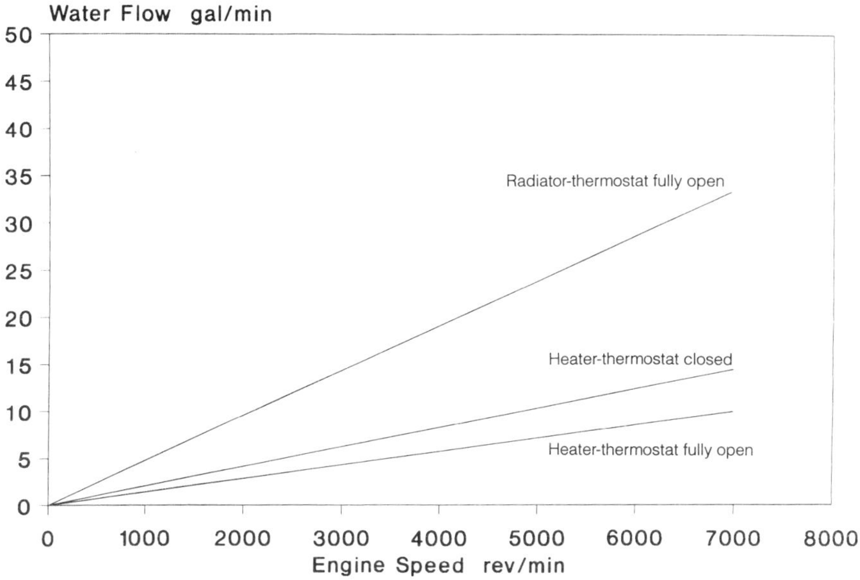

Figure 5. Water Flow in Cooling Circuit

solar heating and conductance through the shell. Two conducting layers are represented: the air/metal and trim interface. Heat loss due to the outflow of air from the vehicle is calculated based on the ventilation rate of fresh air.

The body of the vehicle is represented by a shell model through which heat can be transferred from the interior. Heat energy is removed through convection to the outside at a rate which is dependent on the vehicle speed. A simple relationship between vehicle speed and local air velocity on the shell is used to calculate convective heat transfer from the shell to the exterior.

Average interior temperature is calculated from the net heat input to the air assuming perfect mixing. No allowance is made for the flow of air in the passenger compartment and its effect on local temperature and/or heat transfer distribution. Calculations can start from a pre-defined steady state condition; the rate of change of interior temperature with respect to time can be predicted by applying the energy equations:

$$(MC)pc\frac{dTpc}{dt} = \dot{Q}_{\text{heater}} + \dot{Q}_{\text{occ}} + \dot{Q}_{\text{solar}} - \dot{Q}_{\text{shell}} - \dot{Q}_{\text{vent}}$$

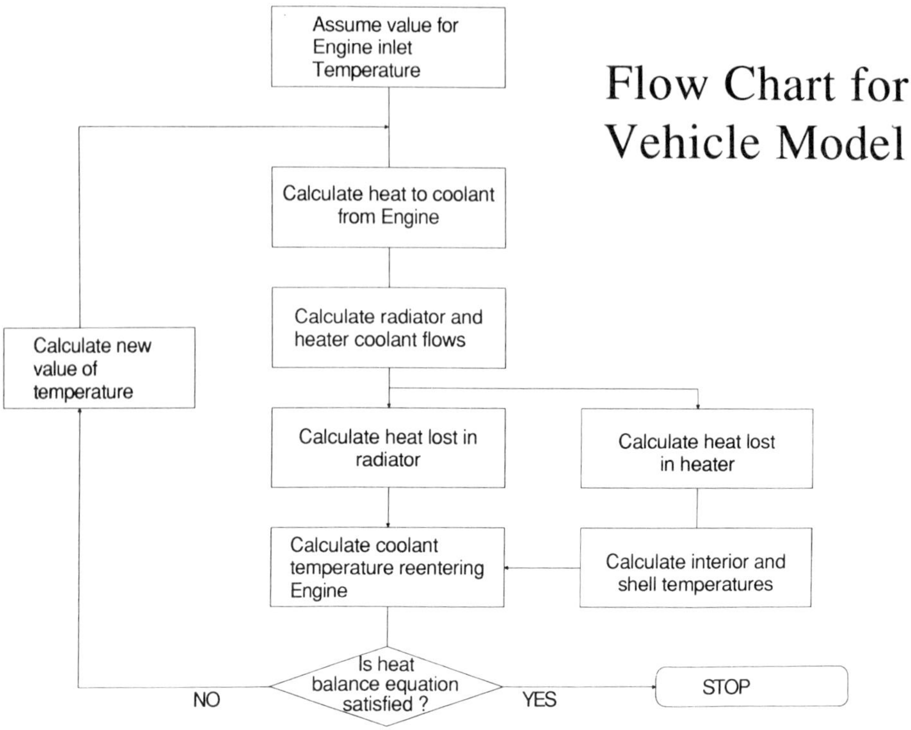

Figure 6. Heat Balance Flow Diagram

where:

$$\dot{Q}_{\text{heater}} = (MC)_{\text{air}}(T_{\text{ao}} - T_{\text{pc}})$$

$$\dot{Q}_{\text{vent}} = (MC)_{\text{air}}(T_{\text{pc}} - T_{\text{a}})$$

$$\dot{Q}_{\text{shell}} = U_{\text{shell}}A(T_{\text{pc}} - T_{\text{shell}})$$

The same principles can be applied to the vehicle shell to model its thermal response:

$$(MC)_{\text{shell}}\frac{dT_{\text{shell}}}{dt} = \dot{Q}_{\text{cond}} - \dot{Q}_{\text{conv}}$$

where:

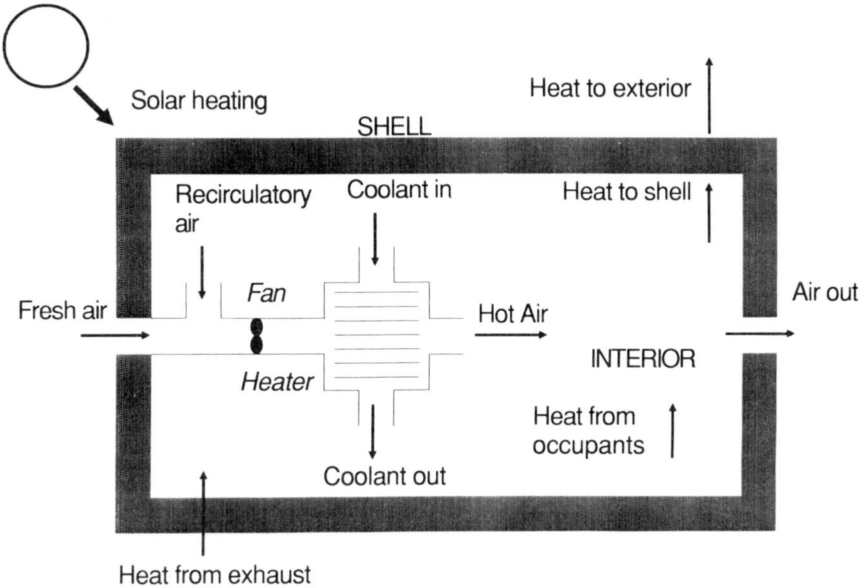

Figure 7. Passenger Compartment Model

$$\dot{Q}_{\rm cond} = U_{\rm cond} A(T_{\rm pc} - T_{\rm shell})$$

$$\dot{Q}_{\rm conv} = U_{\rm conv} A(T_{\rm shell} - T_{\rm a})$$

The above equations can be solved to calculate the temperature variations in the passenger compartment and body shell. Note that the thermal resistance of the body shell is estimated on the basis of the thermal resistivity of the glass, trim and metal.

Simple Euler method is used here to calculate interior temperature with respect in time. This method proved adequate for this sub-model as no sudden changes and/or instabilities were to be modelled; generally a slow temperature response is expected. Typical warm-up times are of the order of 5-20 minutes depending upon weather conditions.

The heat transfer through the bulkhead to the interior is specified by discrete heat flow values; the model will be later replaced by a model which will be based on the temperature difference between the engine and passenger compartments. Heat transfer from the exhaust pipe through the shell into the vehicle is at present considered negligible. The heat

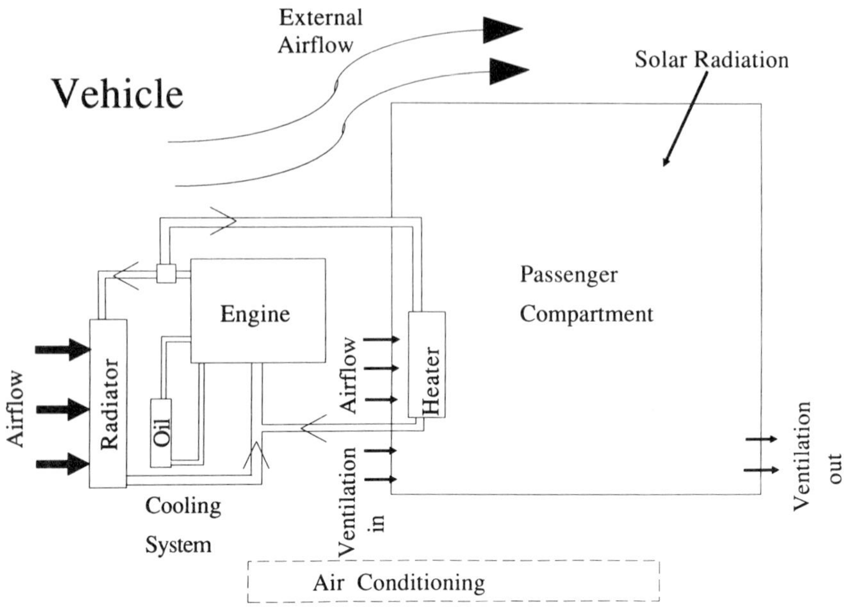

Figure 8. Vehicle Model

output from heater matrix is calculated as a function of supplied coolant temperature, air flow, ambient temperature and the specific dissipation of the matrix.

3.4 Vehicle Model

The vehicle model was based on a Rover 800 vehicle for initial development work. Although this restricted the use of the model as a general design tool, it will be possible to simulate the behaviour of any other vehicle system by replacing all the input data required for representing component characteristics. This approach is however not satisfactory in the long term and it is intended to replace empirical models of common components, such as radiator and heater matrix by analytical models so that the model can be applied to a wider range of components and vehicle systems.

Figure 8 shows the vehicle model and its three main components. As can be seen, it links the engine, cooling system and passenger compartment models in an iterative loop set up to solve for steady state conditions of the overall system.

Calculations are started by using assumed values of the flow rate and temperature of coolant entering the engine; the model then calculates all

Table 1. Uphill Towing Simulation

Coolant into engine = 107.8C
Coolant out of engine = 115.0C
Mean shell temperature = 40. 5C
Interior temperature = 45.7C
Heat dissipated by radiator = 27.0KW
Heat dissipated by heater = 0.0 KW
Heat to coolant = 26.9 KW
Heat to exhaust = 34.5 KW
Heat to oil = 2.3 KW
Other heat = 43.2 KW
Engine power = 41.1 KW

the heat flows to and from individual components and checks the heat balance between them (see Sections 3.1 - 3.3 for details). A steady state solution (representing a balanced condition) is assumed to have been obtained if there is no imbalance in the heat balance equations. In practice, however, an acceptable tolerance/error is defined for each equation as an exact solution may not be possible or may require long computer runs. Depending on the tolerance limit and initial guesses, the model converges after a number of iterations and outputs all the results.

A simple numerical technique is employed to correct successively the numerical values of the variables on which the model iterates. The correction method is based on a technique which used linear interpolation between the results of each two successive iterations. Having calculated a new value, all vehicle model calculations are repeated with this value and all the heat balance compatibility criteria are checked again to find whether an acceptable solution has been reached.

Due to the number of iteration loops and extent of calculations, more efficient numerical techniques should be used to keep the running times short.

4 Validation

Although only the major heat flow components have been combined in the vehicle model to date, performance predictions have shown reasonable results correlating to several wind tunnel tests on actual vehicles, and discrepancies can be accounted for by considering the effect of the effect of the neglected components such as oil coolers, etc.

A typical set of results is shown in Table 1. Figure 9 shows the variation of heat rejection to coolant and lubricant as a function of vehicle speed in

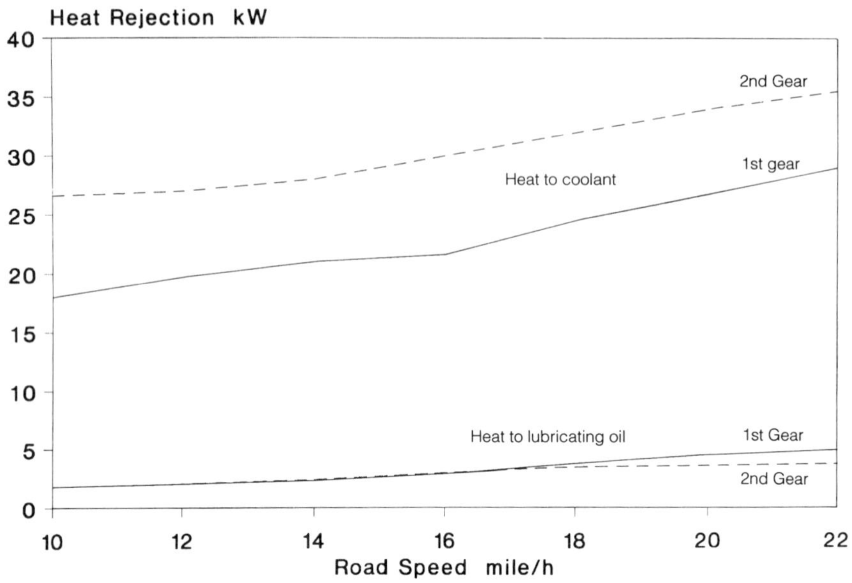

Figure 9. Heat Loss to Coolant and Sump Oil - Uphill Towing

Table 2. Cold Weather Cruising

Coolant into engine = 65.2C
Coolant out of engine = 70.7C
Mean shell temperature = -11.6C
Interior temperature = 29.5C
Heat dissipated by radiator = 0.0 KW
Heat dissipated by heater = 5.9KW
Heat to coolant = 5.9KW
Heat to exhaust = 7.4KW
Heat to oil = .8KW
Other heat = 10.5KW
Engine power = 7.5KW

two gears for an uphill towing situation. The results for the opposite case of low load cold weather cruising are shown in Table 2.

Further validations are required to ensure the integrity of the model and its accuracy. The model has to be also calibrated for other vehicle models.

The model sensitivity to integration step length and error band should also be investigated.

5 Future Work

Air flow through the radiator is, at present, assumed to be directly related to vehicle speed and set at 20% of vehicle speed with uniform distribution over the full radiator face area. A more representative model of air flow to take account of the vehicle installations will be developed. Furthermore, a heat exchanger model is under development which will enable heat dissipation predictions on the basis of the type of construction and the dimensional details. The cooling fan performance and its effect on air flow velocity distribution has to be also modelled to enable more representative assessment of the cooling system and interaction with vehicle aerodynamics.

Heat transfers in the gearbox oil cooler and the engine lubrication system have to be taken into account. Modelling work has already started on a heat transfer model of the lubrication circuit but it needs further refinement before it can be integrated into the vehicle model.

Air conditioning will be included in the model to simulate vehicles fitted with such equipment. The interior and cooling circuit models will need to be modified to allow for the cooling effect of the air conditioning unit as well as the inclusion of the evaporator which is normally packaged close to the radiator affecting its performance.

Work on these and other aspects will be carried out at the University of Warwick Advanced Technology Centre in close collaboration with the Rover Group, over the next year.

6 Conclusions

An empirical model was developed for the design of vehicle heating and cooling systems. The basic idea underlying the model was to define and model the performance characteristics of the vehicle and its components based on experimental data. Simple energy balance equations were employed together with numerical methods to simulate the interaction between vehicle components. Initial simulation runs indicate that the various submodels are programmed properly and that the model provides useful predictions of system behaviour. The convenience of using the model can lead to significant reduction of vehicle and/or its component tests and the model, once fully validated, can be a valuable means of reducing development time and cost.

It should however be emphasised that an air flow analysis of the vehicle and some of its components e.g. radiator should be carried out so that the effects of aerodynamics on heat transfer can be established and modelled without the need of using numerous simplifying assumptions and/or extrapolating from existing empirical data. Despite the present shortcomings, the model shows promise in making the vehicle heating and cooling systems more efficient and should aid the designer in the selection of components and optimisation of the overall system.

7 Acknowledgement

This work was carried out at the Advanced Technology Centre as a research programme funded by the Rover Group. The authors gratefully acknowledge the sponsorship and provision of test facilities by the company and the permission to publish this paper. Thanks are also due to D Brown for the assistance and advice received and to MSc students Messrs. C. Manohar and Y. Babu for their contribution.

Predicting the Flow Characteristics and Aerodynamic Forces of Road Vehicles Using Computational Fluid Dynamics

I.R. Hawkins[1], C.T. Shaw[2] and S. Simcox[1]

Harwell Laboratory[1] and University of Warwick[2]

Abstract This paper describes research into the use of mathematical modelling techniques, which can model the flow of a fluid, to predict the flow of air around a road vehicle. In particular, the qualitative predictions of the flow structure and the quantitative predictions of drag are analysed, for a series of simplified vehicle shapes. Whilst the predictions of flow structure are usually realistic, the predictions of drag are always too high. Various techniques have been used to improve the accuracy of the drag predictions, including the use of higher-order differencing schemes for the advection terms and the adaption of the computational mesh in response to the flow solution.

1 Introduction

When designing a road vehicle, it is essential that the aerodynamic forces and moments on the vehicle can be determined at an early stage in the design process. This enables predictions to be made of the performance, fuel economy and dynamic stability of the vehicle. At present, much of this work is done by placing scale models and full-size mock-ups of a vehicle in a wind tunnel, enabling any necessary shape changes to be made before roadworthy prototypes are built. Such a process is expensive, given that a large number of man-hours is necessary to prepare the physical models, and that complex test facilities have to be used. If mathematical modelling techniques can be developed that provide useful and reliable predictions of the aerodynamic characteristics of a road vehicle, then it is possible that the design process can be influenced quickly and cost-effectively.

Partial differential equations, the Navier-Strokes equations, can be derived [12] that govern the flow of fluids, however, analytical solutions to these equations can only be found for a small number of idealised flows. These are usually laminar flows within simple geometrical boundaries. When considering the external aerodynamics of a road vehicle, not only is the geometry of the vehicle complex, but the flow is also turbulent. Hence, the mathematical models developed for such flows utilise numerical techniques which discretise the governing partial differential equations on small elements or cells of the fluid, and use semi-empirical corrections to model the effects of turbulence [9]. Such techniques form part of the subject, Computational Fluid Dynamics (CFD).

In order that sensible models can be built to predict the flows over vehicles, specialist computer software must be available that takes advantage of the facilities, in terms of both the hardware and software, of modern supercomputers. Using such a combination of hardware and software, the authors have been investigating the suitability of CFD in providing aerodynamic data for use within the vehicle design process. Useful comparisons have to be made between the computed flow fields and physical ones. To do this, three idealised shapes have been modelled for which there is extensive experimental data. These are the MIRA blockage model [4], the Ahmed model [1] and the GM model [15]. The first is a simplified vehicle shape of notchback design, whereas the other two are idealised blunt-nosed, slab-sided bodies with rear shapes of various slant angles.

This research started in 1987 as a collaboration between Jaguar Cars Ltd. and Harwell Laboratory. At that time, Jaguar was investigating the use of CFD for predicting the external flow over vehicle shapes, and had the capability to build the necessary meshes of points to describe the geometry. The first stage was to simulate the flow over the MIRA blockage model using a mesh of some 17,000 cells. Harwell also ran a model with some 140,000 cells, again produced at Jaguar, and the results from these two cases [14] showed that although the predictions of the qualitative features of the flow were correct, the predicted drag figures were far too high. Further work has since been carried out on the two simpler shapes. Different differencing schemes for the advection operator have been tried, together with adaptively redefining the mesh, and the prediction of drag has been improved.

This paper describes the tools that have been used and analyses both the qualitative and quantitative predictions. Finally, the prospects of actually using such techniques in the vehicle design process are assessed.

2 Modelling Tools

2.1 Computer Software

For all the work described here, the central piece of software that has been used is HARWELL-FLOW3D [3]. This uses finite volume techniques to discretise the Navier-Stokes equations. The computational mesh, or grid, is contained within a domain which is topologically cuboid; that is, it is like a pliable brick which can be distorted, but not torn. Within the domain, the mesh consists of a regular structure of topologically cuboid cells. Each cell is specified by the x,y,z locations of its corners, and is identified by the indices, I,J,K in each of the coordinate directions respectively. Since there must be a fixed number of cells in each direction, internal blockages are used to specify the region inside the vehicle surface. This requires extra calculations to be carried out within the solid region, and so there is a computational overhead, but this is offset by the speed-up due to vectorisation of the program.

Traditionally, finite volume CFD programs have used a so-called staggered mesh, where the fluid pressure is stored at the cell centres and the velocities are stored at the cell faces. This was necessary to avoid undesirable numerical effects, but HARWELL-FLOW3D makes use of the algorithm developed by Rhie and Chow [10] to overcome these effects, and so it can store all quantities at the cell centres. Also, contravariant velocity components are used, rather than the local velocity components orthogonal to the local cell directions.

2.2 Computer Hardware

Large computer models are required to give sensible predictions of the flow around vehicle shapes, and so it is difficult to perform the calculations on even mainframe computers. Harwell possesses a Cray 2 supercomputer which has 256 Mwords of directly addressable memory, and achieves a performance in excess of 40 MFLOPS with HARWELL-FLOW3D.

3 Production of Computer Models

3.1 Geometrical Parameters

One of the most difficult tasks in building a mathematical model suitable for the numerical solution of partial differential equations, is the meshing process, where the mesh of points that defines the cells of the domain is created. It is to these cells that the equations are applied to form a discrete system of linearised equations. Usually, the vehicle designer will have some information on the shape of the vehicle, in the form of physical drawings, data on a CAD system or even a scale model. This information defines only the surfaces of the vehicle, whereas the solution of the Navier-Stokes equations demands that the mesh is formed for the whole volume of air around the vehicle.

One way of creating a full volume is to, first, create a mesh of points which are on the surface of the vehicle, and then to use this to create a mesh of the whole volume. A process which is described by Shaw [13]. The surface mesh can be created from CAD data directly, using a simple computer program, if the surface is not too complicated, or from data

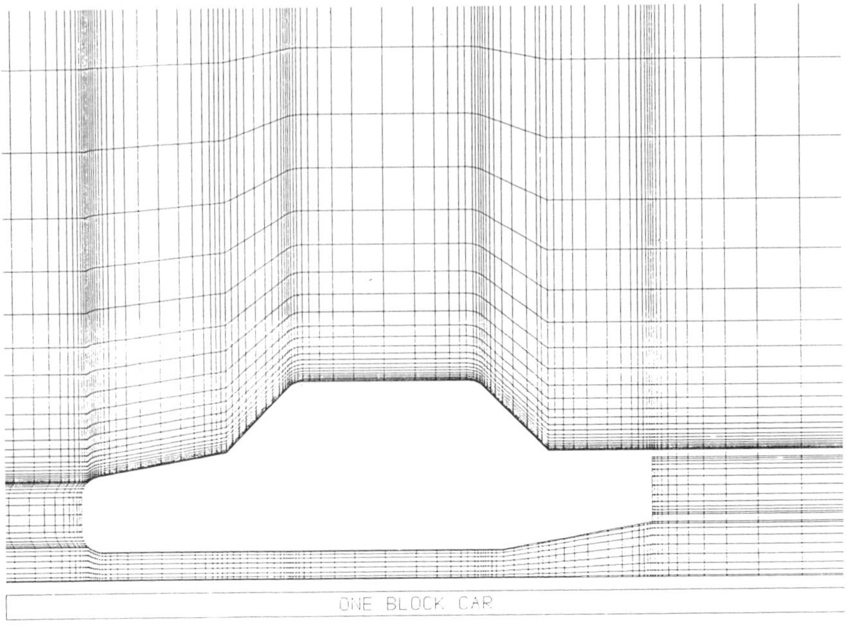

Figure 1. Centreline mesh for the MIRA model

supplied by a digitiser which scans the surface of a model. The points of the vehicle surface mesh are then used as the basis of an interpolation procedure which places points in blocks of air around the vehicle. Figure 1 shows a two-dimensional slice of a mesh for the MIRA model.

With any numerical techniques such as this, the quality of the solution is highly dependent on the mesh used. Smoothness of the mesh is required to avoid large truncation errors and small cell size is required to resolve regions of high gradients. If a flow solution is obtained, quantities such as the pressure gradient can be calculated, and the mesh adapted to place more points in regions of large pressure gradients [7]. Such modifications can be made as often as is deemed necessary.

3.2 Flow Parameters

Having created a suitable mesh of points to describe the geometry of the problem, the flow parameters must be defined. These parameters control the boundary conditions of the problem, the turbulence modelling and numerical solution procedure.

In all the simulations reported here, the standard two-equation $k - \varepsilon$ turbulence model has been used with logarithmic wall functions, set on

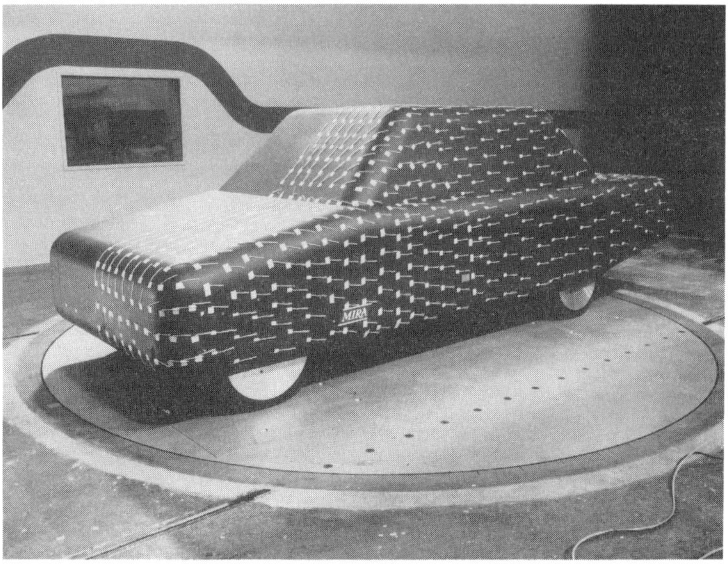

Figure 2. Flow over the front of the MIRA model

the floor of the tunnel and the vehicle surface. The inlet conditions are specified by fixing the value of the velocity at the inlet cells, and the outlet conditions specify that there is no velocity gradient normal to the outlet. Pressure is fixed at one point to remove the singularity in the pressure calculation.

One other feature of the model that can be varied is the way in which the advection terms in the Navier-Stokes equations are discretised. Various schemes have been proposed which eliminate the oscillatory solutions that can occur, at large Reynolds numbers, if central differences are used to model these terms. Several of these schemes are included within HARWELL-FLOW3D, including central and upwind differencing, a hybrid of these two, as well as the less diffuse schemes; HUW [16], QUICK [8], CCCT [6], and CONDIF [11].

4 Qualitative Results

One of the main features of the flow around the MIRA blockage model is the extent of separated flow. Separation takes place in three areas; at the base (rear) of the vehicle, at the top of the rear screen and at the windscreen-bonnet junction. These separation regions can be seen in Figs. 2 and 3. The computer model predicts the first two of these separations,

Figure 3. Flow over the rear of the MIRA model

but not the third, which is very small in extent.

To visualise the massive amounts of data from the computer model, particle tracking is a useful technique. Figures 4 and 5 show particle tracks around the vehicle, calculated using the Remote Interactive Particle-tracking program (RIP) developed at NASA Ames [5]. This program runs simultaneously on the Cray-2 and a Silicon Graphics IRIS workstation, calculating the tracks from the file of the mesh geometry and the velocity data using the Cray and displaying the results as they are calculated, on the IRIS. In the figures the recirculation regions can be seen clearly.

To compare with the tufts in Figs.2 and 3, the shear stress vectors on the vehicle surface can also be plotted, as shown and Figs. 6 and 7. Here, again, the agreement of the predicted flow patterns with those of the experiment is clear.

For the Ahmed shape, two major separations are found on the base of the body, together with a separation on the upper slant surface of the body, for a slant angle of 30°. The computer model for this case predicts the base separations, but does not predict the separation on the slant. A

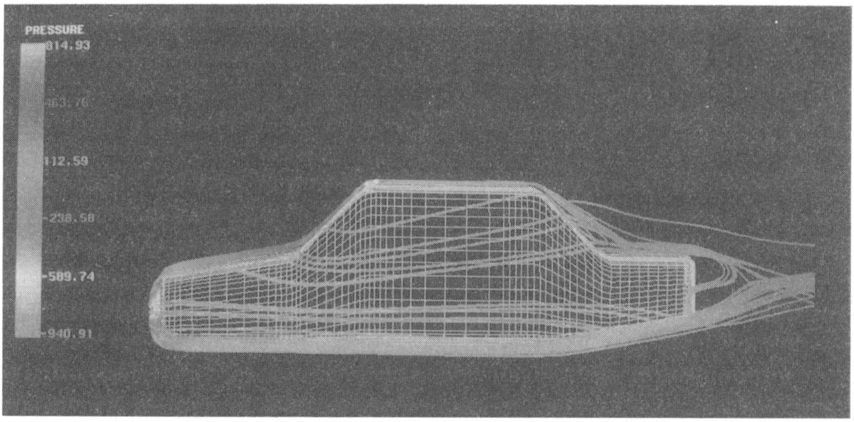

Figure 4. Particle tracks over the MIRA model

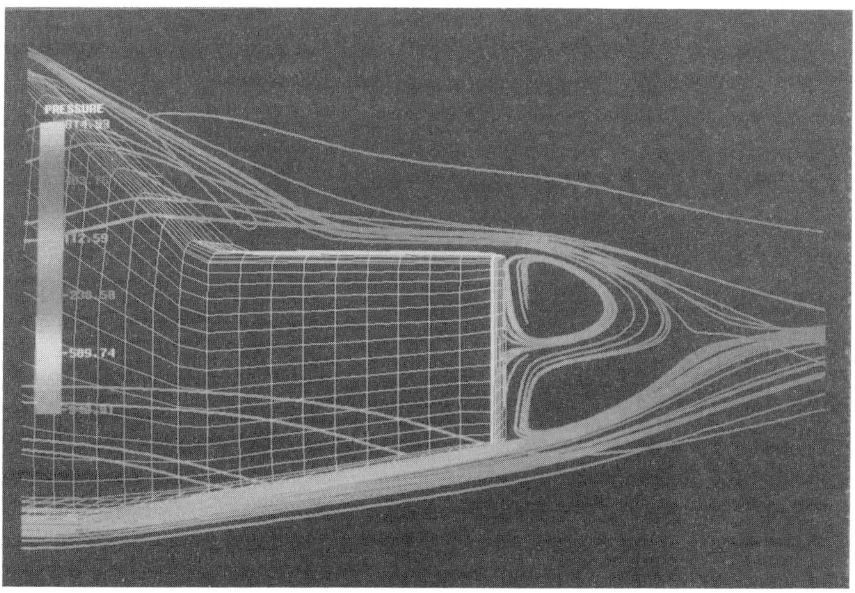

Figure 5. Particle tracks over the rear of the MIRA model

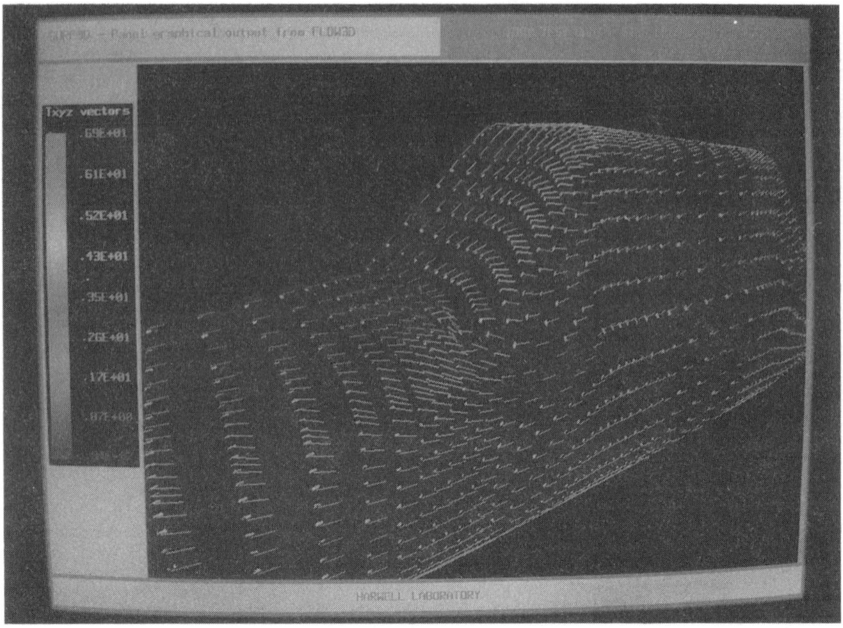

Figure 6. Predicted shear stress vectors on the front on the MIRA model

large vortex is formed on the side of the slant surface, and this is predicted by the model, as shown in Fig. 8. It should be noted that the flow structure at this slant angle was not stable and showed two distinct structures, one giving high drag the other giving low drag.

From these results it is clear that major flow features are predicted by the computer models, but that the more subtle features such as small separation regions are not.

5 Quantitative Results

5.1 Pressure Distributions

For most vehicles, the drag derived from the fluid pressure on the vehicle surface is significant. When calculating the drag coefficient, the choice of the reference pressure is important, as it modifies the component parts of the drag coefficient but not the total drag coefficient. Here, where necessary, the reference pressure has been set to ensure that the pressure coefficient at the front stagnation point on a body is unity. Also, the coefficients have been corrected to allow for wind tunnel blockage as shown by Shaw

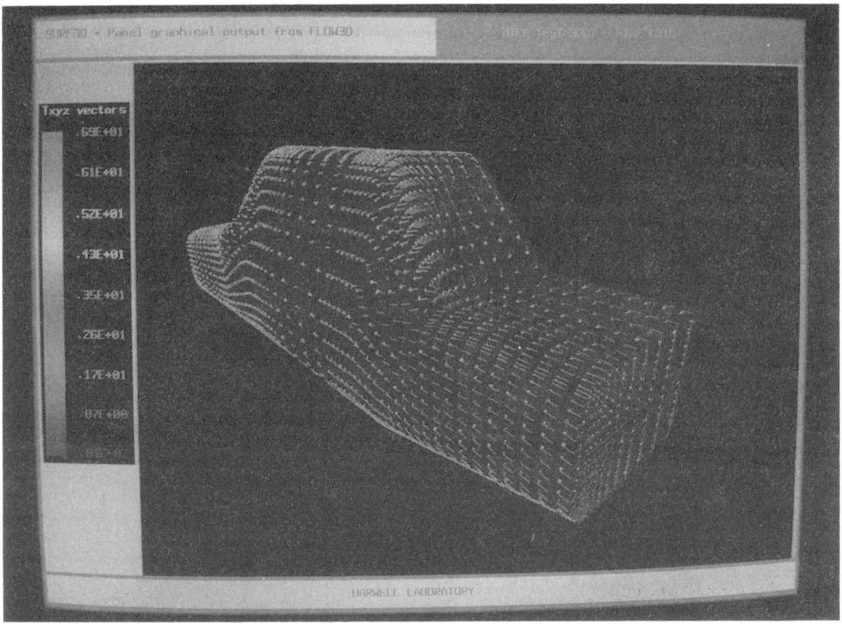

Figure 7. Predicted shear stress vectors on the rear of the MIRA model

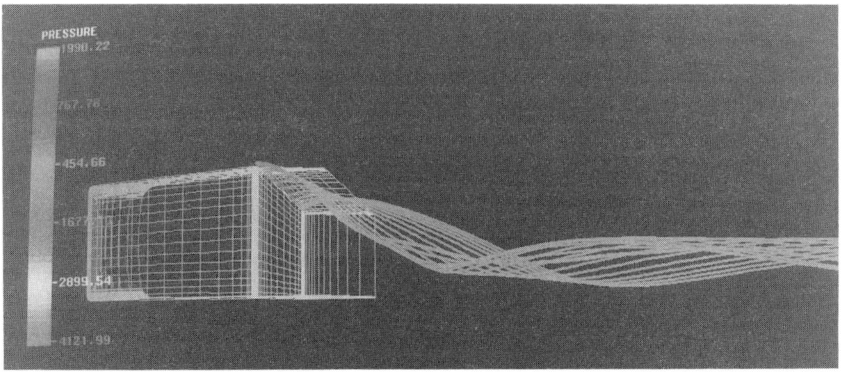

Figure 8. Particle tracks over the rear of the Ahmed model

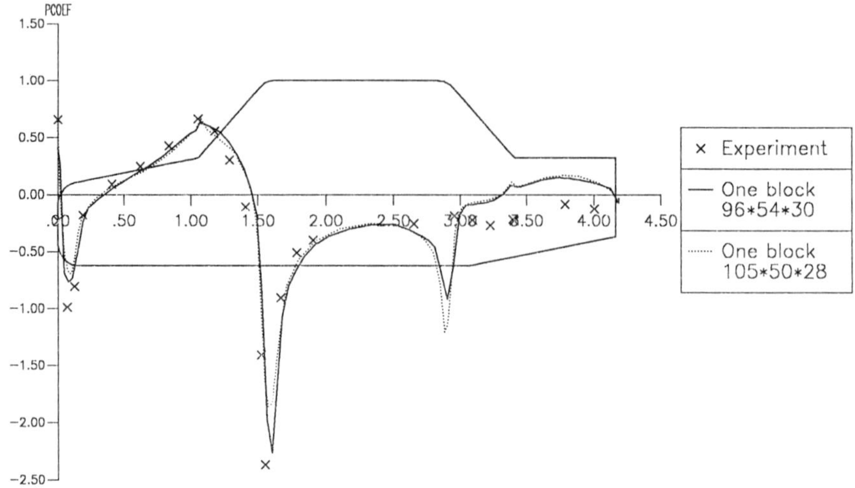

Figure 9. Pressure distributions on the upper surface of the MIRA model

[13].

Figures 9 and 10 show a comparison of the experimental and predicted pressure distributions for the MIRA model. On both surfaces the comparison is very good, excepting that the rear pressure peak that is predicted by the computer model is not evident in the experimental results, as at that location the experimental data is very sparse. Intuitively, this rear pressure peak should be evident in the experimental results, and similar tests on other models show such a peak.

Comparisons of the pressure distributions help in the understanding of the discrepancies in the drag predictions. For example, at the front of the vehicle the pressure suction peaks are not fully developed, and so the predicted drag should be higher than the experimental one. Also, the predicted pressure on the windscreen is higher than the experimental values, and so the predicted drag should, again, be high.

5.2 Vehicle Drag

To give a comparison of the predictions of the vehicle drag, let us first of all look at the Ahmed model. Table 1 gives the predictions of the pressure-derived drag coefficients, together with the experimental values for both the low drag and high drag flow regimes. The coarse mesh has 49 cells down the length of the wind tunnel, 30 in the vertical direction and 16 laterally, for a model of half the vehicle, whereas the fine mesh has 96 x 58 x 30

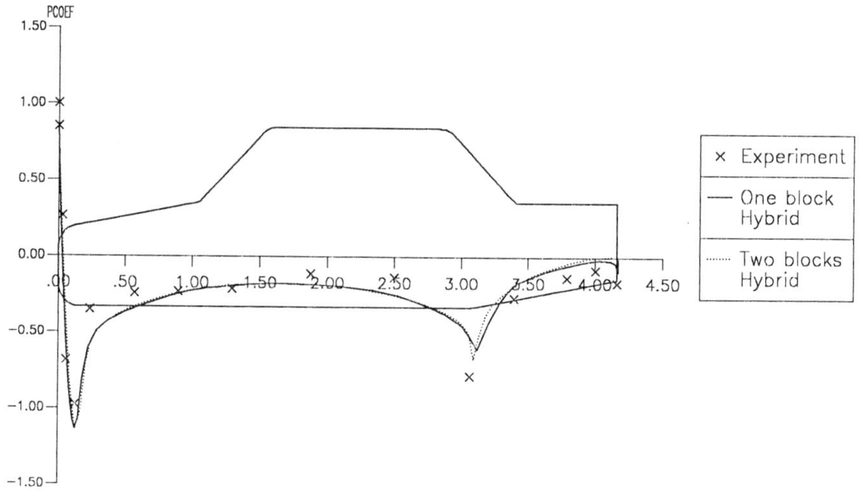

Figure 10. Pressure distributions on the lower surface of the MIRA model

Table 1. Drag Coefficients for the Ahmed Model

Ahmed Model - Drag Coefficients				
Run	Front	Rear	Slant	Total
Hybrid Coarse	0.032	0.284	0.354	0.670
Hybrid Fine	0.181	0.093	0.181	0.455
HUW Fine	0.092	0.067	0.176	0.335
Experiment Low	0.019	0.101	0.089	0.209
Experiment High	0.016	0.092	0.213	0.321

cells in the same directions. The experimental data has been estimated by integrating the pressure on the surfaces of the model.

Essentially the computer model can be seen to predict the high drag flow regime, where the drag is high on the slant surface. Refinement of the mesh has improved the prediction of the drag on the base and the front of the model, but made it worse on the slant. Given the incomplete prediction of the flow structure on the slant, the discrepancy, here, is not surprising.

Table 2. Drag Coefficients and Timing Information for the GM Model

GM Model - Drag Coefficients and Timings			
Run	Total Drag	Iterations	CPU secs
Hybrid Coarse	0.567	269	301
Hybrid Adapted Coarse	0.350	-	-
Hybrid Fine	0.294	-	-
HUW Coarse	0.259	223	283
HUW Adapted Coarse	0.210	-	-
HUW Fine	0.181	465	3269
CONDIF Fine	0.210	-	-
CCCT Fine	0.180	-	-
QUICK Coarse	0.427	-	-
QUICK Fine	0.181	-	-
Experiment Free	0.147	-	-
Experiment Tripped	0.166	-	-

Use of the HUW differencing scheme has improved the drag at the front, but this is still a major source of error.

Table 2 shows values of total pressure-derived drag coefficients for the GM model, together with some sample timing information. Here the coarse mesh has 49 x 30 x 16 cells and the fine mesh has 84 x 58 x 30 cells. The experimental results are the total drag resulting from the pressure distribution on the surface and the viscous shear at the surface, and so these values are higher than the pressure-derived drag would be alone. Looking at the hybrid differencing cases, the improvement in the prediction is clear when the mesh is adapted or refined. With HUW differencing the improvement is even better. Using the other differencing schemes with the fine grid, the drag prediction is between 0.18 and 0.21, which is close to the experimental value when a trip causes a turbulent boundary layer over most of the vehicle surface.

Table 3 gives the results for the MIRA model for meshes of similar density to those for the other two models. The experimental results are the pressure-derived results calculated using the MIRA empirical drag prediction method [2], and the data from the wind tunnel. Only hybrid differencing is reported here, for a coarse mesh and two fine ones, each created using only one block for the vehicle. The 2-block mesh has a fine density of points, and is formed using two blocks for the vehicle body, not one block

Table 3. Drag Coefficients for the MIRA Model

MIRA Model - Drag Coefficients			
Run	Front	Rear	Total
Original Coarse Hybrid	0.340	0.216	0.556
Original Fine Hybrid	0.257	0.164	0.421
New Fine Hybrid	0.235	0.144	0.379
2-Block Hybrid	0.268	0.037	0.305
Experiment	0.000	0.156	0.156

as usual. This has been done to improve the distribution of the mesh points on the vehicle surface, but although it reduces the total drag, it increases the drag adversely at the front. The improvement from the coarse to the fine mesh is clear, as is the improvement from the old fine mesh to the new fine mesh. This is due to a better selection of points at the front of the vehicle.

6 Prospects for the Future

Clearly, the predictions of qualitative flow behaviour are not enough to convince vehicle designers to use CFD. The predictions of drag for various shapes must be accurate, if designers are to place any reliance on the results. At the moment, such a situation has not been reached, but advances are being made. Further work to identify the reasons that small features of the flow are not predicted is needed, as well as work to improve the prediction of drag at the front of the vehicle. Whilst this paper has looked at some ways of improving the predictions, other avenues of investigation also need to be followed, such as the effect of different turbulence models. It is interesting to note that despite the use of the latest mathematical analysis and computer power, considerable artistry is required to obtain sensible results.

7 Conclusions

Computer models can be used to produce predictions of the flow field around a road vehicle. At present, such predictions are qualitatively correct, in most cases and quantitatively correct when considering the pressure distributions. However, they are quantitatively incorrect when considering drag. Some of the errors are due to the lack of refinement of the mesh, especially at the front of the vehicle, and the diffuse nature of the differencing schemes used to discretise the advection terms in the Navier-Stokes equations. If further resources can be devoted to these and other areas, it is very likely that useful results will, one day, be possible using CFD.

References

1. Ahmend, S.R., Ramm, G. and Faltin, G., (1984), Some Salient Features of the Time-Averaged Ground Vehicle Wakes, SAE Technical Paper 84030.

2. Ashley, C. and Carr, (1982), G.W., A Prediction Method for the Aerodynamic Drag of Cars, SAE Technical Paper 82068.

3. Burns, A.D. and Wilkes, N.S., (1987), A Finite Difference Method for the Computation of Fluid Flows in Complex Three Dimensional Geometries, AERE-R 12342, HMSO.

4. Carr, G.W., Motor Industry Research Association, Private Communication.

5. Choi, D. and Levit, C., (1987), An Implementation of a Distributed Interactive Graphics System, Int. J. of Supercomputer Applications, 1, 82-95.

6. Gaskell, P.H., Lau, A.C., (1987), An Assessment of Direct Stress Modelling for Elliptic Turbulent Flows with the Aid of a Non-Diffusive, Boundedness Preserving, Discretization Scheme, In "Numerical Methods in Laminar and Turbulent Flow, Vol. 5" Pineridge Press, Swansea.

7. Hawkins, I.R. and Kightley, J.R., (1989), Grid Adaption for Turbulent Flow Problems, In "Numerical Methods in Laminar and Turbulent Flow, Vol. 6", Ed. Taylor, C., Gresho, P., Sani, R.L. and Hauser, J., Pineridge Press, Swansea, 1651-1661.

8. Leonard, B.P., (1979), A Stable and Accurate Convection Modelling Procedure Based on Quadratic Upstream Interpolation, Computer Methods in Applied Mechanics and Engineering, 9, 59-98.

9. Patankar, S.V. and Spalding, D.B., (1972), A Calculation Procedure for Heat, Mass and Momentum Transfer in Three Dimensional Parabolic Flows, Int. J. Heat Mass Transfer, 15, 1787-1806.

10. Rhie, C.M. and Chow, W.L., (1983), Numerical Study of the Turbulent Flow Past an Airfoil With Trailing Edge Separation, AIAA J, 1527-1532.

11. Runchal, A.K. (1987), CONDIF: A Modified Central-Difference Scheme for Convective Flows, Int. J. Numerical Methods Engineering, 24, 1593-1608.

12. Schlichting, H., (1960), "Boundary Layer Theory", McGraw-Hill, London.

13. Shaw, C.T., (1988), Predicting Vehicle Aerodynamics Using Computational Fluid Dynamics - A User's Perspective, SAE Technical Paper 880455.

14. Shaw, C.T. and Simcox, S., (1988), The Numerical Prediction of the Flow Around a Simplified Vehicle Geometry, In "Supercomputer Computer Applications in Automotive Research and Engineering Development", Ed. C. Marino, 219-231 and 533-539.

15. Sumantran, V. and Hammond, D.C., (December 1987), Experimental Data for the Evaluation of Computational Models of Flow Over Automobile-like Bluff Bodies: Geometries and Experimental Conditions, General Motors Report GMR-6078.

16. Thompson, C.P. and Wilkes, N.S., (1982), Experiments with Higher-Order Finite Difference Formulae, AERE R-10493, HMSO.

8 Acknowledgements

The provision of the experimental data for the MIRA model by MIRA, and the provision of the photographs of the MIRA model by Jaguar are both gratefully acknowledged. Work described in this report was undertaken as part of the Underlying Research Programme of the UKAEA.

Finite-element Analysis of Anisotropic Material Deformation

I. Pillinger[1], P. Hartley[1], C.E.N. Sturgess[1], A. Van Bael[2], P. Van Houtte[2] and E. Aernoudt[2]

University of Birmingham[1] and Katholieke Universiteit Leuven[2]

Abstract This paper describes how Taylor-Bishop-Hill models of polycrystalline aggregates can be incorporated into an elastic-plastic finite-element program in order to simulate the deformation of anisotropic materials.

1 Introduction

Finite-elements have proved to be very valuable in simulating many types of metal deformation process. Most of the finite-element formulations developed so far have assumed that the material is isotropic, that is, its properties (and in particular its yield stress) are the same in all directions. While this approach is adequate in many cases, there are deformation processes, involving polymers, composites or heavily cold-worked metals for example, in which the material is most definitely anisotropic. It is important, therefore, to be able to model anisotropic material deformation using the finite-element method.

Although it is natural, from a macroscopic point of view, to look upon isotropy as being the usual state of metals, on a microscopic level, it is anisotropy that is the norm, since a single crystal exhibits pronounced directionality of properties dependent upon the configuration of the crystal lattice. Metals are made up of many tiny crystals, and it is only when these are randomly orientated (after annealing, for example) that the properties appear to be the same in every direction. During deformation, the crystals in a metal tend to align themselves in preferred directions, thus allowing their anisotropic characteristics to become apparent.

This approach, in which metals and other materials are considered to be aggregates of many individual anisotropic crystals, forms the basis of the anisotropic finite-element formulation developed later in this paper. The emphasis will be on how existing isotropic codes need to be modified to incorporate anisotropy. It is useful, therefore, to start with a review of some of the important principles of a large-deformation finite-element formulation, and to consider how this is usually employed to model isotropic material deformation.

2 Finite-Element Theory

The finite-element theory presented here is necessarily brief. For a more detailed discussion see Rowe et al. (1991).

The starting point for any finite-element analysis of deformation is the discretisation of the domain under investigation into a number of elemental regions inter-connected at nodes. The values of the state variables within each element are approximated by functions that interpolate between the values at the nodes of the element. In this way, the behaviour of a continuous body can be specified, at least approximately, by the values of a finite number of variables. The governing equations that describe plastic deformation are non-linear because they are themselves functions of the current state of deformation. In the work described here, an incremental (up-dated Lagrangean) approach is adopted in which the state variables are assumed to be constant during each finite step. For each increment, minimisation of the total power developed in volume V of an element leads to a set of matrix equations relating $\triangle f_{Im}$, the change in component m of the force at node I of the element during time interval $\triangle t$, to $\triangle d_{Jn}$, component n of the displacement of node J of the element during this time:

$$\triangle f_{Im} = (K^{\epsilon}_{ImJn} + K^{\sigma}_{ImJn} + K^{\phi}_{ImJn})\triangle d_{Jn} \tag{2.1}$$

in which the usual summation convention has been extended to sum over all the nodes of an element when an upper-case subscript is repeated.

The deformation-stiffness matrix K^{ϵ}_{ImJn} is similar to the familiar infinitesimal stiffness matrix:

$$K^{\epsilon}_{ImJn} = \int B_{ijIm} D_{ijkl} B_{klJn} dV \tag{2.2}$$

except that the product of the matrix B_{ijIm} with $\triangle d_{Im}$ is the Linearised Co-Rotational (LCR) increment of strain $\triangle\epsilon_{ij}$ instead of the more usual infinitesimal definition (Pillinger et al., 1986). The elastic-plastic constitutive matrix D_{ijkl} expresses the Jaumann increment of Cauchy stress $\triangle\sigma_{ij}$ in terms of the LCR increment of strain:

$$\triangle\sigma_{ij} = D_{ijkl}\triangle\epsilon_{kl} \tag{2.3}$$

The exact form of D_{ijkl} will depend upon the assumptions made concerning the elastic-plastic behaviour of the material and this will be dealt with in greater detail later.

The stress-increment correction matrix K^{σ}_{ImJn} is required because the expression for power dissipation per unit volume involves the rate of nominal or Piola-Kirchhoff I stress, and not the Jaumann rate of Cauchy stress (Lee, 1976):

$$K^{\sigma}_{ImJn} = \int (N_{I,i}\delta_{mn}\sigma_{ik}N_{j,k} \\ -2B_{ijIm}\delta_{jl}\sigma_{ik}B_{klJn})dV \tag{2.4}$$

In this equation, $N_{I,i}$ denotes the partial derivative of the interpolation function for node I of the element with respect to the direction i, and δ_{ij} is the Kronecker delta.

The constant-dilatation correction matrix K^{ϕ}_{ImJn} is introduced in order to ensure that the effects of incompressible plastic flow do not overconstrain the deformation. (Nagtegaal and de Jong, 1980). The inclusion of this term is equivalent to under-integrating the bulk-strain contribution to the element stiffness:

$$K^{\phi}_{ImJn} = \kappa\left(\frac{1}{V}(\int B_{iiIm}dV)(\int B_{jjJn}dV) - \int B_{iiIm}B_{jjJn}dV\right) \tag{2.5}$$

The bulk modulus k in equation 2.5 is $E/3(1-2v)$, where E is Young's Modulus and v is Poisson's Ratio.

A global stiffness matrix describing the behaviour of the collection of elements is obtained by assembling the individual element matrices defined by equation 2.1. This global matrix is solved in the context of the boundary conditions appropriate to the current disposition of the tool surfaces (Hartley et al., 1988) to obtain the components of displacement of all the nodes of the mesh. The displacement information allows the distribution of strain and strain rate to be calculated throughout the workpiece. The constitutive relationship (equation 2.3) is then used to determine the corresponding changes in stress in the body. Changes in temperature may be calculated by considering heat generation and conduction in the body (Pillinger et al. 1987).

3 Elastic-Plastic Formulation

The theory outlined so far has made no assumptions about the nature of the material undergoing deformation; it has merely been stated that there exists a relationship between changes in the strain tensor and changes in the stress tensor. In order to define this relationship, it is necessary to propose a model for the behaviour of the material during deformation.

Both of the models considered here, the isotropic and the anisotropic, assume that the material behaves in a linear elastic/non-linear plastic manner. This means that an initially-undeformed body may be subjected to small amounts of deformation without permanent change in shape. Moreover, the extent of deformation under such circumstances will always be proportional to the magnitude of the load. More generally, at any stage of the deformation, there are changes that may be made to the stress tensor at a given infinitesimal region of a body for which the corresponding change in strain is completely recoverable and equal to the product of the stress change tensor and a matrix of constant coefficients.

Thus with an elastic-plastic approach, the constitutive matrix D_{ijkl} may take two forms: a matrix of constant elastic coefficients or a matrix of coefficients that are functions of the current deformation state. The deformation model must therefore provide a yield criterion to determine which of these matrices is applicable in any particular instance.

The set of all stress states that may be arrived at by purely elastic stress increments is called the elastic region and is bounded in stress space by a hyper-surface called the yield locus:

$$Y(\sigma_{ij}) = A \tag{3.1}$$

The yield criterion indicates that a stress state is elastic if the value of the function Y for that stress is less than A.

4 Isotropic Model

4.1 Yield criterion

For an isotropic material, the conditions for yield must be independent of the axis system chosen to represent the state of stress. The function describing the yield locus can therefore only be a function of the three invariants of the stress tensor (Hill, 1950). Empirical tests suggest that, for most isotropic materials, the yield locus is independent of the first and third invariants and is a linear function of the second of these. A convenient way of writing the resulting Huber-von Mises criterion is:

$$(\frac{1}{2}\sigma'_{ij}\sigma'_{ij})^{\frac{1}{2}} = \tau_c \tag{4.1}$$

in which σ'_{ij} is the deviatoric stress ($= \sigma_{ij} - \frac{1}{3}\sigma_{kk}$) and τ_c is the current shear yield stress. This is, in general, a function of plastic strain, strain rate and temperature.

4.2 Elastic constitutive relationship

For stress changes within the elastic region the isotropic elastic constitutive matrix $D^{\bar{e}}_{ijkl}$ is used. This is obtained from the generalised form of Hooke's law:

$$D^{\bar{e}}_{ijkl} = 2G(\delta_{ik}\delta_{jl} + \delta_{ij}\delta_{kl}\frac{v}{1-2v}) \tag{4.2}$$

The rigidity modulus G is defined to be $E/2(1+v)$.

4.3 Elastic-plastic constitutive relationship

Whenever materials are plastically deformed well below their re-crystallisation temperature, a certain amount of elastic recovery always occurs on unloading. The total increment of strain may therefore be divided into an elastic part $\Delta\epsilon^e_{ij}$ and a plastic part $\Delta\epsilon^p_{ij}$.

$$\Delta\epsilon_{ij} = \Delta\epsilon^e_{ij} + \Delta\epsilon^p_{ij} \tag{4.3}$$

By definition, $\Delta\epsilon^e_{ij}$ is the strain elastically recovered on removal of the applied stress increment $\Delta\sigma_{ij}$, so:

$$\Delta\sigma_{ij} = D^{\bar{e}}_{ijkl}\Delta\epsilon^e_{kl} = D^{\bar{e}}_{ijkl}(\Delta\epsilon_{kl} - \Delta\epsilon^p_{kl}) \tag{4.4}$$

Since changes in stress within the elastic region have been considered in the previous section, assume without loss of generality that σ_{ij}, the stress at the start of the increment, lies on yield locus. During the increment,the size of the yield locus, determined by the value of τ_c, may change due to strain hardening. If the shear yield stress increases by $\Delta\tau_c$ during the increment, the stress σ^*_{ij} defined by:

$$\sigma^*_{ij} = \left(1 + \frac{\Delta\tau_c}{\tau_c}\right)\sigma_{ij} \tag{4.5}$$

will be on the yield locus at the end of increment. To a first-order approximation, the final stress state will lie on the hyper-plane that is a tangent to the yield locus at stress state σ^*_{ij}. By considering the work done during arbitrary loading cycles, Drucker (1951) has demonstrated that the plastic strain-rate tensor $\dot{\epsilon}^p_{ij}$ is normal to the yield surface at the current point in stress space (figure 1). Thus the actual stress increment $\Delta\sigma_{ij}$ satisfies:

$$\dot{\epsilon}^p_{ij}\Delta\sigma_{ij} = \dot{\epsilon}^p_{ij}\frac{\Delta\tau_c}{\tau_c}\sigma_{ij} \tag{4.6}$$

Define the effective plastic shear strain rate:

$$\dot{\gamma} = (2\dot{\epsilon}^p_{ij}\dot{\epsilon}^p_{ij})^{\frac{1}{2}} \tag{4.7}$$

and the shear stress gradient:

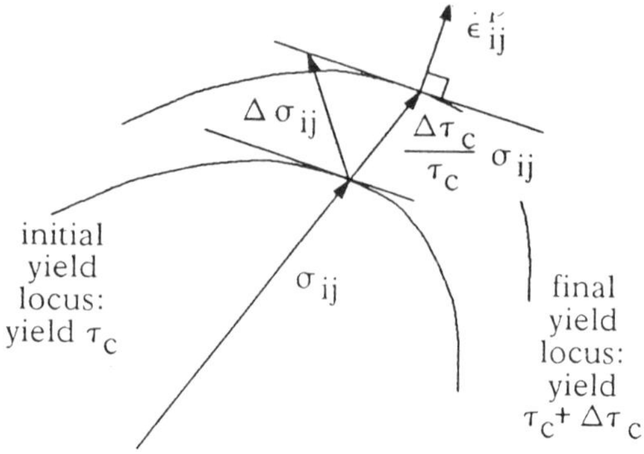

Figure 1. Initial and Final Yield Loci

$$\tau_c' = \frac{d\tau_c}{d\gamma} \tag{4.8}$$

which may determined from a material test. Linearising τ_c' with respect to effective plastic shear strain and with respect to time produces:

$$\Delta\tau_c \simeq \tau_c'\Delta\gamma \simeq \tau_c'\dot{\gamma}\Delta t \tag{4.9}$$

P, the plastic power generated per unit volume may be written in two ways:

$$P = \dot{\epsilon}_{ij}^{p}\sigma_{ij} = \tau_c\dot{\gamma} \tag{4.10}$$

Note that there is no contradiction in using the Cauchy stress in equation 4.10 since this is identical to the nominal stress at the start of the increment. The rates of these two stresses are, of course, not the same.

Equation 4.10 is an expression of the concept that all plastic deformation is a process of shear, and that in an isotropic material there are always slip systems available to allow shear to occur along the direction of maximum shear stress.

Substituting from equations 4.9 and 4.10 into equation 4.6:

$$\dot{\epsilon}_{ij}^{p}\Delta\sigma_{ij} = \tau_c'\dot{\gamma}^2\Delta t \tag{4.11}$$

But from equation 4.4, substituting for the plastic strain increment in terms of plastic strain rate and the time increment:

$$\dot{\epsilon}_{ij}^{p}\Delta\sigma_{ij} = \dot{\epsilon}_{ij}^{p}D_{ijkl}^{\bar{e}}(\Delta\epsilon_{kl} - \dot{\epsilon}_{kl}^{p}\Delta t) \tag{4.12}$$

Equating 4.11 and 4.12:

$$\tau_c' \dot{\gamma}^2 \Delta t = \dot{\epsilon}_{ij}^p D_{ijkl}^{\bar{e}} (\Delta \epsilon_{kl} - \dot{\epsilon}_{kl}^p \Delta t) \tag{4.13}$$

Re-arranging and substituting from equation 4.2:

$$\Delta t = \frac{2\dot{\epsilon}_{ij}^P \Delta \epsilon_{ij}}{\dot{\gamma}^2 (1 + \frac{\tau_c'}{G})} \tag{4.14}$$

Substituting for the time increment into the linearised form of equation 4.4 leads to the result that the isotropic elastic-plastic constitutive matrix $D_{ijkl}^{\bar{e}p}$ for use in yielded increments of deformation is given by:

$$D_{ijkl}^{\bar{e}p} = 2G \left(\delta_{ik}\delta_{jl} + \delta_{ij}\delta_{kl} \frac{v}{1 - 2v} - \frac{2\dot{\epsilon}_{ij}^p \dot{\epsilon}_{kl}^p}{\dot{\gamma}^2 (1 + \frac{\tau_c'}{G})} \right) \tag{4.15}$$

For Huber-von Mises yield, in which the yield surface is a hyper-sphere in deviatoric stress space, the strain-rate tensor is parallel to the deviatoric stress at the start of the increment. Thus:

$$\frac{\dot{\epsilon}_{ij}^p}{\dot{\gamma}} = \frac{\sqrt{3}\sigma_{ij}'}{2Y} \tag{4.16}$$

in which Y is the yield stress in simple tension. If Y' is the slope of the stress-strain curve for simple tension then:

$$\tau_c' = \frac{Y'}{3} \tag{4.17}$$

Substituting from equations 4.16 and 4.17 into equation 4.15 gives the more usual form of the isotropic elastic-plastic constitutive matrix expressed as a function of stress state (Yamada et al., 1968):

$$D_{ijkl}^{\bar{e}p} - 2G \left(\delta_{ik}\delta_{jl} + \delta_{ij}\delta_{kl} \frac{v}{1 - 2v} - \frac{\frac{3}{2}\sigma_{ij}'\sigma_{kl}'}{Y^2 (1 + \frac{Y'}{3G})} \right) \tag{4.18}$$

5 Anisotropic Model

5.1 Yield criterion

The plastic strain-rate tensor is symmetric and, since plastic flow is incompressible, $\dot{\epsilon}_{kk}^p = 0$. Similarly, $\sigma' - \sigma_{ij}'$ is symmetric and by definition $\sigma_{kk}' = 0$. Thus for both tensors there are only 5 independent components. For convenience, define a new pair of five-dimensional vectors, $\dot{\epsilon}_k$ and s_k such that:

$$\dot{e}_1 = \frac{1}{\sqrt{2}}(\dot{\epsilon}^p_{11} - \dot{\epsilon}^p_{22}) \tag{5.1}$$

$$\dot{e}_2 = \sqrt{\frac{3}{2}}(\dot{\epsilon}^p_{11} + \dot{\epsilon}^p_{22}) \tag{5.2}$$

$$\dot{e}_3 = \sqrt{2}\dot{\epsilon}^p_{23} = \sqrt{2}\dot{\epsilon}^p_{32} \tag{5.3}$$

$$\dot{e}_4 = \sqrt{2}\dot{\epsilon}^p_{31} = \sqrt{2}\dot{\epsilon}^p_{13} \tag{5.4}$$

$$\dot{e}_5 = \sqrt{2}\dot{\epsilon}^p_{12} = \sqrt{2}\dot{\epsilon}^p_{21} \tag{5.5}$$

with a similar definition for s_k in terms of σ'_{ij}. It is clear from the above that, for example, $\dot{\epsilon}^p_{ij}\dot{\epsilon}^p_{ij} = \dot{e}_k\dot{e}_k = \| \dot{e} \|$.

Anisotropy is concerned with the variations of yield stress that occur with changing direction of strain rate, independently of its magnitude. It is therefore useful to define the plastic strain-rate mode:

$$a_k = \frac{\dot{e}_k}{\| \dot{e} \|} \tag{5.6}$$

The Taylor-Bishop-Hill theory (Bishop and Hill, 1951), states that the power dissipation per unit volume for a single crystal is:

$$P = \tau_c \| \dot{e} \| M \tag{5.7}$$

where τ_c is now interpreted as the critically-resolved shear stress causing yield on any slip system and M is the Taylor factor. Just as in the isotropic case, τ_c is a function of temperature and the magnitudes of plastic strain and strain rate. The Taylor factor is a function of a_k and the orientation of the crystal, which is usually defined in terms of the 3 Euler angles g_i.

Intuitively, the Taylor factor is a means of relating the macroscopic strain rate to the shear strain rate on the slip systems that are actually active during plastic deformation. Since the Bishop-Hill theory assumes that the shear stress on all such systems is τ_c, equation 5.7 is simply an expression for the power dissipation on active slip systems.

The values for the Taylor factor need only be derived once for any given crystal structure. They can be calculated by a linear programming technique (Van Houtte, 1988) and tabulated for suitable increments in the 4 independent components of a_k and the 3 Euler angles.

For a polycrystalline material, the texture may be described by the orientation distribution function $f(g_i)$ which expresses the volume fraction of crystals, per unit volume of orientation space, that have a given orientation g_i. The values of the orientation distribution function can be measured experimentally for increments in g_i using X-ray diffraction methods.

When applied to a polycrystalline aggregate, equation 5.7 becomes:

$$P = \tau_c \parallel \dot{e} \parallel \oint M f dg = \tau_c \parallel \dot{e} \parallel F \tag{5.8}$$

in which the integration is carried out numerically over the whole of orientation space. The polycrystalline Taylor factor F is a function of plastic strain-rate mode.

Since the power can also be expressed as the product of the macroscopic plastic strain rate and deviatoric stress, the quantity P/τ_c is a potential function for s_k/τ_c.

Thus:

$$\frac{s_k}{\tau_c} = \frac{\delta(\frac{P}{\tau_c})}{\delta \dot{e}_k} = \frac{\delta F}{\delta a_k} + a_k(F - a_l \frac{\delta F}{\delta a_l}) \tag{5.9}$$

The values of F can be calculated for arbitrary a_k, but in order to provide an analytical expression for s_k, the polycrystalline Taylor factor is approximated by a sixth-order series expansion in a_k:

$$F \simeq F_o + F_{ij} a_i a_j + F_{ijkl} a_i a_j a_k a_l + F_{ijklmn} a_i a_j a_k a_l a_m a_n \tag{5.10}$$

with $1 \leq i \leq j \leq k \leq l \leq m \leq n \leq 5$ and the coefficients are calculated to obtain a least-squares fit to the previously-determined values of F. Since the expansion contains only even-order terms, the Bauschinger effect cannot be modelled. An example of a yield locus obtained using a sixth-order approximation to F is shown in figure 2.

Equation 5.9 determines the components of stress at yield for a given strain-rate mode. However, the finite-element program requires the yield criterion in a different form: given a stress tensor, is the material plastically yielding and, if so, what is the direction of plastic strain rate?

If the given stress vector is s_k, suppose $\alpha_o s_k$ is on the yield locus. The stress state is then elastic if $\alpha_o > 1$. For some arbitrary strain rate $\dot{e}_k$, let the yield stress be s_k^*, and let αs_k lie on the hyper-plane which is a tangent to the yield locus (and normal to $\dot{e}_k$) at stress state s_k^* (figure 3). Then:

$$\alpha s_k \dot{e}_k = s_k^* \dot{e}_k = P = \tau_c \parallel \dot{e} \parallel F \tag{5.11}$$

$$\therefore \alpha = \frac{\tau_c \parallel \dot{e} \parallel F}{s_k \dot{e}_k} = \frac{\tau_c F}{s_k a_k} \tag{5.12}$$

a is therefore a function of plastic strain-rate mode only (since τ_c and s_k are assumed constant for the purposes of this exercise). Clearly, the plastic strain-rate mode for the point on the yield locus defined by $\alpha_o s_k$ will be the one that minimises equation 5.12. The minimisation is accomplished

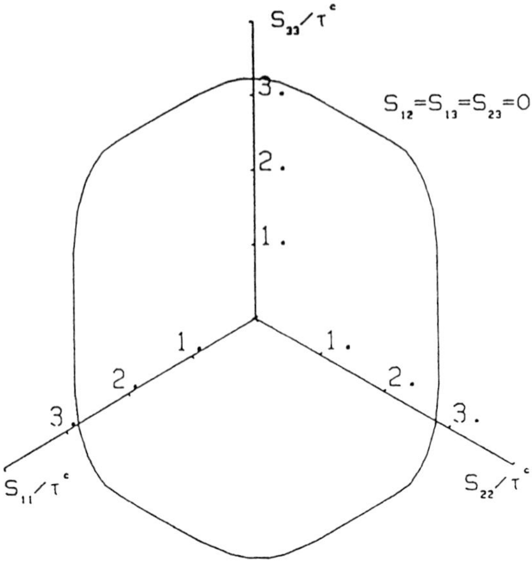

Figure 2. Yield Locus for Cold-Rolled Aluminium Sheet

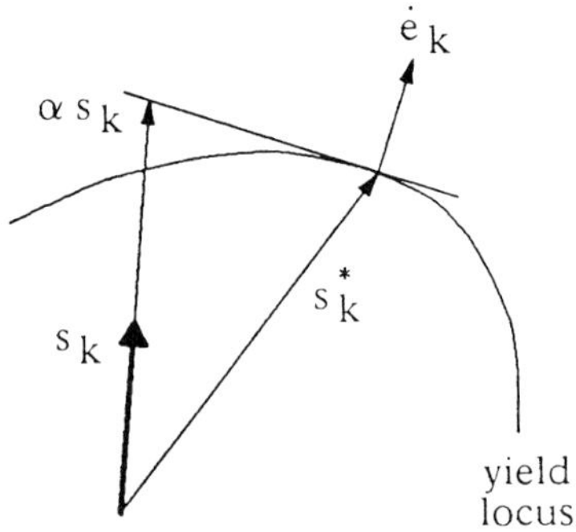

Figure 3. Yield Criterion in Terms of Current Stress State

iteratively by expressing the term on the right-hand side of equation 5.12 as a Taylor expansion about successive approximations to the required value of a_k.

5.2 Elastic constitutive relationship

The anisotropic elastic constitutive matrix D^e_{ijkl} is also known as the anisotropic compliance matrix. In its most general form this contains 21 independent coefficients, but may contain fewer in special cases. For orthotropic materials, for example,there are only 9 anisotropic elastic moduli.

5.3 Elastic-plastic constitutive relationship

The derivation of the anisotropic elastic-plastic constitutive matrix is similar in many respects to that outlined for the isotropic case. In particular, equation 4.6 and figure 1 are equally applicable to isotropic and anisotropic flow, since they do not assume that σ'_{ij} is parallel to $\dot{\epsilon}^p_{ij}$.

The main differences in the anisotropic case are, firstly, that the scalar expression for plastic power dissipation is now given by the right-hand side of equation 5.8:

$$P = \dot{\epsilon}^P_{ij}\sigma_{ij} = \tau_c \parallel \dot{e} \parallel F = \frac{1}{\sqrt{2}}\tau_c\dot{\gamma}F \tag{5.13}$$

and, secondly, that the change in the critically-resolved shear stress is caused by the increase in the effective plastic shear strain along the active slip systems. The latter can be expressed in terms of the macroscopic effective plastic shear strain increment by means of the polycrystalline Taylor factor:

$$\Delta\tau_c \simeq \frac{1}{\sqrt{2}}\tau'_c\dot{\gamma}F\Delta t \tag{5.14}$$

Substituting equations 5.13 and 5.14 into equation 4.6 gives:

$$\dot{\epsilon}^P_{ij}\Delta\sigma_{ij} = \frac{1}{2}\tau'_c(\dot{\gamma}F)^2\Delta t \tag{5.15}$$

Equating this with the anisotropic form of equation 4.12:

$$\frac{1}{2}\tau'_c(\dot{\gamma}F)^2\Delta t = \dot{\epsilon}^p_{ij}D^e_{ijkl}(\Delta\epsilon_{kl} - \dot{\epsilon}^p_{kl}\Delta t) \tag{5.16}$$

$$\therefore \Delta t = \frac{\dot{\epsilon}^p_{ij}D^e_{ijkl}\Delta\epsilon_{kl}}{\frac{1}{2}\tau'_c(\dot{\gamma}F)^2 + \dot{\epsilon}^p_{ij}D^e_{ijkl}\dot{\epsilon}^p_{kl}} \tag{5.17}$$

Hence the anisotropic elastic-plastic constitutive matrix D^{ep}_{ijkl} is given by:

$$D^{ep}_{ijkl} = D^e_{ijkl} - \frac{D^e_{ijpq}\dot{\epsilon}^p_{rs}D^e_{rskl}}{\frac{1}{2}\tau'_c(\dot{\gamma}F)^2 + \dot{\epsilon}^p_{pq}D^e_{pqrs}\dot{\epsilon}^p_{rs}} \tag{5.18}$$

6 Computational Considerations

It is clear from the preceding sections that a large-deformation isotropic elastic-plastic finite-element code can be modified quite easily to make it able to model anisotropic behaviour. The main changes that have to be made are in the code that defines the constitutive relationship and in the sections that test for yield.

There are, however, two practical implications of developing an anisotropic finite-element deformation program.

The first point is that the amount of material property information that has to be input for anisotropic analyses is much greater than for isotropic ones. The sixth-order series expansion of the polycrystalline Taylor factor for a particular crystal structure and material texture contains almost 300 coefficients, and, in general, the texture will vary from element to element. In practice of course, the series coefficients would have to be generated by computer programs anyway, so there is no problem obtaining them in a form suitable for the finite-element program.

The second point is that the anisotropic formulation described here involves much more computation than an isotropic approach. Each calculation of the elastic-plastic constitutive matrix, for example, requires extra multiplications of 6x6 matrices, and the determination of the state of yield involves an iterative search along the yield locus. Since even isotropic simulations of realistic material forming can require inconveniently-large amounts of computing time, use of an anisotropic formulation makes it especially important to develop parallel and vector processing approaches to the finite-element calculation.

7 Future Work

The theory outlined in this paper has considered how an anisotropic finite-element simulation of material deformation can be performed using information about the initial material texture determined from X-ray measurements. However, as the deformation proceeds, the texture will change, so it is important to periodically re-calculate the orientation distribution function for each element. There is, in principle, no difficulty in doing this as the Taylor-Bishop-Hill model provides the theoretical basis for the re-calculation using values of the strain tensor in the elements. The practical implications of such an approach are, once again, greatly increased amounts of computation. Each time the texture is updated, a set of 300 simultaneous linear equation would have to be solved for each element in order to determine the new series expansion coefficients of the polycrystalline Taylor function. As before, the scope for parallelising the calculation is considerable.

References

1. Bishop, J.F.W. and Hill, R., (1951), A theoretical derivation of the plastic properties of a polycrystalline face centred metal. Phil. Mag. 42:1298-1307.

2. Drucker, D.C., (1951), A more fundamental approach to plastic stress-strain relations. Proc. 1st Nat. Cong. on App. Mechanics, p. 487-491.

3. Hartley, P., Pillinger, I. and Sturgess, C.E.N., (1988), Modelling frictional effects in finite-element simulation of metalforming. Proc. Winter Meeting ASME, Chicago.

4. Hill, R., (1950), The mathematical theory of plasticity. Oxford: Clarendon Press.

5. Lee, E.H., (1976), The basis of an elastic-plastic code. SUDAM rep. 76-1, Stanford.

6. Nagtegaal, J.C. and de Jong, J.E., (1980), Some computational aspects of elastic-plastic large strain analysis. In J.T. Oden (ed.), Computational methods in nonlinear mechanics, p. 303-334.

7. Pillinger, I, Hartley, P., Sturgess, C.E.N and Rowe, G.W., (1986), A new linearized definition of strain increment for the finite-element analysis of deformations involving finite rotation. Int. J. Mech. Sci. 28:253-262.

8. Pillinger, I, Hartley, P., Sturgess, C.E.N. and Rowe, G.W., (1987), Thermo-mechanical modelling of metal-forming using the finite-element method. Proc. Int. Conf. on Computational Plasticity, Barcelona, p. 1073-1086.

9. Rowe, G.W. Sturgess, C.E.N., Hartley, P. and Pillinger, I., (1991), Finite-element plasticity and metalforming analysis. Cambridge: CUP.

10. Van Houtte, P., (1988), Textures and Microstructures 8-9: 313-350.

11. Yamada, Y., Yoshimura, N. and Sakurai, T., (1968), Plastic stress-strain matrix and its application for the solution of elastic-plastic problems by the finite element method. Int. J. Mech. Sci. 10:343-354.

Thermal and Mechanical Modelling of Electronic Interconnects for the Automotive Environment

D.C. Whalley and D.J. Chambers

Lucas Automotive, West Midlands

Abstract In the harsh environments automotive electronics are required to endure there is the continued need to ensure a long term reliability of both separable connectors and soldered joints. Extended warranty periods and the growing use of safety critical electronic control systems are placing increasing emphasis on reliability. Evaluation by conventional environmental testing is both time consuming and expensive and there is therefore a need for design tools to optimise interconnect reliability.

The primary reliability hazard in solder joints is low-cycle fatigue due to cyclic thermal stresses. These thermal stresses are caused by both variations in the ambient temperature and in the operating conditions of the vehicle.

In separable connectors two major mechanisms are loss of contact force through thermally activated creep and fretting corrosion due to vibration. Finite element analysis is a powerful tool for the modelling of these complex physical problems and is being used extensively within Lucas Automotive for electronic product design and evaluation:-

1) Transient thermal analysis is used in the modelling of temperature distributions in electronic modules.

2) Plasticity and creep analysis is used to determine the response of solder joints to these thermal loads.

3) Connectors are currently modelled using creep analysis, and dynamic structural analysis is being investigated for the modelling of vibration problems.

Another numerical method being developed is a one dimensional non-linear model for solder creep, for use in the rapid and cost effective modelling of solder joints under different operating conditions.

These analyses have required considerable effort in establishing a database of the relevant material properties particularly where these are non-linear and/or temperature dependent. However, by using modelling methods, significant cost savings have been made over prototype manufacture and testing without any reduction in long term reliability.

1 Background

1.1 The importance of design for reliability

Long term reliability is a crucial factor in the automotive electronic systems of today. Not only are design lives and warranty periods increasing, but there is increasing use of safety critical systems such as antilock braking.

Added to this there are difficult economic constraints on these systems, which place limitations on the packaging technologies that can be adopted.

Design optimisation is usually performed using accelerated environmental testing, but simulating the field failure characteristics of interconnection systems is particularly difficult because of the complex mechanisms involved. This can result in tests being of excessive duration and considerable cost.

There is therefore a need to optimise designs using modelling methods such as those described in this paper.

1.2 The automotive environment

The automotive environment is characterised by extremes of temperature, high levels of shock and vibration, high humidity and exposure to corrosive pollutants such as SO_2, NO_x, Cl_x^- etc - worse in fact than some avionic environments.

These environmental stresses may result in thermal and mechanical fatigue, thermally activated creep, and corrosion.

1.3 Interconnection reliability

The trends in automotive electronics towards higher complexity have significant effects on the technological requirements of the connections between system components. There are now more and more electronic control systems fitted to vehicles which has meant the numbers of interconnections required has increased significantly. The nature of vehicle circuits has also changed. Whereas in the past they were exclusively 12V, and normally passing a few amps, those in electronic circuits can be operating at very much lower voltages and currents ("dry circuits").

Interconnection reliability is now considered by many to be the limiting factor in overall system reliability. Several studies across the electronics industry have shown 50% of all system failures are diagnosed as 'no fault found' (ie no cause for the failure can be detected) and it is probable that a large proportion of these failures are due to undetected interconnection failure. Interconnection problems can also be difficult to diagnose because of their often intermittent nature.

2 Modelling of Solder Joint Fatigue

2.1 Causes of solder joint fatigue

The principal failure mechanism in soft soldered joints is low cycle thermal fatigue (high cycle vibration induced fatigue may also occur, but is considered to be less of a problem). The cause of thermal fatigue is a mismatch in the thermal expansions of component and substrate. This thermal expansion mismatch is caused by either different thermal expansion coefficients of component and substrate or by temperature differences between component and substrate (caused by component power dissipation). The widespread adoption of surface mounted device attachment has exacerbated this problem since the joint strength relies entirely on the very soft solder fillet.

2.2 The Coffin-Manson relation

The isothermal fatigue of near eutectic tin-lead solders has been shown to fit the empirical Coffin-Manson model [1]:-

$$N_f = (\frac{m}{E_p})^{1/n} \tag{2.1}$$

where N_f is the number of cycles to failure, E_p is the plastic strain per cycle, and m and n are empirical constants for the solder alloy in question.

This model is unable to take account of time dependent effects such as corrosion and recrystallisation, and its general applicability to thermal fatigue is in doubt, but until the development of a more sophisticated damage accumulation model it will continue to be widely used.

The plastic strain per cycle (E_p) is determined from mechanical modelling of the problem as described in the next section.

2.3 Mechanical modelling methods

Soft soldered joints in electronic assemblies are mechanically complex structures with non-uniform stress fields and non-linear materials behaviour. Methods for modelling the plastic stresses occurring in solder joints do exist however:-

The simple shear model

This model utilises simplifications of both geometric and materials properties. The model assumes that only shear forces act on the solder (Figure 1), and that all thermal strain is absorbed plastically by the solder (ie the component and substrate have infinite rigidity, and the solder has zero strength). The model is:-

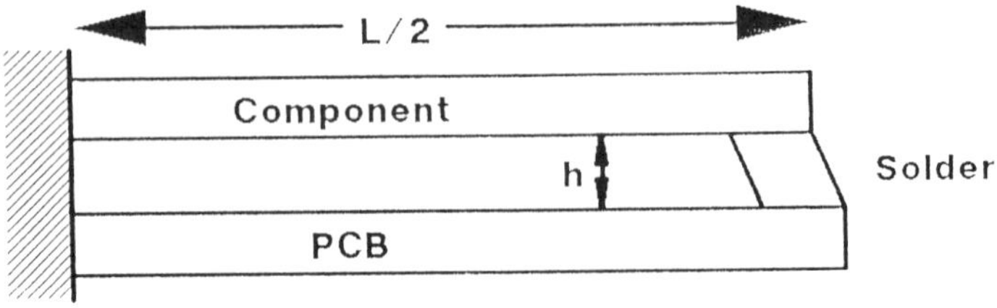

Figure 1. Simple Shear Model of Solder Joint

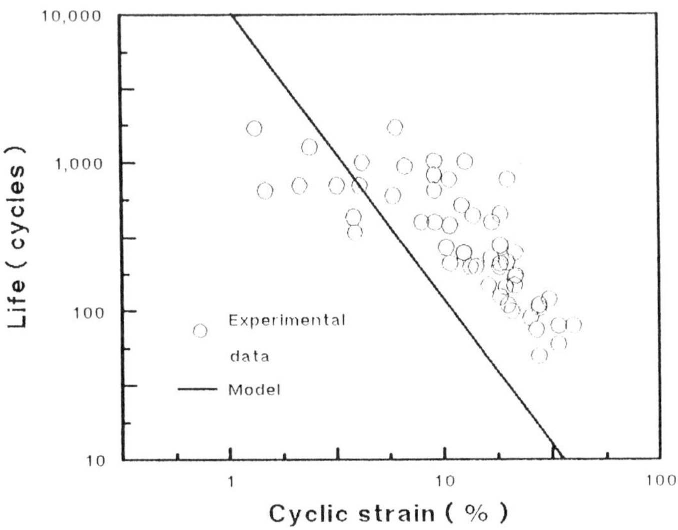

Figure 2. Comparison of Simple Shear Model with Experimental Results

$$E_p = \frac{L(\Delta T_c \alpha_c - \Delta T_p \alpha_p)}{2h} \tag{2.2}$$

where α_c and α_p are the thermal expansion co-efficients of component and PCB, and ΔT_c and ΔT_p are the cyclic changes in temperature of component and PCB.

In Figure 2 this, model is compared with experimental results published by a number of workers in this field [2]. At high cyclic strains the model appears to give a reasonable fit to the data, but at low strain levels/high cyclic lives there is considerable scatter and little confidence may be placed in the model.

Finite element methods

This computer based modelling technique is a powerful analytical tool for solving complex field problems such as stress/strain, thermal flux, magnetism etc. The basis of the technique is to divide the geometry of the problem into discrete regions (elements) of simple shape over which a continuous solution of the field equations may be found. These individual solutions are then assembled into a matrix solution for the whole structure.

The principal system used within Lucas Industries is the Bersafe package developed by the CEGB. This package will not only solve both stress and thermal/thermal stress problems but will also solve non-linear (plasticity and creep) and fracture mechanics problems.

The data required to perform an FE model may be divided into three main areas:-

a) **Geometry** - the shape of the structure must be adequately described although some approximations will inevitably have to be made - particularly in the shape of the solder fillet.

b) **Materials data** - a substantial amount of material properties data must be obtained before any analysis can take place. The principal data used in thermal modelling are the thermal conductivity, specific heat capacity and density. For linear stress analysis the Young's Modulus and Poisson's ratio are required. This data is readily available for common structural materials such as steels, aluminium, copper, glasses etc. In the case of many materials used in microelectronics this information is more scarce and less reliable, especially for proprietary materials and components. Considerable effort has therefore been spent in establishing a database of electronic materials properties to collate what data is available and to assist in identifying cases where actual measurements of these properties have to be made.

In most soldered assemblies the only material to show significant non-linear behaviour (creep) is the solder. Further data is required to enable the modelling of this non-linearity, but fortunately this is available for the commonest solder alloys eg SnPb40.

c) **Loading data** - the individual component power dissipations will usually be known, but the external environmental conditions such as temperature and airflow rates are generally less well defined.

The use of the FE method for modelling solder joints has been reported in several papers eg [3][4][5], but its use has been largely limited to the comparison of stress levels in different designs rather than direct predictions of cyclic strain. This is because elastic and even elastic/plastic models will not accurately determine the solder joint strains which in practice are primarily due to creep.

The cost of computer time has now fallen sufficiently for full cyclic creep modelling to be economically viable, although it is still too expensive and

time consuming for regular use.
Figures 3(a) and (b) show an FE model of a chip capacitor soldered to a PCB. Figure 3(c) shows the stress distribution occurring in the solder joints. The large stresses at the material interfaces correlate well with observed fracture locations in experimental studies. Figure 3(d) shows the stress strain hysteresis loop occurring during a temperature cycle from -55 to 125°C.

State Variable Model

Workers at Cornell University [6] have developed a more sophisticated analytical model for use in solder flow problems. Their solder joint model assumes the simple geometry described in section 2, but describes the solder flow by the differential equation:-

$$GsS(L/h)(\alpha_c - \alpha_p)\dot{T} - \dot{\gamma}p + \tau S\dot{G}s/Gs - \dot{\tau} = 0 \tag{2.3}$$

where Gs is the shear modulus of the solder, S is a measure of the stiffness of the whole assembly, τ is the shear stress in the solder and $\dot{\gamma}p$ the rate of change of plastic strain in the solder with temperature and stress, ie the creep model:-

$$\dot{\gamma}p = f(t, \tau) \tag{2.4}$$

The numerical solution of this equation is much simpler than using a non-linear FE model and allows simple comparison of different situations. By adjusting the value of S the group at Cornell obtained a good correlation with experimental data, but in order to directly predict the plastic strains in new designs a way of estimating the stiffness (S) of any assembly must be developed.
If there is no plastic flow in the solder, equation 3.3 may be rearranged to give:-

$$\Delta\tau = GsS(L/h)(\alpha_c - \alpha_p)\Delta T + \tau S\Delta Gs/Gs \tag{2.5}$$

If a linear FE analysis is performed and gives a stress change $\Delta\tau$ for a small thermal load ΔT then the above may be solved for S. Figure 4 shows a comparison of a full FE analysis with a solution of equation 3.3 using an FE determined value for S. It may be seen that there is a reasonable correlation between these modelling methods, but because of the stress redistribution that occurs in the solder joint there is not a perfect match.

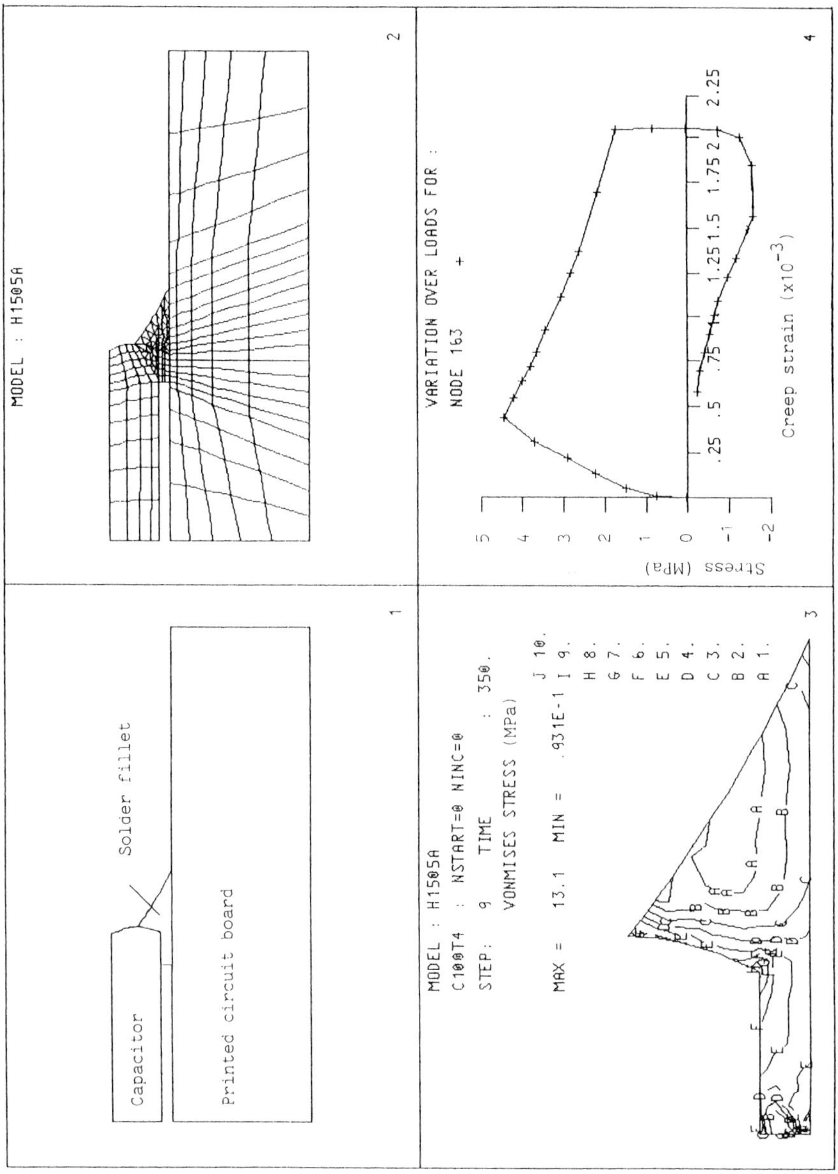

Figure 3. Finite Element Model of Chip Capacitor

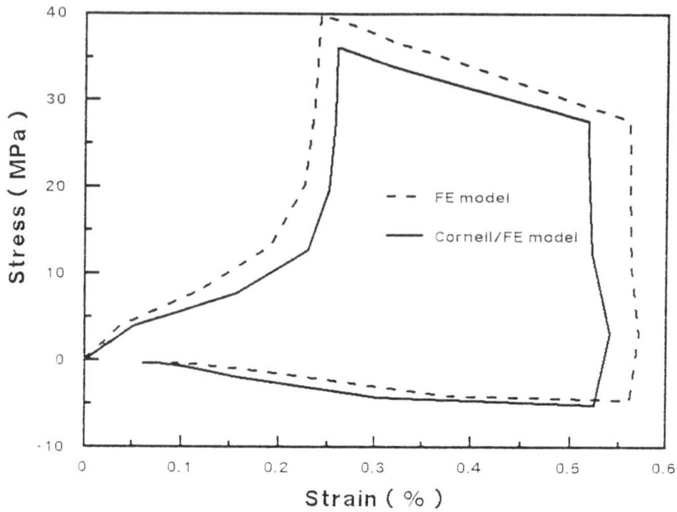

Figure 4. Comparison of Non-Linear FE Model with Cornell and FE Model

3 Modelling of Separable Connections

3.1 Background

Contact theory and reliability

The reliability of a pressure contact can be defined in terms of its ability to maintain a low resistance. The total resistance (Rt) is given by:

$$R_t = R_B + R_C + R_F$$

where R_B, R_C and R_F are the bulk, constriction and film resistances. R_B is the resistance of the metallic contact elements above the contact zone. R_C is the resistance generated by the constricted nature of the contact interface. Actual electrical contact only takes place over a small proportion (typically a few %) of the apparent contact area. These contacts occur as a function of the number of asperities on the metal surface, the elastic/plastic properties of the interfaces, the geometry and the applied load. They are generally termed A-spots and the constriction resistance they cause is termed R_C. These two parameters are relatively unimportant compared to R_F. Contaminant films are always present, even on noble metal contacts, from adsorbed gases in the atmosphere. Provided these films are only fractions of a nm there is generally enough power in the circuit for electrons to tunnel through. However, once these films grow to nm thickness, they

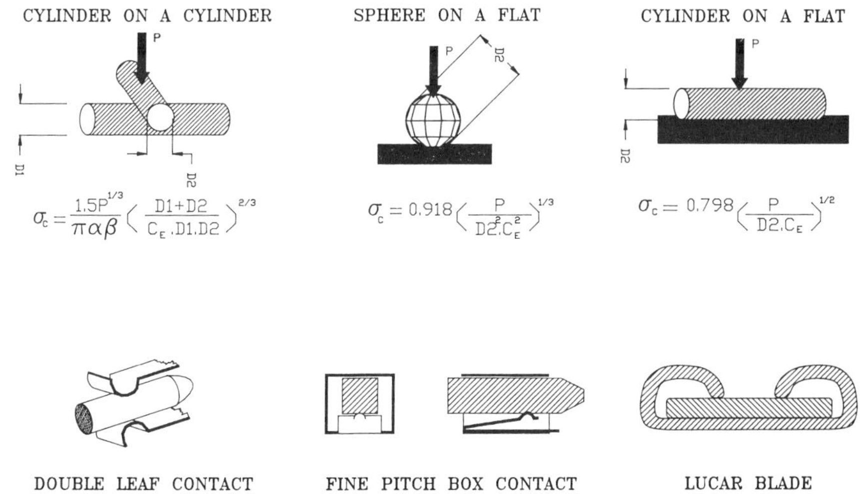

Figure 5. Hertz Stresses for Different Connector Styles

can seriously affect the contact resistance and eventually cause an open circuit. In vehicles such deterioration is caused by Cl^- from road salt and NO_x and SO_2 from exhaust emissions. This failure mechanism is widely thought to contribute to the problems of intermittency and open circuit. Indeed the problem of connector faults is such that throughout Europe they cause around 15% of roadside breakdowns [7].

In order to minimise the effect of corrosive attack at the interface it is desirable that the applied (normal) force is as high as possible without causing wear or excessive insertion and extraction forces. High normal force is in itself however not a sufficient measure of reliability, but by using Hertz stress evaluation a more accurate indicator can be calculated [8].

As an example compare three types of connector (Figure 5). The relevant Hertzian stress equations are [9]:

(a) Sphere on a flat:

$$\sigma_c = 0.918[\frac{P}{D_2^2 C_E^2}]^{1/3} \tag{3.1}$$

where

$$C_E = \frac{1-\nu_1^2}{E_1} + \frac{1-\nu_2^2}{E_2} \tag{3.2}$$

(b) Cylinder on a flat

$$\sigma_c = 0.798[\frac{P}{D_2 C_E}]^{1/2} \tag{3.3}$$

(c) Cylinder on a cylinder

$$\sigma_c = \frac{1.5P^{1/3}}{\pi\alpha\beta}[\frac{D_1 + D_2}{C_E D_1 D_2}]^{2/3} \tag{3.4}$$

α and β depend on $\frac{D_1}{D_2}$, according to the following

D_1/D_2	1	1.5	2	3	4	6	10
$\propto$	.908	1.045	1.158	1.350	1.505	1.767	2.175
β	.908	.799	.732	.651	.602	.544	.481

NB: as D_1/D_2 tends to 1 then the solution tends towards the sphere on a flat.

The resultant Hertz stress for these three types of design are shown in Figure 6. Design (a) is clearly superior to (b) and (c).

Although the actual behaviour of the materials at the interfaces is extremely complex and involves elastic/plastic deformations at contact asperities this method is a useful design tool which has been verified by experiment. These tests have shown that the amount of Hertz stress necessary to ensure stable contact resistance is in excess of 2000Mpa.

In order to perform these calculations it is necessary to determine the normal forces acting on the elements as described in the rest of this section.

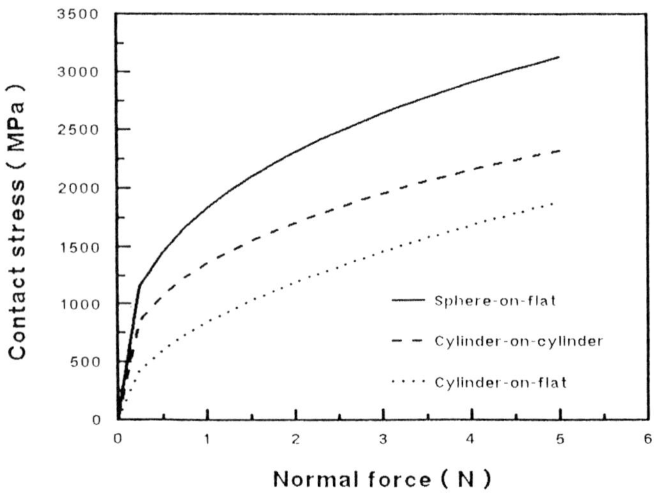

Figure 6. Hertz Stress Against Normal Force for Different Connector Styles

3.2 Analytical methods of determining normal force

Most contact elements can be considered as built-in cantilever beams, where reaction force P is given, from:-

$$P = \frac{Ewt^3\delta}{4l^3} \tag{3.5}$$

where E is Young's modulus, δ is the deflection and w, l and t are the width, length and thickness respectively.

A recommended modification to improve accuracy of this equation is to replace Young's modulus with a "secant" modulus. Since the design stress of a contact is often above the linear portion of the stress-strain curve the modulus can become a function of the stress [10].

In order to achieve greater accuracy of prediction an analytical model for curved end cantilever beams can be used [11]. This more closely resembles modern box-type electronic contacts (Figure 5).

$$P = \frac{Ewt^3\delta}{12r^3[\frac{\lambda^3}{3} + \beta\lambda^2 + 2\lambda(1 - \cos\beta) + \frac{\beta}{2} - \frac{\sin 2\beta}{4}]} \tag{3.6}$$

where $\lambda = L_1/r$

Another approach is to use strain energy methods. Provided the strain induced on insertion is wholly elastic then the work done $U = P\delta/2$.

From [8] the force generated in a cantilever beam including shearing effects is given by:

$$P = \frac{\delta.wt^3}{4}l^3(1 + \frac{3t^2E}{10}l^2G) \tag{3.7}$$

where G is the shear modulus $= \frac{E}{2(1+\gamma)}$.

Techniques of analysis using first principles of strain energy and Castigliano's theory have been reported for analysing non-uniform cantilever contacts [12], but the mathematics become extremely complex. A more user friendly approach is to use finite elements.

Finite element analysis

By using finite element techniques it becomes possible to model the complete connector body, rather than assuming built-in beams. It is also possible to model material orthotropy and non-linearity.

Models can be generated from drawings, but discrepancies in the geometry arise from the forming process eg stress raisers from thinning of the section. These discrepancies can cause significant errors in modelling. An alternative method which can be used when comparing different materials or modifications for an existing design is to digitise sections of actual components.

3.3 Deterioration of contact force with time

The contact force of separable connectors decreases with time by a mechanism analogous to creep. Creep is the time - dependent deformation of materials under stress whereas stress relaxation is the time-dependent decay of stress which converts elastic strain into plastic deformation. The mechanisms of both phenomena are similar in that they depend on dislocation movement and atomic diffusion and as such are highly temperature dependent. The rate of relaxation depends also on the alloy, and to a lesser extent on the grain structure and direction, its condition (rolled to required hardness or annealed down) as well as the level of stress initially applied [13,14].

In order to estimate the stress relaxation behaviour for connector alloys over the operating life of the vehicle it is necessary to extrapolate from limited duration test data. A widely used technique is the Larson-Miller parameter method [15]. This is based on the Arrhenius rate-process equation:-

$$\dot{\epsilon} = Ae^{(\frac{-Q}{RT})} \tag{3.8}$$

where $\dot{\epsilon}$ is the creep rate, A is a constant, Q is the activation energy of the creep process, R is the gas constant and T the absolute temperature.

The Larson-Miller Parameter (P) relates time, temperature and stress in a single relationship:

$$P = T(20 + \log t) \tag{3.9}$$

where t is the exposure time in hours. Plotting log $\Delta\sigma$ against P gives a "master curve" for each alloy. By testing at higher temperature for relatively short periods (typically 1000 hours) the stress drop at lower temperatures over longer times (typically 100,000 hours) can be estimated. Figure 7 shows some master curves for copper based alloys suitable for use as automotive contacts.

This method can be used in conjunction with finite elements as a design tool to ensure optimum material selection for high reliability. As an example after 30,000 hours operation at 75°C (eg for bulkhead mounted electronics) the lucar connector shown in Figure 5 would experience a 30% stress drop if it were made from 70/30 brass (C26000) whereas for C72500 the stress drop would be negligible (Figure 8).

4 Conclusions

Electronic interconnects are of increasing importance to automotive reliability. By using mathematical modelling methods significant improvements

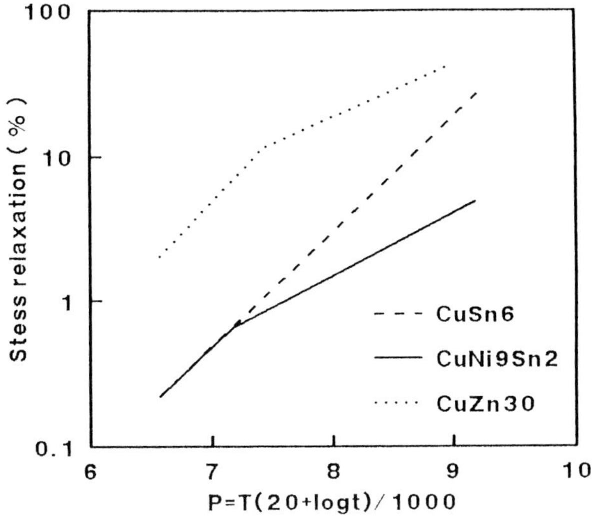

Figure 7. Stress Relaxation Versus Larson-Miller Parameter 14

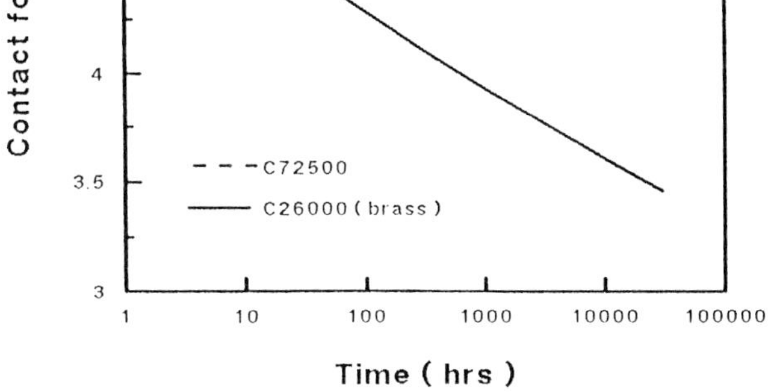

Figure 8. Creep/Normal Force Reduction in a Lucar Connector at 75°C

in reliability may be obtained for reasonable cost. These methods also provide economies in design time and validation testing.

These methods are however dependent on accurate mechanical properties data which may be difficult to obtain for materials of interest in electronic packaging. There is also still a need for the development of better models for such processes as fatigue.

References

1. Solomon, H., (December 1986), Fatigue of 60/40 solder, IEEE transactions on components, hybrids and manufacturing technology CHMT-9 no 4 pp 423-432.

2. Whalley, D. and Chambers, D., (1989), Evaluation of SMD solder joints using finite element analysis, 7th European Hybrid Microelectronics Conference. Hamburg.

3. Clatterbaugh, G. and Charles, H., (1985), Thermomechanical behaviour of soldered interconnects for surface mounting : a comparison of theory and experiment, 35th Electronic Components Conference pp 60-72.

4. Fox, L., Sofia, J. and Shine, M., (September 1987), Investigation of solder fatigue acceleration factors, IEEE Transactions on components hybrids and manufacturing technology, CHMT -8 No 2 pp 275-282.

5. Lau, J., Rice, D. and Avery, P., (September 1987), Elastoplastic analysis of surface-mount solder joints, IEEE Transactions on components, hybrids and manufacturing technology. CHMT-10. No 3 pp 346-357.

6. Sloane, D., Hannula, S. and Li, C., (1985), The effects of service and material variables on the fatigue behaviour of solder joints during the thermal cycle, 35th Electronic Components Conference pp46-51.

7. IEE Colloquium No 1989/58 "Connectors on Vehicles" 13/4/89.

8. Kantner, E. and Hobgood, L., (1989), Hertz stress as an indicator of connector reliability, Connection Technology, pp 14-22, 3/89.

9. Roark, R., and Young, W., (1989), "Formulas for stress and strain". 6th Edition. Chapter 13.

10. Hann, T. and Zarlingo, S., (1985), "Spring designers data package". Olin Brass.

11. Masutani, T. and Mayumi, H., Development of small connectors for automotive wiring harness, SAE 860395.

12. van Kessel, C. and Glashoerster, J., (1983), Design method for slotted beam springs for insulation displacement contacts. IEEE - CHMT - 6 No 4, pp 408-13, 12/83.

13. Horn, K. and Zarlingo, S., (1982), Understanding stress relaxation in copper alloys, 15th Annual Connector Symposium 11/82.

14. Langer, H. and Duerrschnabel, W., (1984), Spannungssrelaxation von Kupferbasis - Legierungen. DVM - Werkstoffpruefung - Tagung, pp 559-576.

15. Larson, F. and Miller, J., A time temperature relationship for rupture and creep stresses. Trans ASME, pp 765-775, 7/52.

Software Support Tools for Durability Assessment and Design of Automotive Components

J. Devlukia

Rover Group Ltd, Warwick

1 Introduction

The automotive industry is fiercely competitive and under continual pressure to reduce design lead times while at the same time improve quality and reliability. It is recognised that the design-to-product cycle of most components can be reduced considerably by shifting the emphasis from development testing and empirical refinement to an approach based on more rigorous mathematical modelling and simulation at early stages of design. A case in point is the design for strength and durability of load bearing components.

However, the task of providing comprehensive analytical tools for fatigue design of automotive components is a daunting challenge. The reasons include the following:

- a) The Spread in service duty resulting from driver and environmental conditions. Many components are exposed to variable amplitude, multiaxial and thermomechanical loadings. Both low cycle and high cycle regimes are covered.

- b) The range of metallic materials and fabrication technologies used in the manufacture of components. These include mostly ferrous and aluminium alloys used in forged, cast and wrought conditions.

- c) The range in surface finish and treatments. The finish varies from as forged or cast to highly polished machined parts. The treatments include thermal (e.g. induction hardening), thermochemical (nitriding, carburising) and mechanical (shotpeening, rolling) etc.

- d) The Variety of joining and fastening techniques including bolts, rivets, spot and fusion welds.

- e) Above all, the components are mass produced without being individually inspected and hence the part-to-part statistical variation or scatter in durability performance may be large.

In addition to the above problems of modelling basic fatigue mechanisms, it has to be noted that the average designer may not possess the necessary expertise to make efficient use of the analytical tools. Apart from a reliance on experimental data (loads, materials etc), fatigue design encompasses a broad range of technologies and engineering disciplines, such as an understanding of material behaviour under cyclic loads, stress analysis of structures, manufacturing considerations etc. It is nonetheless possible to provide support and assistance to the designer via computer software on a CAD Workstation which provides an ideal environment for the integration of the above technologies. A particularly attractive option is that offered by 'knowledge-based' systems which have an ability to encapsulate expert knowledge.

This paper describes the mathematical details of the basic fatigue life prediction algorithms applicable to metallic components. A prototype 'knowlege-based' system to support fatigue design is also described. Some observations on the reliability of life prediction methods are also included.

2 Fatigue Life Prediction Models

Three basic models are in use for fatigue life prediction of automotive components, namely, the Stress-Life ($S-N$) method, the Local Strain ($\epsilon - N$) approach and the Linear Elastic Fracture Mechanics (LEFM) approach. The basic precepts are discussed first, followed by details on how each method deals with specific requirements of the automotive environment.

The $S-N$ method treats nominal stress as the primary damage parameter. The component damage is characterised by a stress-cycle curve representing the number of cycles of fully reversed, linear elastic stress amplitude which the component is able to survive. The $S-N$ curve is defined by the following equation:

$$\frac{\Delta\sigma}{2} = \sigma'_f(2N_f)^2 \tag{2.1}$$

where $\Delta\sigma =$ Stress range
$\sigma'_f =$ Fatigue strength coefficient
$b =$ Fatigue strength intercept
$2N_f =$ Reversals to failure

A fatigue limit is usually associated with such a curve - the limit representing stress amplitude below which no damage is believed to take place. This method is appropriate for high cycle fatigue regime(life>10e5 cycles).

The Local Strain approach is similar, except that the damage process is characterised in terms of a strain amplitude versus cycles curve defined by the following equation:

$$\frac{\Delta\epsilon}{2} = \frac{\sigma'_f}{E}(2N_f)^2 + \epsilon'_f(2N_f)^c \qquad (2.2)$$

where $\sigma'_f =$ Fatigue strength coefficient
$\epsilon'_f =$ Fatigue strength ductility coefficient
$b =$ Fatigue strength exponent; elastic.
$c =$ Fatigue ductility exponent; plastic.
$2N_f =$ Reversals to failure
$E =$ Young's modulus.

The surface strain, rather than stress, is considered to model the non-linear plastic deformation around the critical sites better, and as such, this method is more appropriate when dealing with large strain, low cycle fatigue.

The LEFM approach is based on crack growth. The method is applicable in situations where the plastic zone surrounding the crack tip is relatively small compared to the crack length; typically about 1.0mm. Life calculations are based on the Stress Intensity Factor which is a scaled product of nominal stress and the incipient crack length as shown below:

$$\Delta K = Y.\Delta\sigma(\pi.a)^{1/2} \qquad (2.3)$$

where $\Delta K =$ Stress intensity range
$Y =$ Compliance factor
$a =$ Crack length

The crack growth curve, for rates typically about $10^{-3} - 10^{-6}$mm/cycle, is modelled by the Paris Law as shown below:

$$\frac{da}{dN} = A(\Delta K)^m \qquad (2.4)$$

where $a =$ crack length
$N =$ number of cycles
$K =$ Stress Intensity Factor
$A, m =$ material constants.

The crack growth regime defined by the Paris law is bounded by a lower limit of stress intensity factor (stress intensity threshold), ΔKth, below which the crack advance is negligible and an upper limit, ΔKc above which the crack growth is unstable and catastrophic.

While the above three fatigue life prediction models are simple in principle, complications arise when they are adopted to handle realistic service and component like features. The technical details of these methods are given below with special attention to those factors affecting the analysis of automotive parts.

2.1 Variable Amplitude Loading

For both stress-life and strain-life approaches, the cumulative damage due to a variable amplitude loading history is handled currently via the Miner's rule. This rule performs a linear sum of fractional damage caused by individual cycles (fully reversed stresses or strains. For inelastic deformation each cycle is made up of a stress-strain hysteresis loop). These cycles are extracted from the variable amplitude signal by using the Rainflow method of cycle counting [1]. Thus for a stress spectrum consisting of cycles n_1, n_2 etc. at stress ranges σ_1, σ_2 etc., the respective cycles to failure N_{f_1}, N_{f_2} etc are calculated from Equation (2.1) and the corresponding fractional damage accumulated as shown below:

$$\text{Total calculate Damage} = \sum n_i/N_{fi} \ldots\ldots \text{ Miner's rule.}$$

At failure, this summation equals unity. While open to challenge on its validity and accuracy, there appears to be no reliable alternative which can consistently provide better correlation between experimental and theoretical results on a range of components and materials. The application of Miner's rule is particularly suspect in situation such as welded joints, where a significant crack may be present at the start. Indeed, the current thinking is that the growth characteristics of the so called short cracks (sub granular size cracks) may ultimately control the accuracy of the damage summation rules [2].

The Miner's rule is also applied for cumulative damage summation for strain-life method.

In the LEFM approach, cumulative damage is calculated by integrating the crack growth law from an initial crack size to a final critical size thus

$$N = \int_{a_o}^{a_f} A(\Delta K)^m da \tag{2.5}$$

where $a_o =$ initial crack length
$a_f =$ final crack length

A sequential cycle-by-cycle accumulation of incremental crack extension is needed to calculate the total number of cycles required to extend the crack to a critical size. For block loading (the block being made of a number of cycles of different amplitudes), the extension in crack by successive blocks is not the same since the initial crack length for each block (or even for each cycle within a block) is not the same.

2.2 Mean Stresses.

Apart from fluctuating loads, many automotive components also bear static pre-loads. The methods for dealing with the effects of combined loads on the service lifetime is considered below.

For the $S-N$ method the most widely used method is that proposed by Goodman, who provided an expression for unnotched specimens to calculate zero mean stress ranges, equivalent in damage terms to stress excursions made up of a ranges and a mean.

$$\sigma_a = \sigma_o(1 - (\sigma_m/\sigma_u)^{m1}) \tag{2.6}$$

where $\sigma_a =$ stress amplitude, unnotched
$\sigma_o =$ Allowable zero mean stress amplitude
$\sigma_m =$ mean stress, unnotched
$\sigma_u =$ Ultimate tensile strength
$m1 =$ 1 for Goodman's Rule

The cumulative damage due to the zero mean equivalent stress ranges is calculated via the use of Miner's rule.

For the strain-life approach, the mean stress can be handled by either of the two methods proposed below:

Morrow [3] introduces the mean stress as a correction factor to the fatigue strength coefficient in the strain-life equation as shown

$$\frac{\Delta\epsilon}{2} = \frac{(\sigma'_f - \sigma_m)}{E}(2N_f)^b + \epsilon'_f(2N_f)^c \tag{2.7}$$

However, a preferred method is that proposed by Smith Topper Watson (STW) [4]

$$\sigma\text{max}\cdot\epsilon_a = \frac{\sigma'^2_f}{E}(2N_f)^{2b} + \epsilon'_f\sigma'_f(2N_f)^{b+c} \tag{2.8}$$

The maximum stress at a given strain amplitude is dependent both on the strain amplitude as well as the position on the stress-strain curve at which the material is being worked locally. The latter depends on the preceeding cycle executed. Thus a cycle-by-cycle algorithm is needed to calculate, for each cycle, the strain amplitude and the peak stress associated with that amplitude. This is discussed further in section 2.3 in conjunction with the calculation of local stresses and strains at stress concentration sites from elastic strain history measured remote from the local site.

For the LEFM approach, the mean stresses are found not to have a very significant effect on the crack growth parameters in steels although in other materials such as aluminium alloys a mean stress dependency is reported. The primary effect of mean stresses, however, is to influence the stress intensity threshold, Kth. This effect is modelled as shown below; the equation being credited to Foreman.

$$\frac{da}{dN} = \frac{A(\Delta K)^m}{K1c(1-R) - \Delta K} \tag{2.9}$$

$$\begin{aligned} \text{where } K1c &= \text{ Fracture Toughness} \\ R &= \sigma\text{min}/\sigma\text{max} \end{aligned}$$

Research has also established that fatigue cracks can close at positive applied stress levels because of the plastically yielded material behind the crack tip. This closure reduces the applied stress intensity factor to some smaller effective factor. This closure effect is considered important for crack growth behaviour under variable amplitude loadings. A detailed exposition of the closure effect is outside the scope of this paper.

2.3 Notches

The presence of a notch such as a groove, bolt hole, sectional change etc introduces a stress concentration on the component. Solutions for elastic stress concentration factors, Kt, for a wide variety of engineering situations are tabulated in [5].

For high cycle fatigue the $S - N$ curve is factored down by Kt in the fatigue limit region. The subsequent calculation methodology is the same as that for the unnotched case.

The procedure for dealing with notch effects in the strain-life method is more rigorous. Recourse is made to the so-called Neuber's Rule which equates the elastic-plastic notch root stresses and strain to the nominal elastic stresses.

$$\frac{(Kt.\Delta S)^2}{E} = \Delta\sigma.\Delta\epsilon \tag{2.10}$$

$$\begin{aligned} \text{where } \Delta\sigma, \Delta\epsilon &= \text{ local stress, strain ranges} \\ Kt &= \text{ elastic stress concentration factor} \\ \Delta S &= \text{ nominal elastic stress ranges} \\ E &= \text{ Youngs Modulus} \end{aligned}$$

This is illustrated graphically in Fig.1. The left hand side of Equation (2.10) is equivalent to the triangle ODE. A point A' is found on the cyclic stress-strain curve such that the area of the triangle $OA'B'$ is equal to that of ODE. The point A' defines the notch stress σ, and strain ε. For a cycle-by-cycle analysis, the procedure is repeated with the origin O shifted

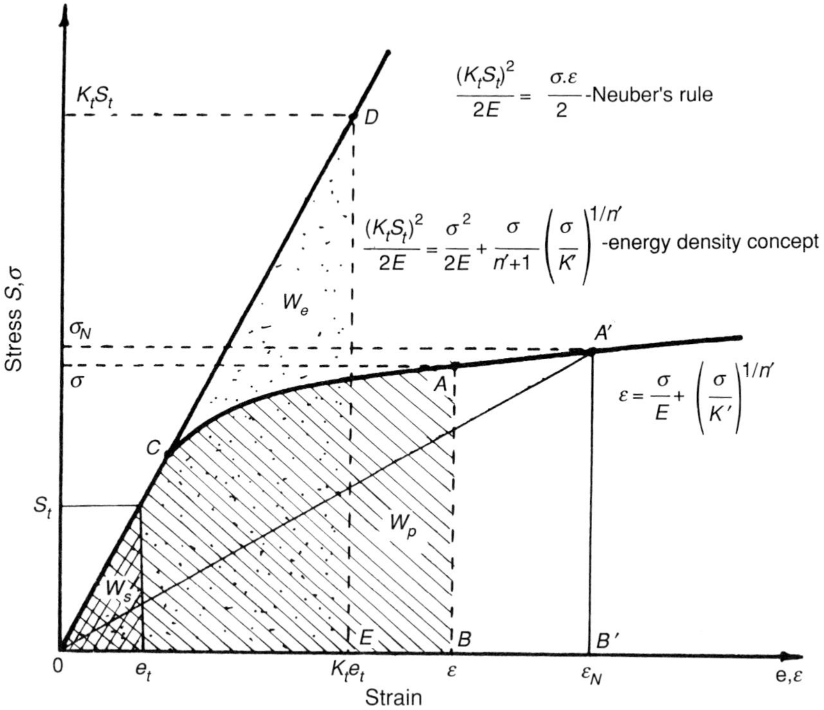

Figure 1. Graphical Interpretation of the Strain Energy Density Concept vs. Neuber's rule

to A'. It is thus possible to identify both the local strain range as well as the maximum stress which are required for cumulative damage using the STW parameter in Equation (2.8). Fig.2 shows a comparison between measured notch strains and Neuber predicted strains under variable amplitude loading.

A phenomenologically rigorous solution to the notch problem is provided by Glinka [6] based on the strain energy density concept. Here, the strain energy represented by the triangle ODE is equated to the area bounded by the cyclic stress-strain curve, the x-axis and the line AB; this area being representative of the deformation strain energy density just ahead of the notch.

$$\frac{(K_t.\Delta S)^2}{E} = \frac{\Delta\sigma^2}{4E} + \frac{\Delta\sigma^2}{n'+1}\left(\frac{\Delta\sigma}{2K'}\right)^{1/n'} \qquad (2.11)$$

where n' = Cyclic strain hardening exponent
K' = Cyclic strength coefficient

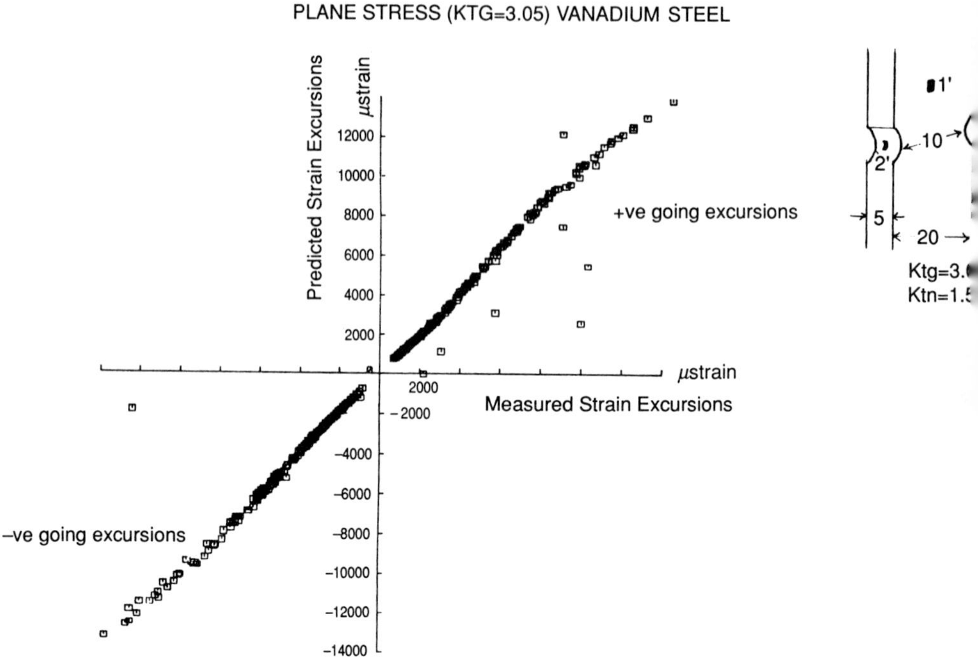

Figure 2. Comparison of Predicted Strain Excursions Based on Neuber's Rule with Measured Local Strain Excursions

This rule uses an explicit form of the cyclic stress-strain curve and thus facilitates the notch calculations even under plane strain conditions by the substitution of the appropriate stress-strain curve.

The LEFM does not appear to be sufficiently developed to cater explicitly for crack growth through a plastically strained region at notch roots. The compliance factor in Equation (2.3) caters for the orientation of the crack and thus partly handles the notch effects but a rigorous definition of the problem is not known to the author.

2.4 Multiaxial Stress States

Complex stress states on automotive components arise through the presence of notches which transform uniaxial loading into multiaxial stresses at notch roots or through combined loading such as the simultaneous presence of torsion and bending. The former, as explained above in section 2.3, is usually dealt with on the premise that the damage can be accounted for by a single dominant stress or strain at the critical site. In any event, the notch stress or strain field under uniaxial loading represents a situation where the ratio of principal stresses remains constant throughout the loading cycle (proportional loading) and hence critical planes (such as the plane of maximum shear stress) remain stationary relative to the component axes. This is relatively simple to analyse compared to the non-proportional case (out-of-phase biaxial loading) where the critical planes are rotating during the loading cycle.

The multiaxial approaches are grouped into three catagories: namely, those dealing with the fatigue limit, the 'equivalent' stress or strain approach for the finite life regime and the fracture mechanics based methods. Examples of these three methods are included below.

Dang Van Criterion for Fatigue Limit

The basis of this criterion is that nucleation of cracks is a microscopic phenomenon, on a scale where the material is no longer homogeneous or isotropic. Thus the local stresses, σ_{ij}, and strains, ϵ_{ij}, can be significantly different from the microscopic stresses and strains usually available to the designers for engineering calculations. The criterion considers the microscopic 'loading' path (locus of σ_{ij} as a function of time) and a 'damage' or 'no-damage' situation identified if this locus transgresses outside a damage line defined by a function $g(\sigma_{ij})$. The explicit form of this function is a linear combination of shear stress, $\tau(t)$, and hydrostic pressure, $p(t)$ expressed by

$$g(\sigma_{ij}(t)) = \tau(t) + a_1.p(t) - b_1 \tag{2.12}$$

$$\text{where}\tau(t) = \frac{\sigma_1(t) - \sigma_3(t)}{2} \quad \text{and} \quad p(t) = \frac{\sigma_1(t) + \sigma_2(t) + \sigma_3(t)}{3}$$

$$\sigma_1 > \sigma_2 > \sigma_3 \quad \text{principal stresses}$$

$$a_1 \text{ and } b_1 \quad \text{material constants}$$

This is illustrated in Fig.3. The endurance domain in the $\tau - p$ plane is delimited by the x-axis and a straight line defined by $g(\sigma_{ij})$. The practical determination of the damage line on the $\tau - p$ plane requires simple uniaxial tension and torsion tests. At least two tests are necessary such as uniaxial tension fatigue limit with two stress ratios ($R = -1$ and $R = 0$, for instances). In addition the fatigue limit in torsion may be helpful. The method is valid both for proportional as well as non-proportional loading although for the latter, the calculation of microscopic stresses is very complicated [7].

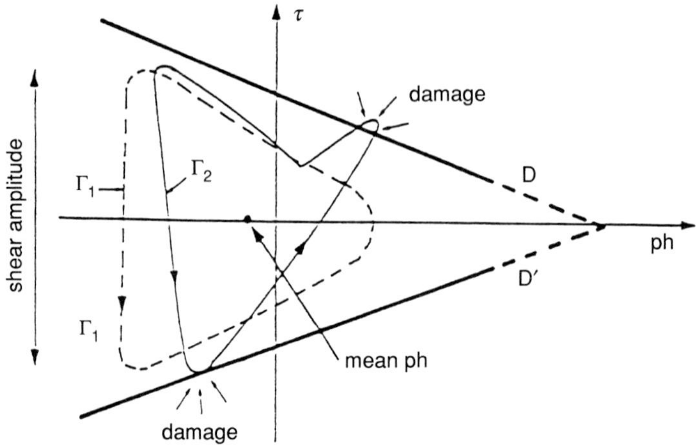

Figure 3. Endurance Domain and Two Typical Local Loading Paths Γ_1 and Γ_2

The Equivalent Strain Approach

This is based on the assumption that the lifetime prediction under multiaxial fatigue can be handled by replacing the uniaxial strains with an equivalent strain calculated from the classical deformation theory. It is thus ideally suited for the local strain approach. The most commonly used are Von Mises or Tresca criterion as listed below:

$$\epsilon_{eq} = \frac{1}{\sqrt{2}(1+\nu)} \{(\epsilon_1-\epsilon_2)^2+(\epsilon_2-\epsilon_3)^2+(\epsilon_3-\epsilon_1)^2\}^{1/2} \quad \ldots \text{Von Mises} \quad (2.13)$$

$$\epsilon_{eq} = \frac{1}{(1+\nu)} \{\epsilon_1 - \epsilon_3\} \quad \ldots Tresca \quad (2.14)$$

This approach is discussed fully in [8]. A typical example of prediction versus measured lives on an automotive component is included in Fig.4.

Crack Growth Approach

The multiaxial application of the crack growth approach is based on a detailed consideration of both Stage I cracking on the 45 degree plane to the component surface and Stage II growth which, under proportional loading, occurs in a direction normal to the axis of greatest principal stress range. The Mode I, Stage II growth law is defined by Equation (2.4). The Mode II, Stage I cracks follow the following rule:

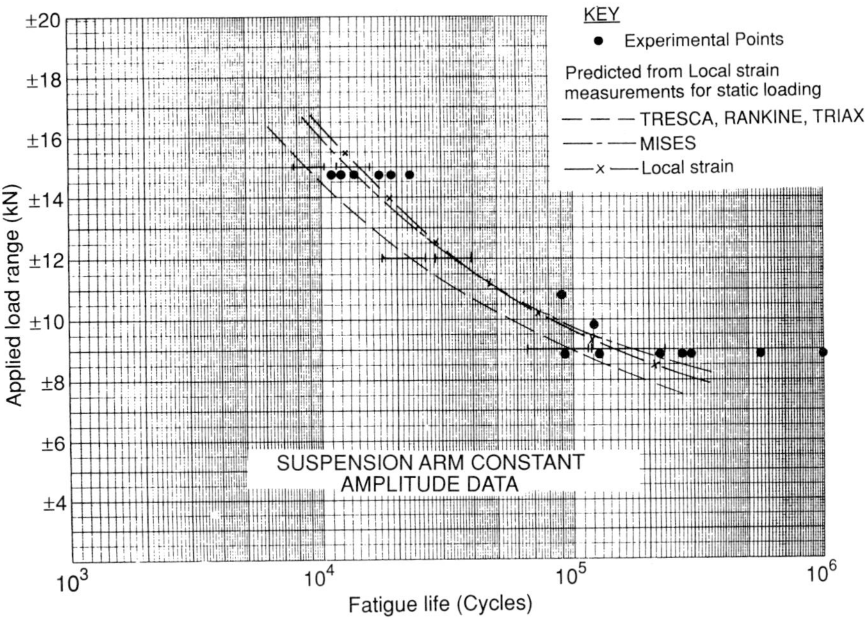

Figure 4. Component Life Prediction (Static Strain Measurements)

$$da/dN = C_s(\Delta\gamma\max)^{ms}(d-a)$$

where C_s, ms are material constants and d is the characteristic dimension for the dominant microstructural barrier to shear mode growth. The maximum shear strain range, σ max, characterises the multiaxial response of shear cracks. Further details of this approach are included in [8].

3 Miscellaneous Effects

Generic models to help with various forms of loadings (except for thermomechanical loading which is outside the scope of this paper) and the effects of stress concentrations arising at notch roots have been outlined above. Additionally, one needs to deal with the durability effects of surface condition of the component and presence of joints.

The conventional method of dealing with the effects of surface roughness is via empirical data available, for example in [9], where the reduction in fatigue strength, Ksf, for various grades of steels is tabulated versus surface roughness parameters. The latter is characterised by 'Rt', the maximum peak-to-valley height of surface roughness. Typical values of Rt for various machining processes such as turning, grinding, milling etc are also tabulated [9]. For lifetime calculations, the fatigue limit is factored by Ksf and the procedure for the $S-N$ method outlined above is used.

A more recent approach is to treat the surface defects as cracks [10] and use the LEFM methodology to calculate the reduction in fatigue strength as illustrated in Fig.5.

The joining techniques used most commonly for the fabrication of sheet metal components, including body shells, consists of spot or fusion welding. Although design codes exist for large, civil engineering structures such as bridges or offshore platforms, see for example [11], they are generally not appropriate for the gauge thickness of the metal used in automotive parts. Nonetheless, in the absence of anything better, recourse is made to these codes in the belief that they represent conservative lifetime estimates and are thus safe.

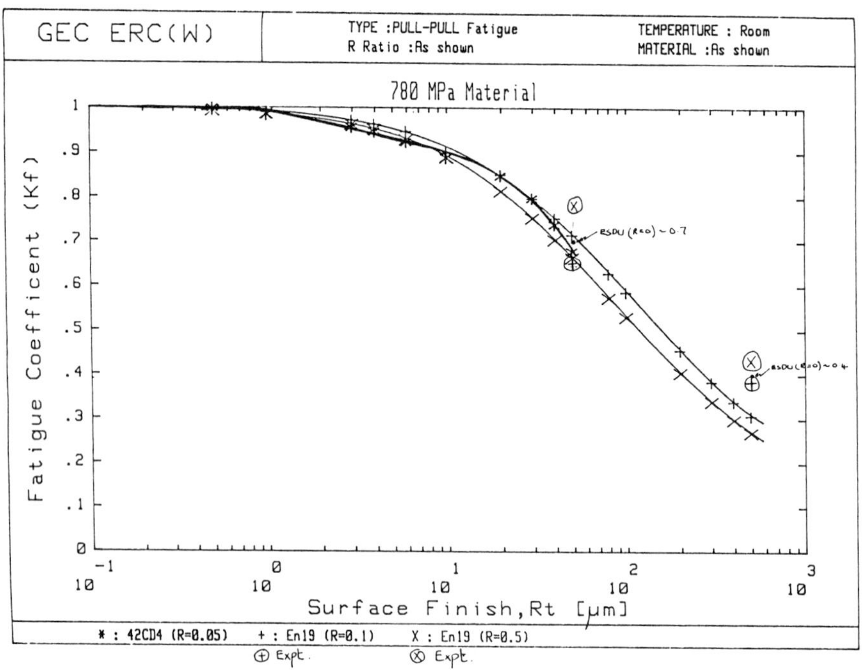

Figure 5. Plot of Reduction in Fatigue Strength as a Function of Roughness for 780MPa Material

4 Reliability of Lifetime Prediction Models

The algorithms described above handle the effects of component loading, geometry, material and surface condition. The inputs to these models are deterministic, usually average or mean values of material strength, duty cycle and surface topography. In turn, the models output a single number representing the estimated lifetime. These deterministic models do not simulate adequately the failure process in service which is a stochastic process yielding a wide scatter (a factor of 10 being typical) in measured lives as well as a poor correlation between measured and predicted lifetimes.

A significant portion of this scatter and lack of correlation is attributable to the statistical variability and uncertainty in the controlling parameters. Probabilistic models are therefore required to handle non-deterministic inputs and provide an output defining the survival probability in service. Such an approach could then handle the component-to-component or batch-to-batch variations and provide a methodology for estimating the 'reliability' of lifetime prediction algorithms. The development of such design methodology based on reliability approaches are well advanced for the offshore and nuclear industries but these do not address the problems of volume production as found in the automotive industry. Hence a major research effort is directed at this problem [12].

5 Strategy for Fatigue Design Support System

The support of fatigue assessment, lifetime improvement or advice on good design practice requires both algorithmic as well as heuristic information. The former has been outlined above, the latter is best handled via the so called 'knowledge-based' computer software technology. In contrast to the conventional computer software, under the 'Knowledge based' approach, the knowledge-base is kept separate from the reasoning mechanisms or 'inference engine', thereby making it easy to update (even by non-programmers) the knowledge-base when more information becomes available without disturbing the reasoning procedure. An integral part of such a system is an interface to a database of engineering information.

An integrated system for fatigue design based on the above philosophy is described in [13]. It is illustrated schematically in Fig.6. Essentially, three separate 'expert' modules are being developed to fulfil the tasks of fatigue assessment, modification and guidance for fatigue prevention. The assessment module includes advice on the selection of the appropriate algorithm for lifetime calculation and processing of input data for these algorithms. For example, the designer is provided with a system to help select an appropriate stress concentration factor, Kt, for critical sites on the component via a pictorial description which idealises component features. This

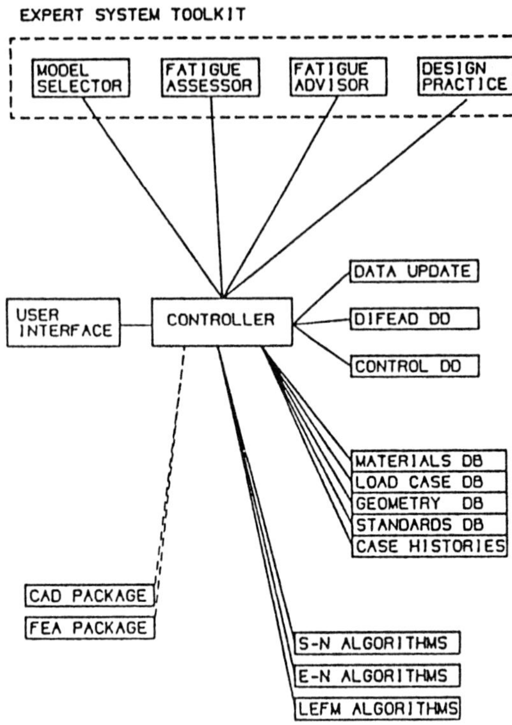

Figure 6. Schematic Illustration of the Expert System for Fatigue Design

information, together with other data from the database, described below, allows a 'rapid but approximate' assessment of the service life of the component. In the event that the design is found inadequate, advice may be sought on the available option for design changes from the fatigue life modification expert module. The changes may include material substitution, treatments or geometric alterations. The third expert on design practice is intended to oversee any obvious errors without recourse to detailed calculations. In addition this module is intended to provide reference to previous, successful design experience.

The success of this approach is critically dependent on the extent and quality of the database which stores information on material performance, service duty cycles, geometric factors etc. In addition, the designer also needs access to company or international standards to ensure that the design conforms to the appropriate sign-off criterion.

The usefulness of the overall scheme depends on its flexibility, hence, all intertask communications and information exchange are handled by a central tool manager. The user access is simplified by making these tools available via a window on a CAD station.

6 Conclusions

The paper has identified the major requirements for software tools to support design for durability of automotive components. Analytical models for fatigue life calculations appropriate for automotive components have been reviewed and the strengths and weaknesses of the $S - N$,the local strain and the LEFM approaches have been pointed out.

It is concluded that a major effort is needed to establish the reliability of the above models when applied to volume produced components where statistical variability introduces large scatter in service lives.

A practical design support system incorporating 'knowledge-based' features has been outlined. Based on a prototype study, this approach appears to offer a powerful medium for the optimisation of the design for durability. Such a strategy offers the real prospect of getting the design 'right-first-time'.

References

1. Sherratt, F., (1982), Fatigue life estimation: A review of traditional methods. Journal of the society of environmental engineering.

2. Miller, K., (1989), A new approach to safe-life and fatigue-life prediction under random loading. Private communication.

3. Wetzel, R., Fatigue under complex loading. SAE publication. Advances in engineering. Vol. 6.

4. Smith, K., Watson, P. and Topper, T., (1970), A stress-strain function for the fatigue of metals. Journal of materials JMLSA.Vol.5. No. 4 pp 767-778.

5. Peterson, R., (1974), Stress Concentration Factors. Wiley Interscience.

6. Glinka, G., (1985), Calculation of inelastic notch-tip strain stress histories under cyclic loading. Engineering Fracture Mechanics. Vol22 No.5.

7. Dang Van, K, Papadopoulos., (1987), Multiaxial Fatigue Failure Criterion: A new approach. Fatigue' 87. Volume II pp 997-1009.

8. Brown, M., (1988), Analysis and design methods in multiaxial fatigue. Advances in Fatigue Science and Technology, Portugal.

9. ESDU Publication No. 74027: The effects of surface roughness on the fatigue limit of steels at zero mean stress.

10. Byrne, P., Taylor, D. and Allen, D., (1989), Fatigue behaviour of rough surfaces: some initial findings. Irish Materials Forum.

11. BS5400: Part 10, Code of practice for fatigue.

12. Devlukia, J., Brite-Euram P-3051-89: Design methodology for durability of mechanical components based on reliability approaches.

13. McMahon, Sims-Williams and Devlukia, J., (1989), A knowledge-based system for fatigue design. IEE conference on exploitable U.K. research for manufacturing industry.

Signal Processing for Fatigue in Both the Time and Frequency Domains

N.W.M. Bishop and F. Sherratt

University of Sheffield and University of Warwick

Abstract The objective of the work in this report was to investigate the effect of finite sample length on various statistical parameters, including fatigue damage. The results from such an investigation are of considerable interest to practising engineers. Work elsewhere [1,2] has shown that the frequency domain approach to fatigue analysis and testing is a viable technique and data presented here will confirm this finding with comparisons on actual time recordings of physical processes. The relative merits of the time and frequency domain approaches are investigated and some interesting and unexpected results concerning the errors involved with predicting fatigue damage from short (or even not so short) lengths of signal, either in the time or frequency domain are presented. The work applies to signals which are stationary and have a Gaussian amplitude probability distribution. Fatigue calculations assume that the Palmgren-Miner cumulative damage hypothesis is valid and that the appropriate cycle count is one of rainflow ranges.

1 Introduction

Time domain techniques are commonly used to predict structural fatigue from data which can only represent one example of any expected service history. Such loading histories may be truly representative of the expected service loading. They will not, however, generally give the expected value of damage. Measurements of damage or other statistical variables computed from samples of the actual loading conditions will have a variance about the expected value. In other words, fatigue damage estimates based on finite lengths of signal taken from a stationary time signal will show scatter. Similarly, fatigue damage estimates taken from (samples of) power spectral densities (PSDs) will also exhibit scatter. In general the design life of a structure or component will be long compared to the sample length on which the fatigue estimate is based. It is likely, then, that the damage actually encountered during service would be nearer to the expected or mean value than to any value derived from a finite sample. The scatter in estimates based on finite length samples then becomes an important factor. The main objective of work reported in this paper was to investigate this scatter.

In particular we are interested in the variance of fatigue predictions using the following two parameters;

(a) a rainflow count on a time sample of duration $L_s = (\Delta_n) * \delta_t$ where Δ_n is the number of points in the sample, and δ_t is the time between successive data points.

(b) rainflow ranges predicted from a PSD of the same time sample of duration L_s.

The frequency domain approach used here will be the one developed by Dirlik [3]. Although this is an empirical approach, it was confirmed later by theoretical work [4].

As well as determining whether to use time or frequency domain techniques, the work presented is relevant to the problem of reliability. In the offshore industry, a considerable amount of work is being done on reliability analysis of offshore platforms [5-9]. Similar discussions have taken place in the automotive industry, concentrating more on the errors involved with the time domain data acquisition process [10-14].

A random signal generator was used in conjunction with various analogue filtering devices for generating the data to be used for this investigation, and this is described in section 2. Section 3 deals with the data acquisition process used to digitise the analogue signals and various errors which result from this process are discussed. Section 4 gives details of various statistical tests such as the 'goodness of fit' normality test which were used to establish that the signals being used were stationary and Gaussian. Section 5 describes techniques for cycle counting from time signals. Section 6 contains a limited discussion on the merits of windowing from a range distribution point of view.

Section 7 presents the computational procedure which was used to check the frequency domain approach against real data. Section 8 details the work on the variance of fatigue predictions from limited sample sizes and section 9 presents the conclusions

2 Signal Generation and Manipulation

To test the validity of any analysis method trials have to be carried out on a record of some sort. In this case the objective was to test the hypothesis that the signal being analysed was random, stationary and Gaussian. Modern systems often use mathematically-generated random noise, filtered in various ways to produce the desired characteristics. In this case it was felt that a technique such as pseudo-random-binary-sequence generation might introduce mathematical bias because of the short lengths of signal being used for computation of the various range probability density functions. Therefore a decision was made to use a signal generator based on shot noise. This uses the arrival of electrons at the electrode of a valve as the primary wide-band source. This was then filtered by analogue methods to produce a variety of time-domain realisations. to do this the broad band random signal produced by the signal generator was split into three and filtered using three pairs of low and high pass filters which had variable cut off frequencies. These signals were then combined using a signal adder device which ensured that the signals were correctly combined. Each signal loop had separate amplification control which was used to adjust the relative magnitudes of certain frequencies being generated. It was therefore possible to generate signals with various spectral shapes, as shown in figure 1.

3 Data Sampling

The signals produced were digitised using a standard desk-top computer acquisition system. Because of the nature of the investigation large samples were needed, from which a number of smaller samples could be extracted. Three sets were generated, each with a maximum anticipated frequency of about 40Hz. Sampling rate is one of the key variables. For the frequency domain, well established theory puts this at the Nyquist rate, but fatigue analysis is less certain about the sampling rate required to accurately characterise a random stress time history. One expression for an upper bound on the error involved in the peak-trough detection process is given by

$$[1 - \cos\left(\frac{\pi}{n}\right)] \times 100$$

where n is the number of samples per highest frequency. At rates of 4, 10, 20 and 25 samples per highest frequency this gives upper bounds for the error of 29%, 4.9%, 1.3% and 0.8% respectively. To get this upper bound below 1% (assuming a highest frequency of 40Hz) requires a sampling rate of about 1000Hz, compared with the Nyquist rate of 80Hz needed for full information in the frequency domain.

A sample length of 1024000 time points was used for each of the three data sets. The memory space restriction imposed on desk top acquisition to RAM tasks meant that acquisition direct to hard disk was required. A software package with a maximum sampling rate of 50000 samples per second and a maximum continuous acquisition data set size restricted only by the disk space (20 megabytes in this case) was used.

4096 intervals were used by the data acquisition package for the quantisation process. This number of levels is well in excess of the required

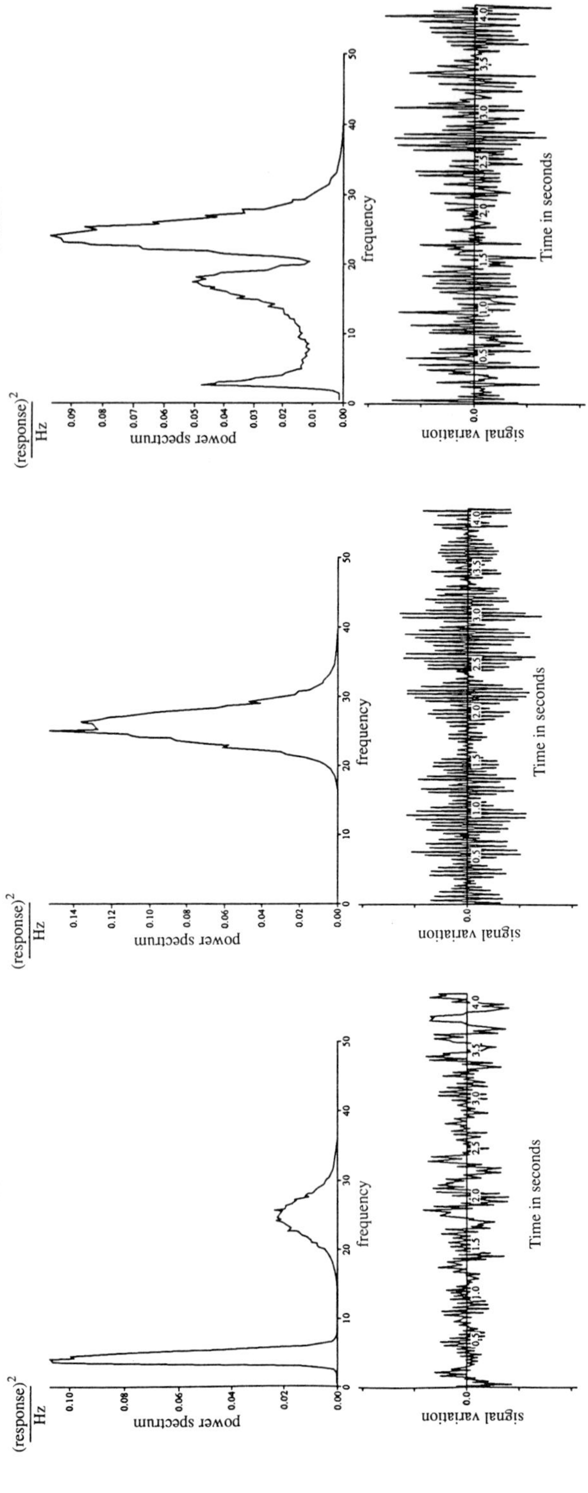

Figure 1. Power Spectral Density plots shown along with the first four seconds of data

number for reasonable accuracy. A second process is described in section 5 where the number of levels is reduced to as low as 32 intervals and the errors then introduced are examined.

4 Data Qualification

Each data set containing 1024000 time points was divided into 16 separate blocks, each containing 64,000 samples. For each block the mean ($\overline{x}$), variance (s^2), minimum, maximum and chi-squared value (χ^2) was computed. The chi-squared value was used in order to test the data for normality using the 'goodness of fit' test. This test was carried out, for each block, on 2000 equally spaced observations (N = 2000).

The method of Bendat and Piersol [15] was used whereby the boundaries of the class intervals were chosen so as to give approximately equal expected frequencies of occurrences (E_i). This expected frequency of occurrence was easily obtained for each class interval by using the computed Gaussian (normal) probability for this interval (obtained using $\overline{x}$ and s^2 from above) multiplied by the number of observations (N). By comparing the observed frequency of occurrence (O_i) with the expected frequency of occurrence (E_i) for each class interval (i) the chi-squared value was computed using;

$$\chi^2 = \sum_{i=1}^{K} \frac{(O_i - E_i)^2}{E_i} \tag{4.1}$$

The number of class intervals, K, was chosen as 43, giving a value of 40 for n, the number of degrees of freedom.

The Reverse Arrangement test was applied to the means and rmss of the 16 groups for each data set in order to test for stationarity. This involved considering the sequence of either mean values or rmss as x_i for $i = 1, 16$ and counting the number of times $x_i < x_j$ for $i < j$.

The results for data set 1 which are typical of the results obtained for the other data sets are given in table 1. Full results for all the data sets are given in reference 16. From the full results we can derive values for the reverse arrangement test applied to the means and variances. These figures are given in table 2. For a 0.05 level of significance we get a value of 55.75 for $\chi^2_{40,0.05}$ from reference 15, page 525, table A.3, and only 2 of the chi-square statistics exceed this value. For these the p-values are given by;

Data set 1 Group 8 $\chi^2 = 56.91 \rho < 0.05$
Group 9 $\chi^2 = 63.05 \rho < 0.025$

It is therefore reasonable to assume normality of the data.

Similarly, for a 0.05 level of significance we get an allowable range of 38 to 81 for the Reverse Arrangement test, from reference 15 page 533, table

Table 1. Qualification tests carried out on data set 1

Group	Mean	Variance	RMS	Min	Max	Chi. Sq
1	2362.62	65542.0	256.012	1518	3215	28.80
2	2365.71	67023.6	258.889	1416	3296	33.50
3	2360.56	69281.6	263.214	1498	3134	37.34
4	2363.54	66137.1	257.171	1566	3290	42.53
5	2365.30	58913.7	242.721	1572	3138	40.05
6	2364.16	61975.7	248.949	1409	3167	30.15
7	2367.81	62995.5	250.989	1603	3438	46.30
8	2362.33	60129.6	245.213	1479	3248	56.91
9	2364.69	63807.4	252.601	1377	3267	63.05
10	2361.70	65358.9	255.654	1493	3456	35.95
11	2366.62	66106.9	257.113	1445	3201	38.18
12	2364.57	65319.5	255.577	1558	3140	42.68
13	2364.49	66249.6	257.390	1479	3177	40.19
14	2365.00	60820.1	246.617	1578	3146	30.77
15	2363.45	60063.4	245.078	1516	3273	39.49
16	2365.88	59489.7	243.905	1472	3266	30.40

Table 2. Analysis of qualification tests carried out on data set 1,2 and 3

	Number of χ^2 values greater than 55.75	Reverse arrangement test on means	Reverse arrangeme test on rms
data set 1	2	51	77
data set 2	1	52	53
data set 3	0	49	52

A.7. All three data sets satisfy this test. Therefore stationarity of the data can be assumed.

5 Cycle Counting from Short Lengths of Time Signal

If one wishes to obtain cycle counts from short samples of data, acquired using standard acquisition packages then four stages of operation are required;

(1) A reduction of the number of intervals used for the first quantisation (data acquisition) process from 4096 to as low as 32 levels.

(2) A peak-trough detection operation resulting in the extraction of only those elements from the original time series which represent alternating peaks and troughs.

(3) A manipulation of the tails of the peak-trough series, followed by a reordering of the series, in order to apply the rainflow range cycle counting process described in (4) below.

(4) Application of the relevant cycle counting routine to determine rainflow ranges.

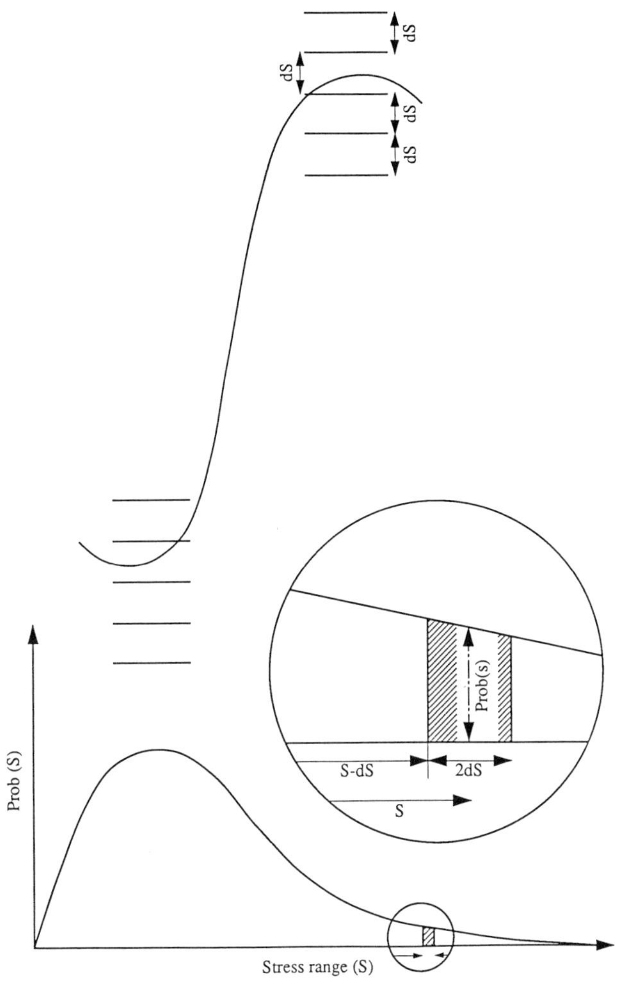

Figure 2. The effect of quantisation of the signal on the range distributions

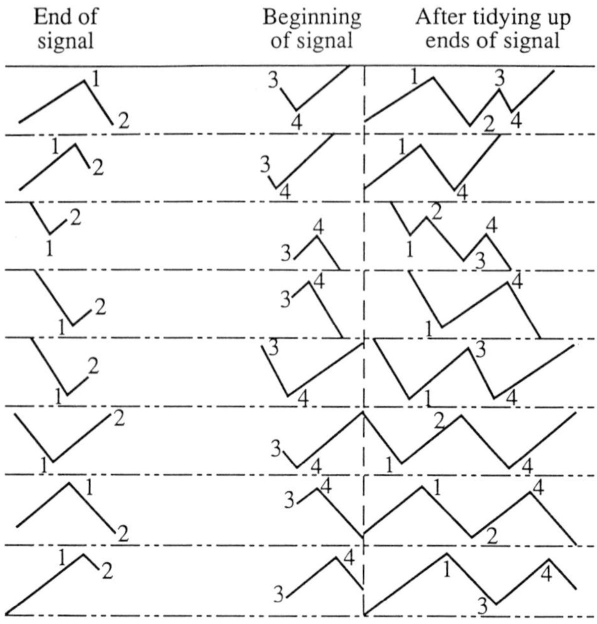

Figure 3. The eighth possible combinations of the signal tails shown along with the most justifiable method of joining the ends together

As mentioned in section 3, errors are introduced into the final fatigue damage results because of the quantisation process. However, because 4096 intervals are used, errors in the data acquisition process can be ignored. But a second quantisation process is required (stage 1 above) before the peak-trough detection process can be applied. Typically, 32, 64, 101 and 192 elements were used for this. Figure 2 highlights this problem. This shows that any range which is counted at a later stage in the analysis, involves two quantisations, one for the peak, and one for the trough. This means that counted ranges of predicted magnitude S, are actually of some magnitude between $S - dS$ and $S + dS$. It is the moments of such rainflow range distributions which are important from a fatigue point of view. Therefore, errors will result because the ranges between $S - dS$ and S will be moved to S and so will be given larger weightings than they should actually have. Similarly, the ranges between S and $S + dS$ will be given smaller weightings than they should have.

Although the error involved with this procedure could be important with certain types of signal, no significant change in damage was observed when the number of intervals was varied between 32 and 192. Therefore for the rest of the work reported in this paper any such error was ignored. A more rigorous study of the problem is planned for the future.

Once the peak-trough detection process had been carried out (stage 2 above) a series of alternating maxima and minima existed to which the application of a rainflow count does not produce a complete distribution because the 'open ended' tails of the peak-trough series would almost certainly result in incomplete closure of all the relevant cycles. Therefore, the last remaining, largest and most important cycles would remain uncounted. The method used by the authors for solving this problem was to consider all peak-trough series as falling into one of eight categories as shown in figure 3. This figure also shows, for each case, the most justifiable methods of joining the end of the signal with the beginning. The series thus created was then reordered to put the signal maximum at the beginning and therefore, also at the end of each signal. A conventional count was then applied.

6 Frequency Domain Analysis of Time Signals

Standard frequency domain methods were used to compute estimates of the PSD. Fast Fourier Transform (FFT) array sizes of between 512 and 4096 were used depending on the available data set size and the number of averages which were used.

Some work was done on the use of windows for improving the range density function estimates. No discernible improvement was observed. This might have been expected since the data being analysed tended to have slowly varying PSD functions with no sharp spikes to which some form of windowing might be applied for improved detection. However, even in such a situation, it is doubtful whether windowing could ever produce improved range density function predictions, since according to the modelling of Dirlik [3] it is the moments of the PSD which completely define these distributions. Because windowing results in bias of the spectral estimates with decreasing scatter, the combined effect on the moments would almost certainly be detrimental.

7 Computational Procedure for Estimating the Fatigue Damage from Both Time and Frequency Domain Information

The following sections describe the computational procedure used for this investigation;

(1) White noise produced by the signal generator was processed using various filters to obtain any desired spectral shape (see figure 1).

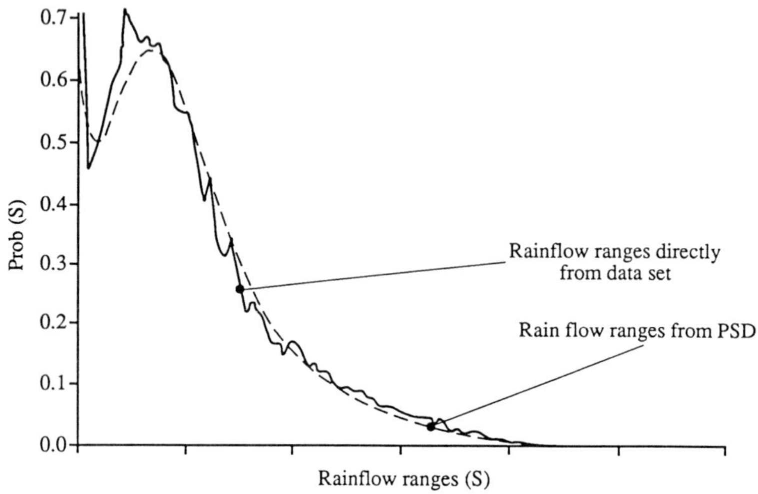

Figure 4. Comparison of rainflow ranges computed directly from full data set with rainflow ranges computed from PSD of full data set using Dirlik's equation

(2) 1024000 points along each time signal were acquired with a sampling interval of 0.001 second, giving 1024 seconds of continuous data. 4096 stress range intervals were used to discretise each sample point. Typical 4 seconds blocks of data are also shown in figure 1.

(3) All of these points were then used (approximately 20 per highest frequency) to extract peaks and troughs and hence, using the methods described in section 5 for manipulating the signal, get the rainflow range density function. For this purpose, 192 stress intervals were used which is well in excess of the required interval resolution needed to give accurate peak-trough extraction.

(4) Every 8th point was used to perform FFTs on the data. This gave a Nyquist frequency of 62.5Hz which was well above the highest frequencies of interest (approx 40Hz). 250 averages were carried out on FFT's of 512 points. Therefore 128000 points were used to obtain the final smoothed spectrum instead of the 1024000 used for the peak/trough count, although the lengths of signal were the same.

(5) The zeroth, first, second and fourth moments of the PSD were computed, and then used with Dirlik's equation to produce the rainflow range density function shown in figure 4 for data set 1. The range

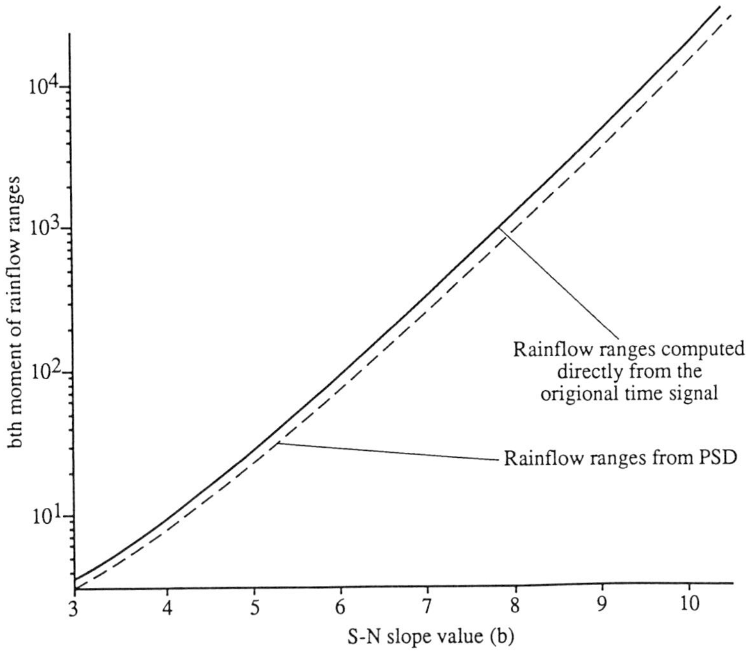

Figure 5. Moments of rainflow ranges shown in Figure 4

function computed directly from the peak/trough count is also shown for comparison.

The prediction using Dirlik's equation is good, particularly at high ranges which are most important for fatigue computations. Figure 5 shows the b_{th} moments of these range functions, where 'b' is the slope of the $S-N$ curve. Further examples of the moments of the rainflow range density function for data sets 2 and 3 are shown in reference 16.

8 The Variance of Fatigue Predictions from Limited Sample Sizes using Both Time and Frequency Domain Methods

When collecting data on site it is important to know the length of data required to obtain a particular accuracy. This is true both for a straightforward peak-trough and then rainflow cycle count and also for the alternative frequency domain approach. In this section results will be presented which highlight the change in the scatter of fatigue predictions for varying sample length L_s with particular reference to the following factors;

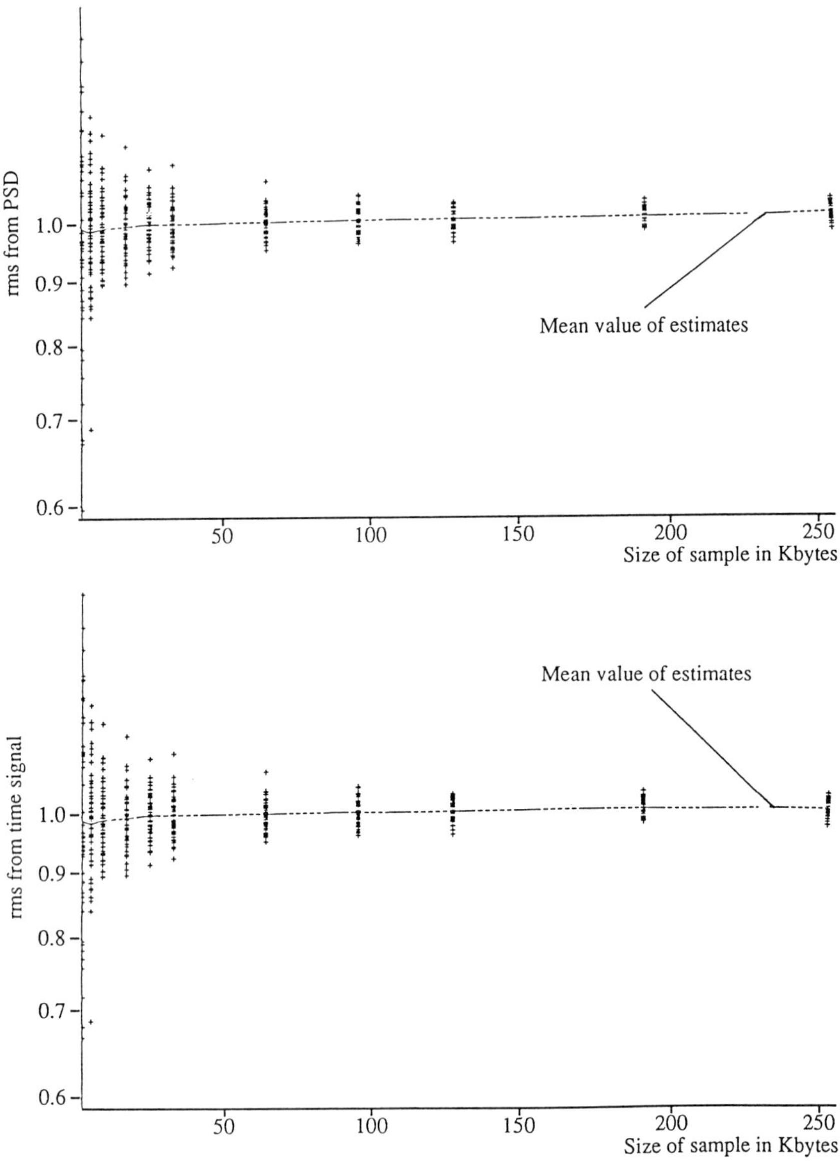

Figure 6. rms (normalised by population rms ... plotted against sample size) computed from (a) PSD of time sample and (b) time sample directly

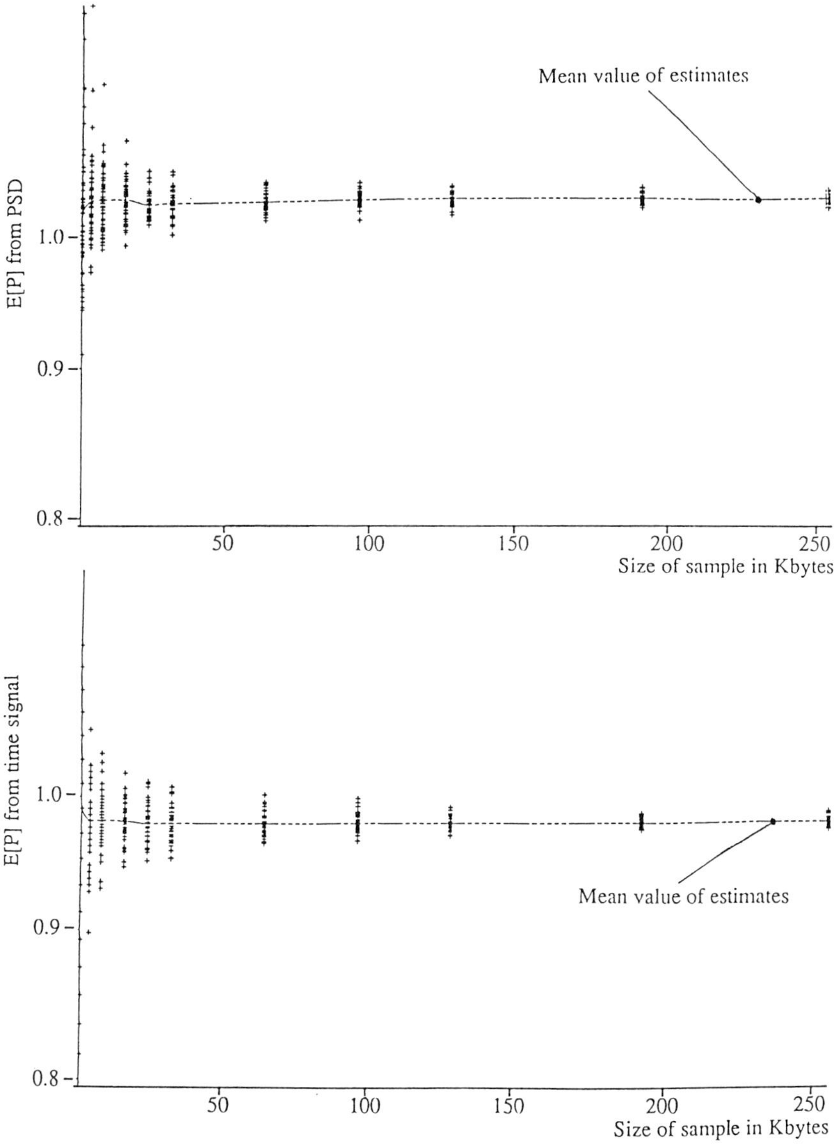

Figure 7. Number of peaks per second (normalised by population number of peaks per second ... plotted against sample size) calculated from (a) PSD of time sample and (b) time sample directly

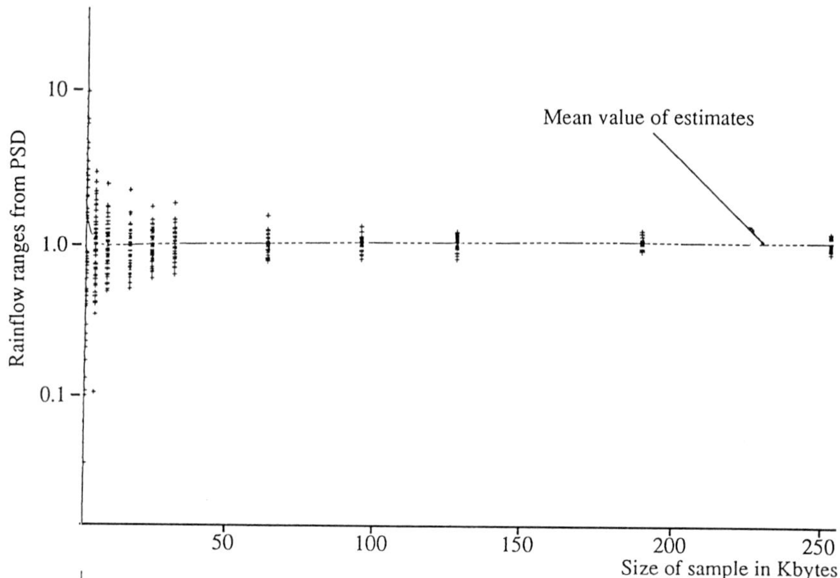

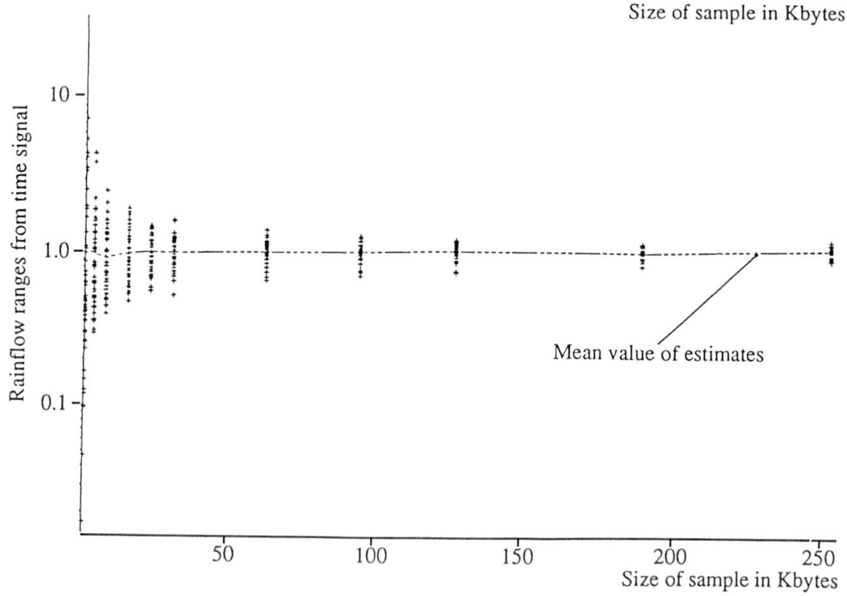

Figure 8. Rainflow range prediction of fatigue (normalised by rainflow range prediction from full time signal ... plotted against sample size) calculated using (a) frequency domain approach (Dirlik's equation) and (b) time domain approach (directly from time signal)

(a) The 6th moment of the rainflow range count of the actual time signal, meaning that the results relate to a typical machined component.

(b) The rms and number of peaks per second from the actual time signal.

(c) Calculation of the PSD from the same length of sample (but with approximately one eighth of the sample rate required) from which the following can be calculated.

(c.i) The 6th moment of the rainflow range density function.

(c.ii) The rms and number of peaks per second.

For the following discussion, the method of calculating the parameters in sections (a) and (b) will be referred to as the 'time domain' approach. The method for sections (c.i) and (c.ii) will be called the 'frequency domain' approach for obvious reasons.

Steps (a) to (c.ii) above were carried out on the data shown in figure 1 for 11 different sample lengths L_s. For each sample length, 50 estimates of the various parameters above were calculated. The results are shown in figures 6(a) to 8(b). All of the plots shown have been normalised by the 'population' values. These are the values computed on the full data sample length (i.e. 1024 seconds).

Figures 6(a) and 6(b) show the reductions in scatter of rms for the frequency and time domain approaches. There is negligible difference between these. The results for the maximum sample length show a scatter of approximately 3% about the mean for both frequency domain and time domain approaches.

The number of peaks per second computed using the frequency and time domain methods are shown in figures 7(a) and 7(b). The time domain method predicts slightly lower rates for the number of peaks per second (approximately 5% less). This is because of the finite interval (192 levels) used for the peak-trough count. If adjacent peaks and troughs fall within the same interval they are ignored by the peak-trough extraction process. This will have little effect on the final fatigue result, however, because the range prediction for the important higher ranges is changed by this reduction in peaks per second to give almost identical fatigue predictions when the number of peaks per second is multiplied by the range distribution moment. The scatter about the mean values for the maximum sample length is approximately 1%.

Figures 8(a) and 8(b) show the comparison of scatter between the time and frequency domain predictions for rainflow ranges. The PSD approach appears to show a slight improvement in scatter but this is not conclusive. However, the important point is that the PSD approach is at least as good as the time domain approach.

These results also show that for a limited sample length (one can never obtain an infinite sample) a factor of safety should be applied to the result to allow for unconservative scatter in the prediction. This applies to both time and frequency domain computations. For example, with figures 8(a) and 8(b) for the moments of the rainflow ranges, the factor of safety would be 16% at 250 kbyte samples (256 seconds), 30% at 64 kbyte samples (66 seconds) and 40% at 32 kbytes (33 seconds).

9 Conclusions

(1) The frequency domain approach to rainflow cycle counting has been shown to be a viable analysis technique. The acquisition rates and numbers of samples required are significantly less than the alternative time domain approach and this may result in significantly less computational time and storage space for on line fatigue analysis tests. Factors like the acquisition rate may be significant for high frequency fatigue work [17].

(2) Scatter in the fatigue predictions has been shown to reduce with sample size as would be expected. The results for the frequency domain approach indicate that it is at least as good as the time domain prediction in terms of scatter. Plots of this scatter show the need for a factor of safety to be applied to any fatigue prediction which is based on a limited sample size. This applies both to time domain calculations and frequency domain solutions where a probabilistic approach is being applied.

References

1. Bishop, N.W.M. and Sherratt, F., (1989) Fatigue life prediction from power spectral density data. Part 1, Traditional approaches, Environmental Engineering, 2, No. 1, pp. 11-29.

2. Bishop, N.W.M., and Sherratt, F., (1989) Fatigue life prediction from power spectral density data. Part 2, Recent developments, Environmental Engineering, 2, No.2.

3. Dirlik, T., (1985) Application of computers in Fatigue Analysis, University of Warwick PhD Thesis.

4. Bishop, N.W.M., and Sherratt, F., (1989) A theoretical solution for the prediction of rainflow ranges from PSD's, Fatigue and Fracture of Engineering Materials and Structures, 13, No. 4.

5. Martindale, S.G., (1982) Reliability-based progressive fatigue collapse, presented at the ASCE Structures '82 Congress and Exposition, New Orleans, LA, October 25-27.

6. Committee on Reliability of Offshore Structures of the Committee on Structural Safety and Reliability of the Structural Division (1983), Application of reliability method in design and analysis of offshore platforms, Journal of Structural Engineering, 109, No. 10.

7. Anderson, W.D., et al., (1982) Reliability procedure for fixed offshore platforms, Journal of the Structural Division, ASCE 108, No. ST11.

8. Vanmarcke, E., (1983) Risk assessment for offshore structures: A review, Journal of Structural Engineering, ASCE 109, No.2.

9. Bea, R.G., et al., (1984) Decision analysis approach to offshore platform design, Journal of Structural Engineering, 110, No.1.

10. Donaldson, K.H., (1982) Field data classification and analysis techniques, paper presented at the SAE Automotive Engineering Congress and Exposition, Detriot.

11. Grenier, G.C., et al., (1987) Integration of fatigue analysis and laboratory simulation requirements in the field data acquisition process, paper presented at the Earthquake Industry Conference, Peoria, Illinois, April 7-9.

12. Grenier, G.C, (1985) Sample rate expansion of random data for peak value determination, MTS internal report.

13. Grenier, G.C, (1986) Sines, sampling, Shannon and Nyquist alas, the alias, A^2, the anti-alias, at random, "Whew"*, MTS internal report.

14. Grenier, G.C, (1986) Fatigue damage vs. sample rate, MTS Internal report.

15. Bendat, J.S., and Piersol, A.G., (1986) Random data, analysis and measurement procedures (2nd edition), John Wiley and Sons, Inc., New York.

16. Bishop, N.W.M., (1988) The use of frequency domain parameters to predict structural fatigue, PhD thesis, Warwick University.

17. Stanzl, S.E., (1986) Fatigue testing at ultrasonic frequencies, Journal of the Society of Environmental Engineers, Vol. 25-1, pp. 11-16.